T0190411

Lecture Notes in Computer Science 12853

More information about this subseries at http://www.springer.com/series/7408

Ibrahim Habli · Mark Sujan ·
Simos Gerasimou · Erwin Schoitsch ·
Friedemann Bitsch (Eds.)

Computer Safety, Reliability, and Security

SAFECOMP 2021 Workshops

DECSoS, MAPSOD, DepDevOps, USDAI, and WAISE
York, UK, September 7, 2021
Proceedings

Editors
Ibrahim Habli
University of York
York, UK

Simos Gerasimou
University of York
York, UK

Friedemann Bitsch
Thales Deutschland GmbH
Ditzingen, Germany

Mark Sujan
Human Factors Everywhere Ltd.
Woking, UK

Erwin Schoitsch
AIT Austrian Institute of Technology GmbH
Vienna, Austria

ISSN 0302-9743 ISSN 1611-3349 (electronic)
Lecture Notes in Computer Science
ISBN 978-3-030-83905-5 ISBN 978-3-030-83906-2 (eBook)
https://doi.org/10.1007/978-3-030-83906-2

LNCS Sublibrary: SL2 – Programming and Software Engineering

This Springer imprint is published by the registered company Springer Nature Switzerland AG
The registered company address is: Gewerbestrasse 11, 6330 Cham, Switzerland

Preface

The SAFECOMP workshop day has for many years preceded the SAFECOMP conference, attracting additional participants. The SAFECOMP workshops have become more attractive since they started generating their own proceedings in the Springer LNCS series (Springer LNCS vol. 12853, the book in your hands; the main conference proceedings are published in LNCS vol. 12852). This meant adhering to Springer's guidelines, i.e., the respective international Program Committee of each workshop had to make sure that at least three independent reviewers reviewed the papers carefully. The selection criteria were different from those for the main conference since authors were encouraged to submit workshop papers, i.e., on work in progress and potentially controversial topics. In total, 26 regular papers (out of 34) were accepted. Additional to the accepted papers were a few invited talks in USDAI, DepDevOps, and WAISE. All workshops included an introduction written by the chairs. The workshops were organized as online (virtual) events because of uncertainties around COVID-19 restrictions.

Four of the five workshops are sequels to earlier workshops, one is new in topic and organizing committee:

- DECSoS 2021 – 16th Workshop on Dependable Smart Embedded and Cyber-Physical Systems and Systems-of-Systems, chaired by Erwin Schoitsch and Amund Skavhaug, and supported by ERCIM, EWICS, and European Horizon 2020/ECSEL JU projects.
- WAISE 2021 – Fourth International Workshop on Artificial Intelligence Safety Engineering, chaired by Orlando Avila-García, Mauricio Castillo-Effen, Chih-Hong Cheng, Zakaria Chihani, and Simos Gerasimou.
- DepDevOps 2021 - Second International Workshop on Dependable Development-Operation Continuum Methods for Dependable Cyber-Physical Systems, chaired by Miren Illarramendi, Haris Isakovic, Aitor Arrieta, and Irune Agirre.
- USDAI 2021 – Second International Workshop on Underpinnings for Safe Distributed AI, chaired by Morten Larsen. This workshop was organized around one invited keynote and an online panel.
- MAPSOD 2021 – First International Workshop on Multi-concern Assurance Practices in Software Design, chaired by Brahim Hamid, Jason Jaskolka, and Sahar Kokaly.

The workshops provide a truly international platform for academia and industry.

It has been a pleasure to work with the SAFECOMP chair, John McDermid, the workshop co-chair, Simos Gerasimou, the publication chair, Friedemann Bitsch, the program co-chairs, Ibrahim Habli and Mark Sujan, the workshop chairs, the Program Committees, and the authors. Thank you all for your good cooperation and excellent work!

September 2021 Erwin Schoitsch

Organization

EWICS TC7 Chair

Francesca Saglietti	University of Erlangen-Nuremberg, Germany

General Chair

John McDermid	University of York, UK

Program Co-chairs

Ibrahim Habli	University of York, UK
Mark Sujan	Human Factors Everywhere, UK

General Workshop Co-chairs

Simos Gerasimou	University of York, UK
Erwin Schoitsch	AIT Austrian Institute of Technology, Austria

Publication Chair

Friedemann Bitsch	Thales Deutschland GmbH, Germany

Local Organizing Committee

Dawn Forrester	University of York, UK
Sarah Heathwood	University of York, UK
Alex King	University of York, UK

Industry Chair

Simon Burton	Fraunhofer IKS, Germany

Workshop Chairs

DECSoS 2021

Erwin Schoitsch	AIT Austrian Institute of Technology, Austria
Amund Skavhaug	Norwegian University of Science and Technology, Norway

DepDevOps 2021

Haris Isakovic	TU Wien, Austria
Miren Illarramendi	Mondragon University, Spain
Aitor Arrieta	Mondragon University, Spain
Irune Agirre	IKERLAN, Spain

MAPSOD 2021

Jason Jaskolka	Carleton University, Canada
Brahim Hamid	University of Toulouse II, France
Sahar Kokaly	General Motors, Canada

USDAI 2021

Morten Larsen	AnyWi Technologies, The Netherlands

WAISE 2021

Orlando Avila-García	Arquimea Reserch Center, Spain
Mauricio Castillo-Effen	Lockheed Martin, USA
Chih-Hong Cheng	DENSO, Germany
Zakaria Chihani	CEA LIST, France
Simos Gerasimou	University of York, UK

Gold Sponsor

Intel

Supporting Institutions

European Workshop on Industrial Computer Systems – Reliability, Safety and Security

University of York

Assuring Autonomy International Programme

ASSURING
AUTONOMY
INTERNATIONAL PROGRAMME

Human Factors Everywhere Ltd

Austrian Institute of Technology

Thales Deutschland GmbH

THALES

Lecture Notes in Computer Science (LNCS)

Chartered Institute of Ergonomics & Human Factors

Chartered Institute
of Ergonomics
& Human Factors

European Training Network for Safer Autonomous Systems

Safety-Critical Systems Club

Safety-Critical
Systems Club

European Network of Clubs for Reliability and Safety of Software-Intensive Systems

European
Network of
Clubs for
REliability and
Safety of
Software

German Computer Society

Informationstechnische Gesellschaft

ITG

Electronic Components and Systems for European Leadership - Austria

ARTEMIS Industry Association

Verband Österreichischer Software Industrie

Austrian Computer Society

European Research Consortium for Informatics and Mathematics

Contents

4th International Workshop on Artificial Intelligence Safety Engineering (WAISE 2021)

16th International ERCIM/EWICS/ARTEMIS Workshop on Dependable Smart Embedded Cyber-Physical Systems and Systems-of-Systems (DECSoS 2021)

16th International Workshop on Dependable Smart Cyber-Physical Systems and Systems-of-Systems (DECSoS 2021)

European Research and Innovation Projects in the Field of Dependable Cyber-Physical Systems and Systems-of-Systems (supported by EWICS TC7, ERCIM, and Horizon2020/ECSEL projects)

Erwin Schoitsch[1], Amund Skavhaug[2]

[1] Center for Digital Safety & Security, AIT Austrian Institute of Technology GmbH, Vienna, Austria
Erwin.Schoitsch@ait.ac.at
[2] Department of Mechanical and Industrial Engineering, NTNU (The Norwegian University of Science and Technology), Trondheim, Norway
Amund.Skavhaug@ntnu.no

1 Introduction

The DECSoS workshop at SAFECOMP 2021 followed in the tradition of the workshops held since 2006. In the past, it focussed on the conventional type of "dependable embedded systems", covering all dependability aspects as defined by Avizienis, Lapries, Kopetz, Voges, and others in IFIP WG 10.4. To put more emphasis on the relationship to physics, mechatronics and the notion of interaction with an unpredictable environment, massive deployment, and highly interconnected systems of different type, the terminology changed to "cyber-physical systems" (CPS) and "systems-of-systems" (SoS). The new megatrend of the "Internet of Things" (IoT), as super-infrastructure for CPS as things, added a new dimension with enormous challenges. "Intelligence" as a new ability of systems and components leads to a new paradigm, "smart systems". Collaboration and co-operation of these systems with each other and humans, and the interplay of safety, cybersecurity, privacy, and reliability, together with cognitive decision making, are leading to new challenges in verification, validation, and certification/qualification, as these systems operate in an unpredictable environment and are open, adaptive, and even (partly) autonomous. Examples are the smart power grid, highly automated transport systems, advanced manufacturing systems ("Industry 4.0"), mobile co-operating autonomous vehicles and robotic systems, smart health care, and smart buildings up to smart cities.

Society depends more and more on CPS and SoS - thus it is important to consider trustworthiness including dependability (safety, reliability, availability, security, maintainability, etc.), privacy, resilience, robustness, and sustainability, together with ethical aspects in a holistic manner. These are targeted research areas in Horizon 2020 and public-private partnerships such as the Electronic Components and Systems for European Leadership Joint Undertaking (ECSEL JU), which integrated the former Advanced Research and Technology for Embedded Intelligent Systems (ARTEMIS), ENIAC, and EPoSS efforts as "private partners". The public parts are the EC and the national public authorities of the participating member states. Funding comes from the EC and the national public authorities ("tri-partite funding": EC, member states, project partners). Besides ECSEL, other Joint Technology Initiatives (JTIs), who organize their own research and innovation agenda and manage their work as separate legal entities according to Article 187 of the Lisbon Treaty, are the Innovative Medicines Initiative (IMI), Fuel Cells and Hydrogen (FCH), Clean Sky, Bio-Based Industries, Shift2Rail, and Single European Sky Air Traffic Management Research (SESAR).

Besides these Joint Undertakings there are many other so-called contractual PPPs, where funding is completely from the EC (via the Horizon 2020 program), but the work program and strategy are developed together with a private partner association, e.g. Robotics cPPP SPARC with euRobotics as the private partner. Others are, e.g., Factories of the Future (FoF), Energy-efficient Buildings (EeB), Sustainable Process Industry (SPIRE), European Green Vehicles Initiative (EGVI), Photonics, High Performance Computing (HPC), Advanced 5G Networks for the Future Internet (5G), the Big Data Value PPP, and the cPPP for Cybersecurity Industrial Research and Innovation.

The end of the Horizon 2020 Program and the current PPPs coincides with the end of the current EU budget period. The landscape of PPPs will be updated in the context of the next EC Research Program "HORIZON Europe" (2021–2027), where reorganized JUs are planned (e.g. Electronic Components and Systems, Key Digital Technologies (ECS-KDT) as successor to ECSEL, including additional key themes like photonics and software, advanced computing technologies, bio-sensors and flexible electronics), besides new PPPs. Due to the COVID-19 crisis and other negotiations within the EC, the new programs are delayed at the time of the writing of this text.

2 ECSEL: The European Cyber-Physical Systems Initiative

Some ECSEL projects which have "co-hosted" the workshop, in supporting partners by funding the research, finished in Autumn last year (delayed by the COVID-19 crisis which influenced negatively the full achievements of some of the co-operative demonstrators because it was not possible to work together at the demonstrators' locations) and this year some projects have also been extended to Autumn (see reports in last year's Springer SAFECOMP 2020 Workshop Proceedings, LNCS volume 12235). This year, mainly H2020/ECSEL projects and a few nationally funded projects "co-hosted" the DECSoS workshop via contributions from supporting partners, and some of the reports covered contributions from more than one project, achieving synergetic effects. The contributors were as follows:

- SECREDAS ("Product Security for Cross Domain Reliable Dependable Automated Systems", https://www.ecsel.eu/projects/secredas), a member of the ECSEL Lighthouse cluster Mobility.E. This project, being close to finalization of the planned work, set up two dedicated sessions in DECSoS 2021 with five papers, using the chance for compact dissemination of results. Some of the work presented also received funding from national authorities (the Czech Republic, Austria, and the Dutch National Research Council).

- AFarCloud ("Aggregated Farming in the Cloud", http://www.afarcloud.eu/), a member of the ECSEL Lighthouse cluster Industry4.E, contributed via two presentations to the DECSoS workshop.
- SMARTEST2, a project supported by the German Federal Ministry of Economic Affairs and Energy (BMWi) .
- IoT4CPS, a project within the "ICT of the Future" program of the Austrian Research Promotion Agency (FFG) and COMET K2 Competence Centers (Virtual Vehicle Research GmbH, which is also active in SECREDAS).
- National funding programs, e.g. from Sweden (KTH), the Czech Republic (together with SECREDAS), and the Dutch National Research Council (project INTERSECT).

Results of these projects were partially reported in presentations at the DECSoS-workshop. Some presentations referred to work done within companies or institutes, not particular public project funding.

Other important ECSEL projects in the context of DECSoS are the two large ECSEL "Lighthouse" projects for Mobility.E and for Industry4.E, which aim at providing synergies by cooperation with a group of related European projects in their area of interest, including the following:

- iDev40 ("Integrated Development 4.0", https://www.ecsel.eu/projects/idev40), contributing to ECSEL Lighthouse cluster Industry4.E.

New H2020/ECSEL projects which started recently, and may be reported about next year at this workshop or SAFECOMP 2022 include the following:

- Comp4Drones (Framework of key enabling technologies for safe and autonomous drones' applications, https://artemis-ia.eu/project/180-COMP4DRONES.html; started October 2019).
- AI4CSM (Automotive Intelligence for Connected, Shared Mobility, https://www.ait.ac.at/en/research-topics/dependable-systems-engineering/projects/ai4csm; https://ai4csm.automotive.oth-aw.de/).

This list is of course not complete, it considers only projects which provided main contributions to the papers of the authors and co-authors directly or indirectly via the work within their organizations and companies. Short descriptions of the projects, partners, structure, and technical goals and objectives are described on the project and the ECSEL websites, see also the acknowledgements at the end of this introduction and https://www.ecsel.eu/projects.

3 This Year's Workshop

DECSoS 2021 provided some insight into an interesting set of topics to enable fruitful discussions. The mixture of topics was, hopefully, well balanced, with a certain focus on multi-concern assurance issues (cybersecurity and safety, privacy, and co-engineering), on safety and security analysis, and on critical systems development, validation, and applications, with a specific focus on autonomous driving systems and smart farming. Presentations were mainly based on the ECSEL, Horizon 2020, and nationally funded projects mentioned above, and on industrial developments of partners' companies and universities. The following overview highlights the projects which, at least partially, funded work presented.

The session started with an introduction and overview to the DECSoS workshop, setting the European research and innovation scene and the co-hosting projects and organizations: ERCIM, EWICS and ECSEL (Austria).

The first session on Dependable Autonomous Driving Systems comprised two presentations:

(1) Dependable Integration Concepts for Human-Centric AI-based Systems, by Georg Macher, Eric Armengaud, Davide Bacciu, Jürgen Dobaj, Majid Dzambic, Matthias Seidl, and Omar Veledar.
Adaptive, cloud-based and AI-based, systems are increasingly used in the context of safety-related and safety-critical systems. This is a huge challenge for dependability engineers involved in developing and assessing trustworthiness of such systems, requiring adaptation of engineering processes and methods, particularly for autonomous applications and in the automotive domain. This is highlighted in this paper. The paper is based on work in the H2020 project TEACHING (nr. 871385).

(2) Rule-Based Threat Analysis and Mitigation for the Automotive Domain, by Abdelkader Magdy Shaaban, Stefan Jaksic, Omar Veledar, Thomas Mauthner, Edin Arnautovic, and Christoph Schmittner.
Cybersecurity is a severe challenge and risk factor with, potentially, a huge impact on safety, particularly in the automotive domain of highly automated/autonomous vehicles. The potential threats must be identified and mitigated to guarantee a flawless operation. This paper presents a novel approach to identify security vulnerabilities in automotive architectures and automatically propose mitigation strategies using rule-based reasoning. The rules, encoded in ontologies, enable establishing clear relationships in the vast combinatorial space of possible security threats and related assets, security measures, and security requirements from the relevant standards. The approach is evaluated on a mixed-criticality platform, typically used to develop autonomous driving (AD) features. Main contributions came from the ECSEL project AFarCloud (nr. 783221) and the "ICT for the Future" project IoT4CPS (nr. 6112792) of the Austrian Research Promotion Agency (FFG).

The first SECREDAS session covered Safety/Security/Privacy Systems Co-Engineering aspects with three papers:

(1) Guideline for Architectural Safety, Security and Privacy Implementations Using Design Patterns: SECREDAS Approach, by Nadja Marko, Joaquim Maria Castella Triginer, Christoph Striecks, Tobias Braun, Reinhard Schwarz, Stefan Marksteiner, Alexandr Vasenev, Joerg Kemmerich, Hayk Hamazaryan, Lijun Shan, and Claire Loiseaux.
Vehicle systems engineering is experiencing new challenges with vehicle electrification, advanced driving systems, and connected vehicles. System complexity is extended, adding vehicle-to-everything (V2X) communication systems, which provide remote communication services that collect, store, and manipulate confidential data. The impact on safety, security, and privacy (SSP) of these new advanced technological systems requires the implementation of new processes during their development phase. This paper proposes rules for introducing new architectural SSP technology elements in a product under development using design patterns.
The paper was developed under the ECSEL project SECREDAS (nr. 783119) by Virtual Vehicle Research GmbH (ViF) and partners from Austria, Germany, France, and the Netherlands; ViF is a COMET K2 Competence Center funded by the Austrian Research Promotion Agency (FFG) and other Austrian governmental institutions.
(2) Structured Traceability of Security and Privacy Principles for Designing Safe Auto-mated Systems, by Behnam Asadi Khashooei, Alexandr Vasenev, and Hasan Alper Kocademir.
There is, in a connected world, no safety without security. With many diverse components addressing different needs, it is hard to trace and ensure the contributions of components to the overall security of systems. Principles, as high-level statements, can be used to reason how components contribute to security (and privacy) needs. This would help to design systems and products by aligning security and privacy concerns. The structure proposed in this positioning paper helps to make traceable links from stakeholders to specific technologies and system components. It aims at informing holistic discussions and reasoning on security approaches with stakeholders involved in the system development process.
This research was carried out as part of SECREDAS (ECSEL nr. 783119) and the Dutch National Research Council project INTERSECT (nr. NWA.1162.18.301).
(3) Synchronization of an Automotive Multi-Concern Development Process, by Martin Skoglund, Fredrik Warg, Hans Hansson, and Sasikumar Punnekkat.
Methods for autonomous driving systems design, validation, and assurance are still in their initial stages and an established practice for how to work with several complementary standards simultaneously is still lacking. The paper presents a unified chart describing the processes, artefacts, and activities for three road vehicle standards addressing different concerns: ISO 26262 - functional safety, ISO 21448 - safety of the intended functionality, and ISO 21434 – cybersecurity engineering. The need to ensure alignment between the concerns is addressed with a synchronization structure regarding content and timing (SECREDAS project).

The second SECREDAS session was dedicated to Critical System Development and Validation:

(1) Offline Access to a Vehicle via PKI-based Authentication, by Jakub Arm, Petr Fiedler, and Ondrej Bastan.
Various approaches and ideas are emerging on how to ensure the access in use cases reflecting car rental services, car sharing, and fleet management, where the process of assigning car access to individual users is dynamic and yet must be secure. This paper shows that this challenge can be resolved by a combination of the PKI technology and an access management system. A vehicle key validation process was implemented into an embedded platform (ESP32) and the real-time parameters of this process were measured to evaluate the user experience. The results indicated that user experience is not worsened by the entry delays arising from the limited computing power of embedded platforms, even when using key lengths that meet the 2020 NIST recommendations for systems deployed until 2030 and beyond.
This work was funded within SECREDAS (ECSEL nr. 783119) and many parts of the technology were supported by several projects of the Central European Institute of Technology (CEITEC) core facilities (Czech Republic).
(2) HEIFU - Hexa Exterior Intelligent Flying Unit, by Dário Pedro, Pedro Lousã, Álvaro Ramos, João Matos-Carvalho, Fábio Azevedo, and Luis Campos.
The increasing use of UAVs led to more stringent EC regulations. In this article HEIFU is presented, a class 3 hexa-copter UAV that can carry up to an 8kg payload (having a maximum take-off weight (MTOW) of 15 kg and a wingspan of 1.5 m), targeting applications that could greatly profit from having fully automated missions. Inside, an AI engine was installed so that the UAV could be trained to fly, following a pre-determined mission, but also to detect obstacles in real-time so that it can accomplish its task without incidents. A sample use case of HEIFU is also presented, facilitating the temporal replication of an autonomous agricultural mission application.
The partners took part in the SECREDAS project to ensure, and finally increase, the safety, security, and privacy of the automated vehicle.

The last session "Software and Application Security – Testing, Modeling and Evaluation" included three software and application-oriented security papers covering particular aspects of security checking and improvement in software, railway interlocking safety impact, and of secure communication in the context of agricultural automation:

(1) Testing for IT Security: A Guided Search Pattern for Exploitable Vulnerability Classes, by Andreas Neubaum, Loui Al Sardy, Marc Spisländer, Francesca Saglietti, and Yves Biener.
This article presents a generic structured approach supporting the detection of exploitable software vulnerabilities of given type. Its applicability is illustrated for two weakness types: buffer overflowing and race conditions.
This project, SMARTEST2, was funded by the German Federal Ministry of Economic Affairs and Energy (BMWi) (project nr. 1501600C9).

(2) Formal Modeling of the Impact of Cyberattacks on Railway Safety, by Ehsan Poorhadi, Elena Troubitsyna, and György Dán.
This work was done by KTH, Royal Institute of Technology, Sweden.
Modern railway signaling extensively relies on wireless communication technologies based on standardized protocols and is shared with other users. As a result, it has an increased attack surface and is more likely to become the target of cyber attacks that can result in loss of availability and in safety incidents. While formal modeling of safety properties has a well-established methodology in the railway domain, the consideration of security vulnerabilities and the related threats lacks a framework that would allow a formal treatment. In this paper is developed a modeling framework for the analysis of the potential of security vulnerabilities to jeopardize safety in communications-based train control for railway signaling, focusing on the recently introduced moving block system. A a refinement-based approach enabling a structured and rigorous analysis of the impact of security on system safety is proposed.
This work was done by KTH, Royal Institute of Technology, Sweden.
(3) LoRaWAN with HSM (Hardware Secure Module) as a Security Improvement for Agriculture Applications - Evaluation, by Reinhard Kloibhofer, Erwin Kristen, and Luca Davoli.
Digital transformation in the agricultural domain requires continuously monitoring of environmental data and recording of all work parameters which are used for decision making and in-time missions. To guarantee data security and protection of sensor nodes, a security improvement concept around LoRaWAN communication using HSM, as presented in DECSoS 2020, is now evaluated as a project result (ECSEL project AFarCloud).

4 Acknowledgements

As chairpersons of the DECSoS workshop, we want to thank all authors and contributors who submitted their work, Friedemann Bitsch, the SAFECOMP publication chair, Simos Gerasimou as general workshop-co-chair, John McDermid, the general chair of SAFECOMP 2021, the program chairs Ibrahim Habli and Mark Sujan, and the members of the International Program Committee who enabled a fair evaluation through reviews and considerable improvements in many cases. We want to express our thanks to the SAFECOMP organizers, who provided us the opportunity to organize the workshop at SAFECOMP 2021 as an online event, despite the still existing uncertainties of the CoVID-19 crisis. In particular, we want to thank the EC and national public funding authorities who made the work in the research projects possible. We do not want to forget the continued support of our companies and organizations, of ERCIM, the European Research Consortium for Informatics and Mathematics with its Working Group on Dependable Embedded Software-intensive Systems, and EWICS, the creator and main sponsor of SAFECOMP, with its chair Francesca Saglietti and the sub- groups, who always helped us to learn from their networks.

We hope that all participants will benefit from the workshop, enjoy the conference, and join us again in the future!

Part of the work presented in the workshop received funding from the EC (H2020/ECSEL Joint Undertaking), various partners, and the National Funding Authorities ("tri-partite") through the projects SECREDAS (nr. 783119), iDev40 (nr. 783163), AfarCloud (nr. 783221), Comp4Drones (nr. 826610), ARROWHEAD Tools (nr. 826452), and AI4CSM (nr. 101007326). Other EC funded projects are, e.g., in Horizon 2020 such as the project TEACHING (nr. 871385). Some projects received national funding, e.g. SMARTEST2 from the German Federal Ministry for Economic Affairs and Energy (BMWi) (nr. 1501600C), IoT4CPS from "ICT for Future" (FFG, BMK Austria), the COMET K2 Program (Virtual Vehicle Research GmbH, Austria), INTERSECT (Dutch National Research Council, Grant NWA.1162.18.301), and CEITEC, which received funding from several Czech Research projects and the Swedish government (KTH).

International Program Committee

Dependable Integration Concepts for Human-Centric AI-Based Systems

Georg Macher[1(✉)], Siranush Akarmazyan[8], Eric Armengaud[5], Davide Bacciu[2], Calogero Calandra[10], Herbert Danzinger[5], Patrizio Dazzi[4], Charalampos Davalas[3], Maria Carmela De Gennaro[10], Angela Dimitriou[8], Juergen Dobaj[1], Maid Dzambic[1], Lorenzo Giraudi[6], Sylvain Girbal[7], Dimitrios Michail[3], Roberta Peroglio[6], Rosaria Potenza[10], Farank Pourdanesh[10], Matthias Seidl[1], Christos Sardianos[3], Konstantinos Tserpes[3], Jakob Valtl[9], Iraklis Varlamis[3], and Omar Veledar[5]

[1] Graz University of Technology, Graz, Austria
georg.macher@tugraz.at
[2] University of Pisa, Pisa, Italy
[3] Harokopio University of Athens, Moschato, Greece
[4] Institute of Information Science and Technologies (ISTI), CNR, Pisa, Italy
[5] AVL List GmbH, Graz, Austria
[6] Ideas & Motion, Turin, Italy
[7] Thales Research and Technology, Palaiseau, France
[8] Information Technology for Market Leadership, Athens, Greece
[9] Infineon Technologies AG, Munich, Germany
[10] Marelli Europe S.p.A, Turin, Italy

Abstract. The rising demand for adaptive, cloud-based and AI-based systems is calling for an upgrade of the associated dependability concepts. That demands instantiation of dependability-orientated processes and methods to cover the whole life cycle. However, a common solution is not in sight yet That is especially evident for continuously learning AI and/or dynamic runtime-based approaches. This work focuses on engineering methods and design patterns that support the development of dependable AI-based autonomous systems. The emphasis on the human-centric aspect leverages users' physiological, emotional, and cognitive state for the adaptation and optimisation of autonomous applications. We present the related body of knowledge of the TEACHING project and several automotive domain regulation activities and industrial working groups. We also consider the dependable architectural concepts and their applicability to different scenarios to ensure the dependability of evolving AI-based Cyber-Physical Systems of Systems (CPSoS) in the automotive domain. The paper shines the light on potential paths for dependable integration of AI-based systems into the automotive domain through identified analysis methods and targets.

Keywords: AI · Dependable systems · CPSoS · Dependability

Supported by the H2020 project *TEACHING* (n. 871385) - www.teaching-h2020.eu.

I. Habli et al. (Eds.): SAFECOMP 2021 Workshops, LNCS 12853, pp. 11–23, 2021.
https://doi.org/10.1007/978-3-030-83906-2_1

1 Introduction

A comprehensive set of methods, tools, and engineering approaches has evolved in the past decades to ensure the correctness of operation and to affirm trust in automotive systems. However, new challenges are exposed through the embrace of non-deterministic components and their no strict correctness characteristics by dependable systems. Several questions arise concerning dependability and standard compliance, including process and technical engineering aspects.

For dependable system integration, different challenges are linked to deterministic and non-deterministic functions. As a deterministic function assures always the same output for a given input, it is possible to predict and determine system behaviour under all considered circumstances. That assumption is the basis for the construction of sufficiently safe products and state of the art safety argumentation based on evidence for appropriate process engineering during design, development, implementation, and testing.

In contrast, as non-deterministic functions deliver different outputs to the same inputs at different runs, traditional processes and engineering approaches do not guarantee accurate prediction of the system behaviour under all considered circumstances. Hence, it is not possible to pledge necessary evidence for appropriate safety assurance.

TEACHING project tackles the specified issue while focusing on autonomous applications running in distributed and highly heterogeneous environments. It emphasises the relationship between Artificial Intelligence (AI), humans and CPSoS by leveraging human perception for adaptation and optimisation of autonomous applications. The resulting human-centric systems leverages the physiological, emotional, and cognitive state of the users for the adaptation and optimisation of autonomous applications. The implementation is based on the structuring of a distributed, embedded and federated learning system, which is reinforced by methods that improve system dependability. The results are exploited in the automotive and avionics domains. Both domains pose an autonomous challenge with high dependability needs for the system with the human in the loop.

This paper focuses on the automotive sector, which is confronted by four main trends: electrification, ADAS and Autonomous Driving (AD), connected vehicles and diverse mobility [1]. The successful response to these trends depends on openness to changes, skills to execute the same and dedicated implementation [22]. That is especially the case for AD, which relies on smart environment sensing and complex decision making supported by CPSoS. The complexity of autonomous decision making induces the need for embedding AI algorithms. Such algorithms must mimic low-level cognitive skills to enable machines to use available data and generate appropriate decisions [22].

Machine Learning (ML) models enable dynamic extraction of knowledge from historic data to anticipate the effect of actions, plans and interactions within the cyber and the physical realm and provide adaptability to effectively handle human interactions. Hence, ML is a key enabling service providing fundamental adaptation primitives and mechanisms for applications running on the CPSoS [3].

Nevertheless, the neural-based empowerment of the CPSoS requires addressing compelling challenges related to the dependability of Neural Networks (NN).

The apparent poor dependability of AI in critical decision-making environments is one of the key causes of the low level of acceptance of and trust towards new technologies. Thus, there is a need to demonstrate and inform the community of reliability approaches for AI and their benefits.

This paper offers dependability perspectives to consider different application case of non-deterministic systems as described in Sect. 3, which is also one key factor of the TEACHING project. Prior to that, we consider the related work and standardisation activities in Sect. 2. The paper is concluded with the key findings and outlook for the TEACHING project in the reported context.

2 Related Work and Regulation Activity Overview

Automotive regulations and working group activities are summarised in this section, which also includes synopsis of consequentially resulting challenges for dependable AI-based systems in the autonomous automotive context. These regulation activities form the framework, in which the proposed approaches need to sustain and provide evidence, as depicted in Fig. 1. European manufacturers comply with regulations provided by the United Nations Economic Commission for Europe ('UNECE') [21], which is the legal basis for uniform type approval regulations. The UNECE Regulations contain provisions for (a) administrative procedures for granting type approvals, (b) performance-oriented test requirements, (c) conformity of production, and (d) mutual recognition of type approvals. The regulations related to the automotive sector come from UNECE world forum for harmonisation of vehicle regulations (WP.29).[1]

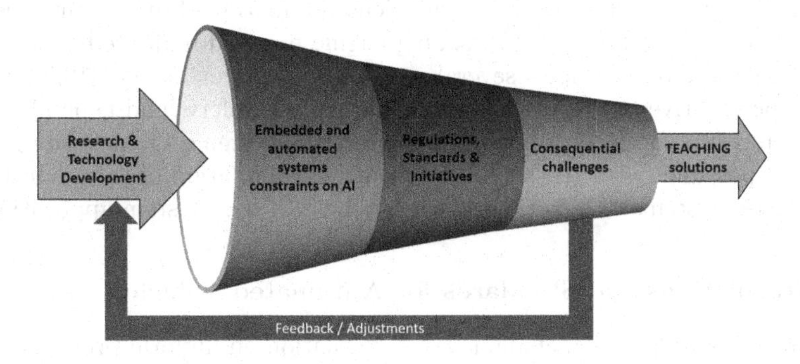

Fig. 1. Regulatory and standardisation constraints on development.

[1] https://unece.org/wp29-introduction.

2.1 Research Related Work

The development of Artificial Intelligence (AI) throughout the years has introduced new concepts and methods for solving complex technical problems. The benefits of AI and machine learning were recognised by many industrial areas, which started to utilise it for their applications. However, ist utilisation in dependable systems brings new challenges into play. Although standardisation bodies from different fields have already started to consider the integration of AI-based systems in the context of safety-critical applications, this topic is still in an early phase [23] and solutions focusing on the dependability of AI-based systems are seldom. G. Montano et al. [17] proposes a novel naturalistic decision-making support system for complex fault management procedures on board modern aircrafts. The framework is responsible for generating applicable configurations at run-time by using sensor data and autonomously generating effective decision support information for the pilot. The authors showed that instead of the constraint programming paradigm, AI could be effectively utilised for analysing the system and supporting the pilot with decision-making. Nevertheless, an evaluation of the quality constraints of the dependability features is not given. In the work of [14] different methods for uncertainty estimation on metrics which were designed to give more insights on the performance concerning safety-critical applications were described.

In [4] explainability is mentioned as the heart of Trustworthy AI and thus the guarantee for developing AI systems aimed at mission-critical (including safety) applications. The authors focus on approaches with humans keeping the responsibility for the decisions, but relying on machine aids.

The nn-dependability-kit [5] is an open-source toolbox to support safety engineering of NNS for autonomous driving systems. The rationale behind this is a GSN structured approach to argue the quality of NNs. The tool also includes dependability metrics for indicating sufficient elimination of uncertainties in the product life cycle and a formal reasoning engine to avoid undesired behaviours.

Besides these publications, several survey and overview papers [10,15,18,19] provide perspectives and descriptions of the AI and safety landscape [9]. However, there is no common approach to protect the systems against wrong decisions and possible harm to the environment, determination of safety measures for AI-based systems or generic pattern for the AI-based system applications.

2.2 Regulations and Standards for Automated Vehicles

The UNECE regulations include new UN regulation on uniform provisions concerning the approval of vehicles with regards to cyber-security and cyber-security management system (UNECE R 155 2021- the final phase of approval), uniform provisions concerning the approval of vehicles with regards to software update and software update management system (UNECE R 156 2021 – under development), and regulations event data recorder (UNECE WP.29 GRVA – 2020 not frozen). There is a multitude of other regulations dealing with more specific parts of the automated vehicle (e.g. Automated Lane Keeping System (ALKS), Advanced Emergency Braking System (AEBS)).

Standards embody the specific topic's global agreed state of the art, within a particular domain. They are not legally binding, but they offer the agreed design and development practices. The following standards are specific to autonomous vehicle design, development, and testing. This is not an exhaustive list for the whole domain, but a fair representation of specific standards to be considered for autonomous vehicle development and safety and cybersecurity functions.

The most prominent automotive standard is ISO 26262 [11] intended for safety-related systems that include one or more electrical and/or electronic (E/E) components. This document addresses possible hazards caused by malfunctioning behaviour of safety-related E/E systems, including the interaction of these systems. The included framework is intended to be used to integrate functional safety activities into a company-specific development framework.

From the perspective of process engineering, the non-deterministic system behaviour is addressed by new standards, such as SotIF (Safety Of The Intended Functionality) [13]. SotIF is a technical product safety standard with a focus on how to specify, develop, verify and validate an intended functionality to be considered sufficiently safe.

The absence of unreasonable risk due to hazards resulting from functional insufficiency of the intended functionality or by reasonably foreseeable misuse by persons is referred to as the Safety Of The Intended Functionality (SotIF). ISO PAS 21448 enhances ISO 26262 and is applied to intended functionality where proper situational awareness is critical to safety. These are situations that are derived from complex sensors and processing algorithms; especially emergency intervention systems and systems with levels of automation 1 to 5 on the OICA/SAE standard J3016 automation scale.

The third key automotive standard is ISO SAE DIS 21434 - Road vehicles - cybersecurity engineering [12]. It replaces the SAE J3061 - Cybersecurity Guidebook for Cyber Physical Vehicle Systems, provides guidelines for the organisation management of cybersecurity (CSMS) and performs operative cybersecurity activities for automotive product development. It is accompanied by ISO DTR 4804 – Road Vehicles, Safety and cyber-security for Automated Driving Systems Design, Verification and Validation. This document provides recommendations and guidance on steps for developing and validating automated driving systems based on basic safety principles derived from worldwide applicable publications. These principles provide a foundation for deriving a baseline for the overall safety requirements and activities necessary for the different automated driving functions including human factors as well as the verification and validation methods for automated driving systems focused on vehicles with level 3 and level 4 features according to SAE J3016:2018. ISO/WD PAS 5112 Road vehicles - Guidelines for auditing cybersecurity engineering and VDA - Automotive Cyber Security Management System Audit provide guidelines on how to perform cybersecurity audits and to evaluate the compliance to CSMS defined in the UNECE Reg 155.

2.3 Regulations and Standards for AI-Based Systems

Aside from multiple ethics guidelines for AI-based systems, which are out of scope of this work, the ethics guidelines for trustworthy AI by an EU Independent High Level Expert Group on Artificial Intelligence highlight the need for AI systems to be human-centric.

Also, in the context of AI-based systems, UNECE WP.29 released a first informal document, WP.29-175-21, about artificial intelligence and vehicle regulation. This work connects AI to two automotive-specific applications: (a) HMI enhancements for infotainment and vehicle management and (b) development of self-driving functionalities (building on HD maps, surrounding detection using sensor data fused with deep learning algorithms, and driving policies for automated driving using deep learning). Currently, there are no established UNECE regulations specifically for AI-based systems.

Additionally, in the last two years, the European Commission has been actively studying AI and its impact on citizens' lives. The European Commission created an independent group of high-level experts for AI. The European Commission released a set of guidelines for AI-based systems and a white paper on AI [20] to create a unique 'ecosystem of trust'. AI technologies may present new safety risks for users when they are embedded in products and services. A lack of clear safety provisions tackling these risks may, in addition to risks for the individuals concerned, create legal uncertainty for businesses that are marketing their products involving AI in the EU.

The new EU regulatory framework would apply to products and services relying on AI. To that aim, the intended regulatory framework will be defined following a risk-based approach. A risk-based approach requires clear criteria to differentiate between the different AI applications, concerning the question of whether they are 'high-risk' or not.

Conformity assessment is needed to verify and ensure compliance of certain mandatory requirements, which address high risks. The prior conformity assessment could include procedures for testing, inspection or certification, as well as checks of the algorithms and data sets used in the development phase.

2.4 Consequential Challenges

Automated Driving: Minimising the potential for fault propagation and limiting complexity requires safety-related systems to include dependable and function-specific encapsulated systems. However, the large number of intercommunicating nodes of ADSs limits the ordinary applicability of functional safety. ADSs require new approaches to real-time fault tolerance and reasoning about the consequences of faults because the fault tolerance of ADSs is unlikely to be efficiently solved solely as a software problem due to the need to coordinate complex integrative system comprised of hardware, software and physical elements.

Connected Vehicle End2End Safety: New security risks may be exposed, opening the opportunity for automated remote attacks on vehicle fleets through increased interlacing of automotive systems with networks (e.g. V2X), new features like autonomous driving, and online software updates. Remote cyberattacks can directly affect vehicles' safety-related functions. Hence, a combined approach is needed for safety and cybersecurity analysis.

Safety of the Intended Functionality: The hazard analysis and risk assessment are followed by triggering condition analysis in line with the safety goals. It would be useful to describe which is the most suitable method to define the SotIF requirements to discover the weaknesses of the system design and reduce Area 3 to an acceptable level already to the first phase of system development without waiting for driving tests, simulation, endurance testing, etc.

Dependability Engineering Methods for AI-Based Systems: The goal is to manage and evaluate the risk posed by inadequate performance of the NNs. Considering a huge encasement in the number of advanced vehicle functionalities, an acceptable safety level for the road vehicles requires the avoidance of unreasonable risk caused by every hazard associated with the intended functionality and its implementation, especially those due to performance limitations.

However, for the systems, which rely on sensing the external or internal environment, potential hazardous behaviour caused by the intended functionality or performance limitation of the fault-free system is not adequately addressed in the ISO 26262. Example of such limitations includes ML algorithms and AI-based system. Therefore, when developing a safety-critical AI, the safety case is used as a key tool for determining safety requirements to encapsulate all safety arguments for the AI. That safety case is based on a SotIF standard [13], demonstrating that all necessary safety measures are appropriately applied for AI. So, both ISO 262626 and SotIF are addressed in parallel to evaluate potential risks which can affect vehicle safety. Combining these two dependability domains will result in the definition of a safe function and mean that weaknesses of the technologies have been considered (SotIF) and that possible E/E faults can be controlled by the system or by other measures (ISO 26262).

3 Conceptual Approaches for Ensuring Dependability of AI-Based Systems

This section presents four conceptual approaches for ensuring dependability features (e.g., safety or security) of AI-based systems for different view points. The conceptualisation is a step closer to identifying key integrated process engineering approaches that support the development of dependable products that rely on non-deterministic algorithms for different application cases. Table 1 provides an overview of the concepts, drawbacks and benefits.

Table 1. Overview of concepts, main intention, drawbacks, and benefits

Concept	Main intention	Benefits	Drawbacks
A	Support of operator	Human takes decisions, traditional safety measures guarantee system safety	System operation not autonomously, decision-making mechanism must be qualified adequately
B	Selection of policy	Policy-based decision making ensures deterministic system behaviour, traditional safety measures guarantee system safety, autonomous system operation possible	Only restricted AI algorithm capabilities due finite set of policies
C	Taking critical decisions in supervised manner	Comparison with deterministic supervisor system, monitor meets classic safety requirements, less restricted application of AI, autonomous operation of system	AI limited by monitor functionalities, two nearly equally sophisticated systems needed, resource usage increased, synchronisation mechanism required
D	AI to monitor system and enhance dependability	Conventional system works without AI intervention, AI acts as monitor, AI increases realiability of conventional system	AI must learn normal vs. abnormal behaviour of system, AI must not reduce the system reliability in any case

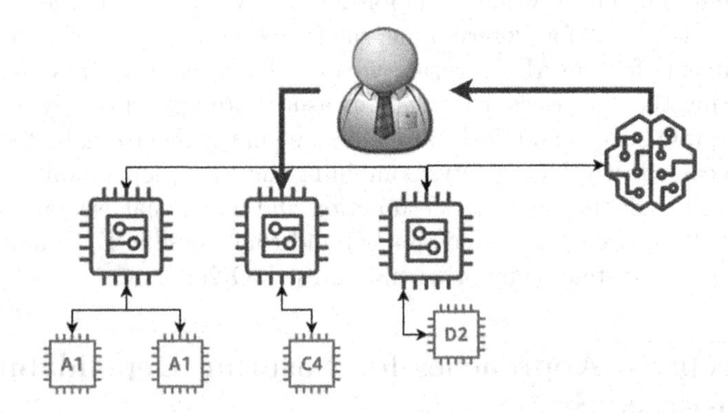

Fig. 2. Conceptual approach A for ensuring dependability of AI based systems.

3.1 Concept A: Human-in-the-Loop

This concept (depicted in Fig. 2) uses AI-based system to observe and analyse specific tasks or components and recommend human-readable actions. As a 'safe' decision gate, the human decides whether the AI recommendations should be applied and if so, then how they should be executed. The presence of the human in the loop enables application of traditional safety measures to warrant system safety. The analysis of complex situations and tasks is transferred to the AI algorithm, which frees up human resources, otherwise dedicated to the analysis.

The implication of potentially wrong decision being made by the AI algorithm (i.e., detection or non-detection of a critical situation) can potential violate the system safety, but are monitored by the human. Thus, the system does not operate autonomously because human intervention is required. Kesuma et al. [8] proposed a kit that utilises AI for data anomaly detection. If AI detects unexpected signal behavior, a human is notified. The AI can enhance the monitoring system by observing many sensor signals and signalling the operators if an anomaly is detected at any stage.

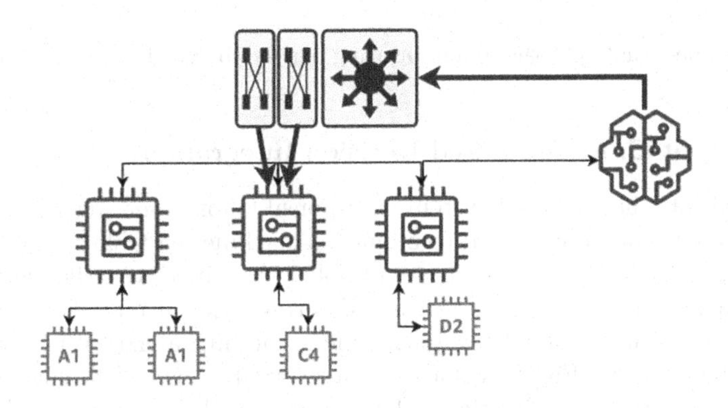

Fig. 3. Conceptual approach B for ensuring dependability of AI based systems.

3.2 Concept B: Policy-Based Decision Integration

This concept (Fig. 3) makes the AI-based system responsible for observing and analysing specific tasks or components and recommends machine-readable actions that can be translated into a finite set of policies and objectives. These policies and objectives are then used to influence the set-point generation of the safety-critical system domain. The finite set of policies and objectives (shown as two hard-wired icons in Fig. 3) can be analysed for safety, and traditional safety techniques can be applied to guarantee system safety. Thus, the system operates autonomously because no human intervention is required to integrate the actions recommended by the AI algorithm and the policy-based approach can be implemented in a resource-efficient manner. A wrong decision by the AI algorithm (i.e., detection or non-detection of a critical situation) could not violate system safety, since the defined policies and transition between the policies have to be intrinsically safe. Since the set of possible actions is limited to a finite number of policies and actions, the AI algorithm's capabilities might be restricted by this limitation, but the AI algorithm itself is not considered to be a safety-critical component.

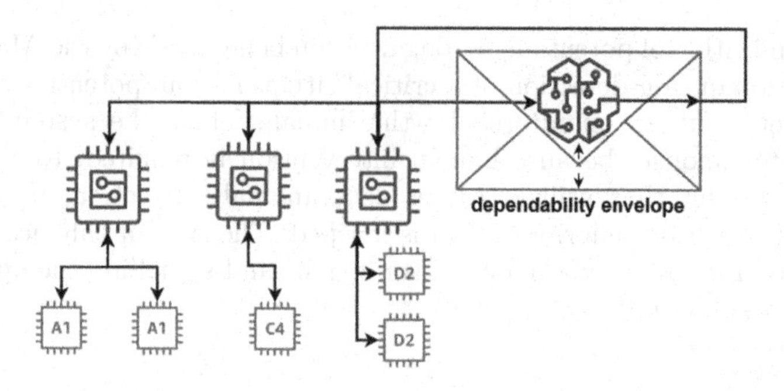

Fig. 4. Conceptual approach C for ensuring dependability of AI based systems.

3.3 Concept C: Model-Based Decision Integration

In this concept, the AI-based system is responsible for observing and analysing specific tasks or components and recommends machine-readable actions. Instead of mapping these actions to a finite set of policies or objectives, the model-based integration approach compares the non-deterministic output of the AI-based system with the output of a deterministic model running along with the AI-based system. The concept (Fig. 4) has also been referred to as 'safety envelope' [16]. The AI-based and deterministic models are designed for the same objectives, while the deterministic model is also designed to meet classic safety systems requirements. Hence, the deterministic model can be used to validate the AI-based system's output to ensure system safety.

The concept envisages AI-based system as a replacement of the human driver in fully autonomous vehicles. In such a use case, the system assumes responsibility for perception and interpretation of the vehicle environment for calculating the input values for the setpoint generator in every specific driving situation. Since the generated inputs have a critical impact on system safety, the vehicle vendor (i.e., the original equipment manufacturer) must therefore guarantee that AI-generated inputs do not violate system safety. That is problematic since the non-deterministic nature of AI-based systems makes them unverifiable with current state-of-the-art safety methods and standards.

The advantage of the system running the deterministic model is that it can be analysed for safety, and traditional safety techniques can be applied to guaranty system safety. As the deterministic model is less restrictive than the policy-based approach, the AI algorithm's capabilities are less restricted. The system operates autonomously as no human intervention is required to integrate the actions recommended by the AI algorithm. Thus, the AI algorithm's wrong decision does not violate system safety. Hence, the AI algorithm is not considered to be a safety-critical component.

The major drawback of this concept is dictated by the limitation of the deterministic model, which might restrict the capabilities of the AI algorithm. This means that two nearly equally sophisticated systems must be developed and the two (possibly) resource-intensive systems must be executed side-by-side in a synchronous manner.

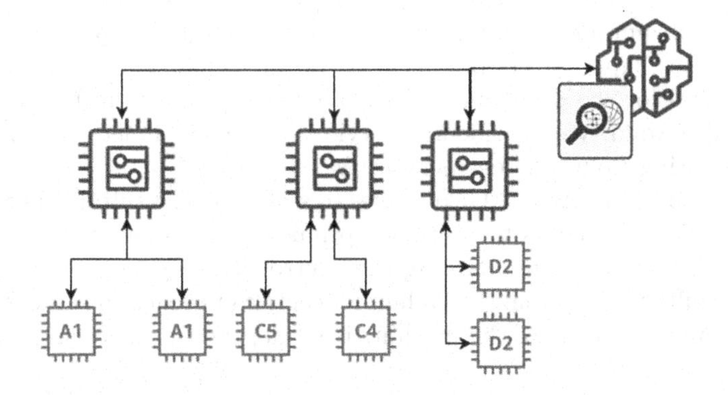

Fig. 5. Conceptual approach D for ensuring dependability of AI based systems.

3.4 Concept D: AI-Based System for Ensuring Dependability

This approach (Fig. 5) inverts assumptions for the application case. The AI-based algorithm is not seen as a potential source of harm for the dependability of the CPSoS but as an intelligent monitoring unit. It is used to monitor the conventional system. The AI algorithm learns the normal/expected system behaviour under real operating conditions without influencing the functionality and dependability of the system itself unless the system violates its specification. Consequently, AI algorithm enhances the dependability of systems through monitoring and learning the behaviour of a (dependable) system under observation. If AI detects abnormal behaviour, countermeasures are either recommended or automatically triggered. Equally, the same is valid for the real-time guarantees of the system. Furthermore, the reliability of the system under observation including the monitoring architecture shall not be lower than the system reliability of the system under observation on its own. Also, the source code of the system under observation shall not be modified by the monitoring architecture [7]. This concept is used in cyber-security applications for anomaly based network intrusion detection [2] or dynamic honeypots [6].

4 Conclusion and Outlook

The assurance of dependability, especially considering novel run-time adaptive AI-based approaches in the automotive domain, is still an open issue that lacks standard solutions for industrialisation. However, there is a necessity to establish the means of delivering a convincing and explicit affirmation that the systems under development are at the appropriate maturity level. To shine a light on possible paths for the dependability of AI systems in the automotive world, this paper presents (I) the body of knowledge of the TEACHING project and related regulatory activities, and (II) four perspectives on dependability architecture concepts and patterns for the adoption of continuously learning AI-based systems into dependable automotive applications. The presented conceptual dependability perspectives support the identification of key integrated process engineering approaches for the development of dependable products in the automotive domain and beyond. We provide an overview of the advantages and drawbacks that are resulting from different perspectives.

In that respect, TEACHING continues to seek the most suitable perspectives and the balance of benefits to continue optimising driving automation applications. Consequently, we are adapting the control strategy to the human in the loop, hence fulfiling the promise of a human-centric approach to driving automation where improvements of the machine itself depend on the human state.

Acknowledgment. The presented work is partially supported by TEACHING, a project funded by the EU Horizon 2020 research and innovation programme under GA n. 871385.

References

1. Armengaud, E., Peischl, B., Priller, P., Veledar, O.: Automotive meets ICT—enabling the shift of value creation supported by European R&D. In: Langheim, J. (ed.) Electronic Components and Systems for Automotive Applications. LNM, pp. 45–55. Springer, Cham (2019). https://doi.org/10.1007/978-3-030-14156-1_4
2. Ashok Kumar, D., Venugopalan, S.: Intrusion detection systems: a review. Int. J. Adv. Res. Comput. Sci. **8**, 356–370 (2017)
3. Bacciu, D., Chessa, S., Gallicchio, C., Micheli, A.: On the need of machine learning as a service for the Internet of Things. In: Proceedings of the 1st International Conference on Internet of Things and Machine Learning, IML 2017. Association for Computing Machinery, New York (2017)
4. Chatila, R., et al.: Trustworthy AI. In: Braunschweig, B., Ghallab, M. (eds.) Reflections on Artificial Intelligence for Humanity. LNCS (LNAI), vol. 12600, pp. 13–39. Springer, Cham (2021). https://doi.org/10.1007/978-3-030-69128-8_2
5. Cheng, C.-H., Huang, C.-H., Nuehrenberg, G.: nn-dependability-kit: engineering neural networks for safety-critical autonomous driving systems. In: Proceedings of Workshop on Artificial Intelligence Safety, SafeAI 2020 (2019)
6. Chowdhary, V., Tongaonkar, A., Chiueh, T.-C.: Towards automatic learning of valid services for honeypots, pp. 469–470 (2004)
7. Goodloe, A., Pike, L.: Monitoring distributed real-time systems: a survey and future directions. Communications in Computer and Information Science. National Aeronautics and Space Administration (2010)
8. Kesuma, H. et al.: Artificial intelligence implementation on voice command and sensor anomaly detection for enhancing human habitation in space mission. In: 2019 9th International Conference on Recent Advances in Space Technologies (2019)
9. Herníndez-Orallo, J.: AI safety landscape from short-term specific system engineering to long-term artificial general intelligence. In: 2020 50th Annual IEEE/IFIP International Conference on Dependable Systems and Networks Workshops (2020)
10. Hinrichs, T., Buth, B.: Can AI-based components be part of dependable systems? In: 2020 IEEE Intelligent Vehicles Symposium (IV), pp. 226–231 (2020)
11. ISO - International Organization for Standardization: ISO 26262 Road vehicles Functional Safety Part 1–10 (2011)

12. ISO - International Organization for Standardization: ISO/SAE CD 21434 road vehicles - cybersecurity engineering (under development)
13. ISO - International Organization for Standardization: ISO/WD PAS 21448 road vehicles - safety of the intended functionality (work-in-progress)
14. Henne, M., et al.: Benchmarking uncertainty estimation methods for deep learning with safety-related metrics. In: Invited paper - 2019 IEEE/ACM International Conference on Computer-Aided Design (ICCAD), pp. 83–90 (2020)
15. Ma, Y., Wang, Z., Yang, H., Yang, L.: Artificial intelligence applications in the development of autonomous vehicles: a survey. IEEE/CAA J. Autom. Sin. **7**(2), 315–329 (2020)
16. Macher, G., Druml, N., Veledar, O., Reckenzaun, J.: Safety and security aspects of fail-operational urban surround perceptION (FUSION). In: Papadopoulos, Y., Aslansefat, K., Katsaros, P., Bozzano, M. (eds.) IMBSA 2019. LNCS, vol. 11842, pp. 286–300. Springer, Cham (2019). https://doi.org/10.1007/978-3-030-32872-6_19
17. Montano, G., Mcdermid, J.: Effective naturalistic decision support for dynamic reconfiguration onboard modern aircraft (2012)
18. Houben, S., et al.: Inspect, understand, overcome: a survey of practical methods for AI safety. CoRR arXiv:abs/2104.14235 (2021)
19. Martínez-Fernández, S., et al.: Software engineering for AI-based systems: a survey. CoRR arXiv:abs/2105.01984 (2021)
20. The European Commission: White Paper on Artificial Intelligence: A European Approach to Excellence and Trust. European Commission (2020)
21. UNECE: Task force on Cyber Security and (OTA) software updates (CS/OTA). https://wiki.unece.org/pages/viewpage.action?pageId=40829521. Accessed 7 Sep 2019
22. Veledar, O.: New Business Models to Realise Benefits of the IoT Technology Within the Automotive Industry. WU Executive Academy, Vienna (2019)
23. Wahlster, W., Winterhalter, C.: Deutsche normungsroadmap künstliche intelligenz (2020)

Rule-Based Threat Analysis and Mitigation for the Automotive Domain

Abdelkader Magdy Shaaban[1], Stefan Jaksic[1(✉)], Omar Veledar[2], Thomas Mauthner[2], Edin Arnautovic[3], and Christoph Schmittner[1]

[1] AIT Austrian Institute of Technology GmbH, Vienna, Austria
{abdelkader.shaaban,stefan.jaksic,christoph.schmittner}@ait.ac.at

[2] AVL List GmbH, Graz, Austria
{omar.veledar,thomas.mauthner}@avl.com

[3] TTTech Computertechnik AG, Vienna, Austria
edin.arnautovic@tttech.com

Abstract. Cybersecurity is given a prominent role in curbing risks encountered by novel technologies, specifically the case in the automotive domain, where the possibility of cyberattacks impacts vehicle operation and safety. The potential threats must be identified and mitigated to guarantee the flawless operation of the safety-critical systems. This paper presents a novel approach to identify security vulnerabilities in automotive architectures and automatically propose mitigation strategies using rule-based reasoning. The rules, encoded in ontologies, enable establishing clear relationships in the vast combinatorial space of possible security threats and related assets, security measures, and security requirements from the relevant standards. We evaluate our approach on a mixed-criticality platform, typically used to develop Autonomous Driving (AD) features, and provide a generalized threat model that serves as a baseline for threat analysis of proprietary AD architectures.

Keywords: Security · Threat analysis · Ontology · Automated driving

1 Introduction

As a crucial automotive trend, connectivity is the key driver of the inevitable digital transformation [1]. This trend is further increased by technological developments, evolving customer needs and consolidating legislation. The resulting dilemma between evolution and extinction is forcing the adjustments to secure the existence [2]. Hence, from an organizational perspective, connectivity is an existential topic and no longer a matter of choice. It goes hand in hand with an important trend of Advanced Driver Assistance Systems (ADAS) and Autonomous Driving (AD), as indicated by the Society of Automotive Engineers [3]. The affiliated IoT solutions aid the dependable operation of systems in settings that, if manipulated, could result in fatal consequences. Security arises as a crucial aspect of sustainable deployment, as it is one of the key stumbling blocks

I. Habli et al. (Eds.): SAFECOMP 2021 Workshops, LNCS 12853, pp. 24–38, 2021.
https://doi.org/10.1007/978-3-030-83906-2_2

for widened user acceptance of driving automation. Hence, the autonomous systems must provide adequate security features that prevent cyberattacks and implement resilience to an ongoing attack.

Ensuring system security begins in the design phase by conducting a systematic analysis of potential security threats, using a method called threat modelling [4]. In dedicated tools such as ThreatGet [5] or Microsoft Visio [6] one defines an abstract model of threats to identify potential risks and security measures to prevent the attack. However, suppose the system does get compromised despite the security measures. In that case, one needs to define mitigation according to the security requirements typically defined by a standard (i.e., IEC 62443 [7]) or Common Criteria (CC) [8].

Unfortunately, the challenge arises when one tries to determine the applicable security mitigation. Proper mitigation would need to satisfy several constraints: (1) it needs to address a concrete threat related to the asset; (2) it must prescribe the security measure; and (3) it must comply with the security requirement addressing the threat. Satisfying these constraints properly is a demanding task due to its complexity and the level of expert knowledge needed. Currently, this task typically involves manual mapping between different entities (threats, assets, requirements, and measures), making it a non-formalized process, prone to inconsistencies due to ad hoc decisions by an expert. Consequently, it remains unclear which methodology should be pursued to track threats and determine relevant security requirements for threat mitigation.

As a solution, we introduce a rule-based approach that combines a threat analysis method and an automated inference to define an applicable set of security mitigation. After applying ThreatGet to identify potential threats, we use inference to determine security measures between potential threats and security requirements. This approach, which builds a mapping strategy, deduces a set of security requirements from a local knowledge base to identify which potential threats to address. A selected set of security requirements from the Common Criteria (CC) [8] is used to build up the knowledge base. Several classes, subclasses, data types, and data properties are specified for expressing domain knowledge of cybersecurity data to map cyber-threats to the most appropriate security requirements in the form of an ontology. The incorporation of the ontology into this work enhances the proposed technique by allowing greater scalability of the existing knowledge base. Finally, we demonstrate the benefits of our approach on an industrial, automotive case study: the *mixed-criticality platform* for developing ADAS. In this example, we check our approach for a particular vehicular asset. However, the proposed rule-based approach can be utilized on a bigger scale for selecting and prioritizing a set of security requirements for various vehicle assets. We identify two main contributions of this paper: (1) leveraging ontologies for establishing traceability in complex domains of security threats, assets, measures, and requirements, and (2) we provide a generalized threat model of a mixed-criticality platform which can serve as a foundation for security analysis of proprietary automotive architectures.

This paper is organized as follows; Sect. 1.1 presents related topics of this research. The structure of the rule-based approach is discussed in Sect. 2 and the architectural aspects of autonomous driving in Sect. 3. Threat modeling for the ADAS system is discussed in Sect. 4. The description among multiple entities in the ontology model is considered in Sect. 5. Section 6 discusses the findings of this proposed approach before the paper ends with a conclusion.

1.1 Related Work

The cybersecurity attacks on connected devices are well-documented in literature [9,10]. Protecting vehicles from such attacks is crucial [11] for improving user acceptance. Thus, it is necessary to apply security mechanisms to prevent cyber-attacks [12]. Threat modeling is an integral part of identifying potential threats and specifying corresponding security mitigation [13,14]. One threat modeling method is based on existing tools for different phases of the automotive development life cycle, such as concept phase, product development, production, and operation [15]. It is important to note that threat modeling integrates well into the development life-cycle based on security standards [16] such as IEC 62443 [7] standard for industrial security or the oncoming ISO/SAE DIS 21434 Automotive Cybersecurity Standard [17].

There are related approaches in threat modeling, as discussed in [18]. Authors in [19] presented a method for security requirements engineering of security based on ontology. Also, [19] presented a general security ontology for requirements engineering. An approach for security requirements verification and validation in the automotive domain is introduced in [20]. This approach utilizes reasoning rules to test the correctness of the applied security requirements and check whether any of these requirements are fulfilled. Our rule-based approach introduces a new practical approach for managing security issues and suggests an applicable set of security requirements that can handle and address them.

2 Structure of the Rule-Based Approach

Security weaknesses and requirements must be identified and clarified to develop a secure system infrastructure. It is essential to identify the most appropriate security mitigations to address security vulnerabilities in a system design. A protection profile (PP) is a method to establish a clear relationship and promote traceability between security requirements, security objectives, and threats. It represents a **Security Objectives Rationale** (i.e., *Threats*⟶Security Objectives), which defines a tracking path between threats and security objectives. Also, it defines **Security Requirements Rationale** (i.e., *Security Requirements*⟶Security Objectives) to represent a tracking of security requirements addresses with relevant security objective. Security requirements are defined in groups, where each group corresponds to a security objective. There has to be at least one objective for every security requirement, as discussed in [21]. However, it is unclear which methodological approach should

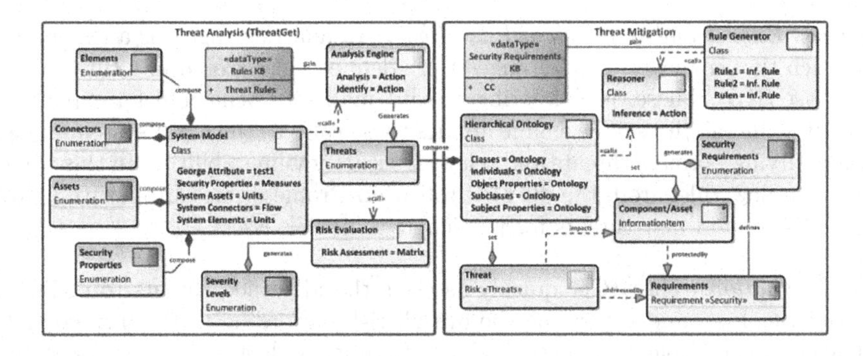

Fig. 1. The metamodel of the rule-based approach

be followed to track threats and relevant security requirements to achieve the main security objectives.

We tackle this problem by introducing a rule-based approach consisting of two phases: threat analysis and mitigation. The threat analysis discovers all security weaknesses in a system model and extracts potential threats and the relevant units affected by the threats. Based on this information, we build a knowledge representation of threats and affected units in the form of an ontology. The underlying semantic of the ontology allows to understand relationships between multiple ontology entities and creates a clear tracking path from threats to security requirements, thus promoting traceability. Figure 1 illustrates the metamodel structure of the proposed approach.

2.1 Threat Analysis with ThreatGet

The threat analysis identifies and investigates all expected negative actions that impact a system design or a part of it (i.e., component/asset). The threat model is required for defining security issues at early system development stages, define security requirements, and create secure applications [4]. For those reasons, the Austrian Institute of Technology (AIT)[1] developed ThreatGet [5], a threat modeling plug-in for Enterprise Architect (EA) [22]. ThreatGet analyzes threat models to perform threat and risk analysis [23] by leveraging domain knowledge encoded in an internal threat database, regularly updated with the latest threat data [5].

System Modeling. In this process, the system design is defined and structured entirely with all relevant security assumptions that need to be a part of a system design. ThreatGet offers a wide range of components and security properties that can be used to perform security analysis for a broad palette of scenarios.

[1] https://www.ait.ac.at/en/.

Analysis Engine and Rules Knowledge Base. An analysis engine is a rule-based approach that manages a wide range of pre-defined rules stored in a rule knowledge base (KB). These rules describe threats' malicious behaviour for exploiting security vulnerabilities. The engine applies these rules to the input system model for identifying possible threats and relevant security vulnerabilities. In this work, we create new rules from the "OTP Protection Profile of an Automotive Gateway" [21], to add new threats for particular vehicular assets.

Risk Evaluation. ThreatGet quickly assesses the identified threats to estimate the severity level and determine the overall risk, according to the likelihood of threats and the impact parameter values. Severity and risk are then stored in a risk matrix that allows a user to understand and analyze these values. As a result of this process, the tool returns the value of severity, as defined by ThreatGet (e.g., low, medium, high, and critical) [24].

2.2 Inferring Threat Mitigation

The threat mitigation phase takes the threat analysis results as input and infers a set of security measures with properties of security requirements. This phase relies on hierarchical ontologies, which encode relationships on different entities such as security threats, assets, measures, and requirements.

Hierarchical Ontology. In this process, we translate the outcomes of the threat analysis into an ontology form to clarify the main security issues in a system design and determine appropriate security requirements. A set of relationships are built among ontology entities to represent a clear understanding of affected units, exploited vulnerabilities, and identified potential threats.

Security Requirements KB. We build an ontology knowledge representation (KB) for selected security requirements based on the common criteria (CC). These requirements are used to handle existing security issues identified in the threat analysis process. By obtaining both threat and requirement ontology, we have the necessary ingredients for applying the rule generator [25].

Rule Generator. In this phase, we generate a set of inference rules based on Semantic Web Rule Language (SWRL) [26] according to the security properties exploited by threats. Each vehicular component/asset has a set of security properties defined as a set of protection mechanisms from different cyberattacks. These properties are defined as security vulnerabilities once a threat exploits them. Therefore, the rule generator determines the exploitation to define the common security issues in vehicular components/assets that match security measures from applicable security requirements.

Reasoner. Infers a group of security requirements according to the rule generator's logical rules. We use SWRLAPI [27] in this work; the API has a SWRLRuleEngine interface with a getOWLReasoner method [28]. The SWRLAPI is

Java-based for maintaining OWL-based SWRL rules and the (Semantic Query-Enhanced Web Rule Language) SQWRL [29] query statements. It uses OWLAPI to manage the OWL ontologies [27]. SWRLAPI provides a set of interfaces for creating rule-driven applications.

Mapping Algorithm. This process builds relationships between components/assets, threats, and security requirements to establish full traceability from threats to security requirements. The selected security requirements are defined after multiple properties are fulfilled, and the chosen ones are the most appropriate set of requirements to address existing security issues.

3 Architectural Aspects of Autonomous Driving

The intermediate step towards fully automating AD systems is the ADAS, which are designed to either eliminate the possibility of a safety-critical event or, in the case of its occurrence, to minimize the impact. Figure 2 depicts a high-level AD System Architecture mapped to a concrete hardware architecture by allocating functions to hardware components. Sensor data (radars, Time-Of-Flight (TOF) cameras, LIDARs, etc.) are fused to create a (static and dynamic) model of the environment. Control algorithms implemented in such a system compute driving strategy to control steering, braking, and powertrain. Additional ADAS functions such as Automated Emergency Braking (AEB), lane assistance, and surround are also commonly supported.

In addition, modern automotive connected systems must guarantee dependable functionality for high-performance cyber-physical systems. Such requirements carry a potentially conflicting undertone in the sense that the development drivers are pushing for the extremely powerful performance, while the dependability aspect is creating a limiting wrapper around this solution-seeking computational system. This is evident in the available systems-on-chip, which are predominantly either highly specialized and offer high computing performance (e.g., with multi-core, multi CPUs on a single chip, GPUs), or highly focused on compliance with the relevant safety standards (e.g., Lockstep CPU cores with clock delay, safety management unit, clock and voltage monitors). In order to offer both high-performance and safety features to applications, the necessary feature of a hardware platform for AD is called the *mixed criticality*. Mixed-criticality systems execute applications with different criticality levels and provide guarantees that the applications of different criticality do not interfere.

ADAS development methods assume several approaches, including Model-Based Design, Driving Simulators, and Virtual Platforms for simulating the environment on different levels [30]. However, these development methods either abstract away the complexity of HW/SW interaction or offer only limited simulation speed, limiting their usability. To overcome these limitations, ADAS are typically developed on a dedicated *mixed-criticality* HW platform [31,32], which enable testing of ADAS at sufficient speed and level of detail early on in the development process. One of the contributions of this paper is the generalized threat

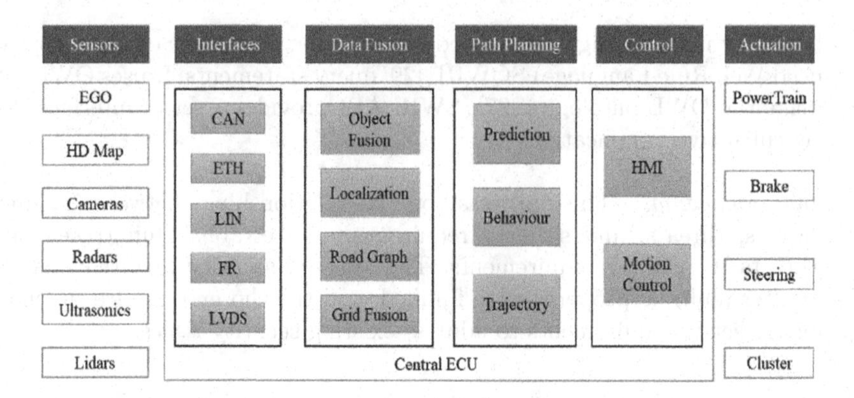

Fig. 2. General automated driving architecture

model of a *mixed-criticality* platform for rapid prototyping of ADAS, which we hope will serve practitioners as a baseline threat model for their proprietary AD architectures.

4 Case Study: ADAS Development Platform

In this work, we define a threat model to encapsulate the security properties of classes of protocols, interfaces, and hardware blocks found in mixed-criticality platforms. This way, we build a general yet sufficiently verbose model which serves as a basis for further specification and extension for a target ADAS. In this model, we introduce the following building blocks:

- COMM_IF: This interface represents the main communication interface from the mixed-criticality platform to the vehicle. As concrete instances of this generic interface, we foresee CAN, Ethernet, and FlexRay interface.
- DEBUG_IF: The ADAS developer uses the debug interface for debugging purposes. Depending on the concrete ADAS function, we can instantiate UART, USB, JTAG, and similar interfaces. The common properties of such interfaces are encapsulated in the current DEBUG_IF element.
- VIDEO_IF: This element represents a bidirectional video interface. On one hand, the video data can be sent from the camera to the Performance CPU to be processed or analyzed, typically over an automotive-grade Gigabit Multimedia Serial Link (GMSL) interface or similar. For example, we can have data from the front view or surround-view camera or a driving monitoring camera. The video output can display the processed or augmented video on a vehicle dashboard or infotainment system, i.e., over an FPD-Link.
- The included sensors are the most relevant ones for monitoring the external environment: Lidar, Radar, and Camera-based sensors. Additionally, we include ECUs that control the vehicle's actuators, i.e., braking and steering.
- The model contains a set of assets that are considered valuable targets for attackers [21], namely:

- HSM2Gateway: Data that exchange between the gateway and the hardware security module.
- Firmware: A set of instructions for managing the control of the Target of Evaluation (TOE).
- Firmware Update: An updated version of firmware to keep the TOE up-to-date with the latest changes.
- Configuration Data: A set of files containing data for configuring the transmission/reception characteristics and type of messages.
- IVN Message: Information transmitted over the In-Vehicle Network interface (IVN).
- ITS Message: Any compliant message exchange between the Intelligent Transport System (ITS) and other TOEs or ITS-Stations.

Each vehicular unit defined in ThreatGet has a set of security measures used as a set of protective mechanisms against different attack scenarios. These properties can be managed and set by the system architect to provide a set of risk mitigation actions against potential threats. For example, a Deterministic Ethernet unit shall provide authentication security mechanisms to mitigate spoofing attacks from unreliable or untrusted source units. In addition, some cyber attacks could be initiated through internal interfaces of the vehicle, such as the debug interface, as is depicted in Fig. 3. An attacker could load malicious code through interfaces. Therefore, some security mechanisms need to be applied, such as input validation to check the correctness of the incoming data to validate the data is out of any malicious content. For instance, the V2X unit has a collection of security properties identified and managed by ThreatGet to represent a set of security measures that should be part of that unit to make a defensive mechanism against various cyberattacks. The absence of these security properties would lead to an increase in cyber-attack incidents. V2X should have access control, authentication, authorization, low communication latency, reliable communication, input validation, the security of data integrity, and secure booting. These protection properties are considered some of the security mechanisms that shall be identified within the V2X unit to establish a protective mechanism against alternative cyberattacks. This will be demonstrated in the next chapter, hence, illustrating the impact of security properties in increasing/decreasing cyberattack incidents.

We have already mentioned the diverging requirements of safety and performance in modern autonomous vehicles and that we can reconcile these conflicting requirements by implementing our system on a mixed-criticality platform. In our threat model, we introduce typical representative components of such a platform: Safety CPU or a microcontroller, a Performance CPU as well an internal communication switch, in this case, based on Deterministic Ethernet.

Safety CPU is typically a Lockstep CPU with additional features to detect runtime faults such as error-correcting code and self-checking monitors to ensure lockstep operation. To provide increased reliability, a lockstep CPU has one (or more) redundant unit which performs the same task on the same data. These CPUs are favorably designed and proven to satisfy safety standards such as

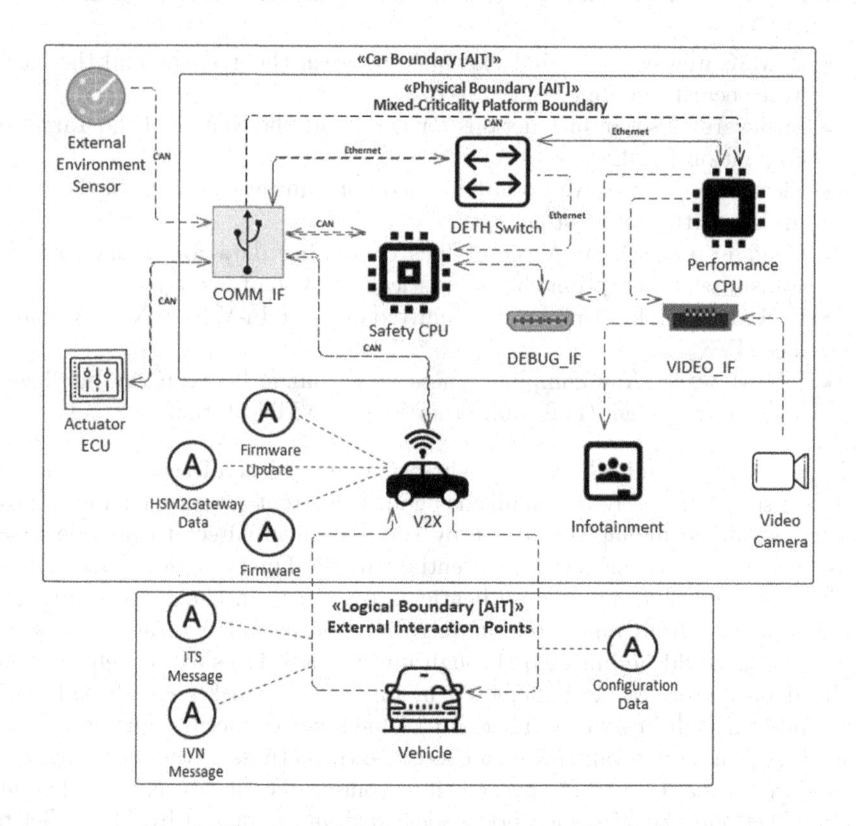

Fig. 3. Threat modeling of a mixed-critically platform

ISO 26262 [33]. Performance CPU can be any dedicated CPU hardware for data processing, i.e., a DSP core or a GPU. DETH Switch is implementing Time-Scheduled Ethernet Standard (Deterministic Ethernet), which provides communication latency guarantees to the standard IEEE 802.3 Ethernet. This way, it is possible to transmit larger amounts of data than using a traditional CAN bus and support the safe integration of dedicated components running safety-critical applications (i.e., Safety CPU).

5 Establishing Relationships with Ontologies

Figure 4 depicts four different elements of our ontology: *integrity* as a security property, *T19_1* as a threat resulting from our threat analysis, *FDP_SDI_2* a security requirement from Protection Profile and an *IVN_Message* which is the asset of our model.

Tracing security issues in a system model to determine applicable security requirements for addressing potential risks is considered quite challenging. It takes much time and effort by an expert. Therefore, the proposed approach automates this process by defining and creating relationships among multiple

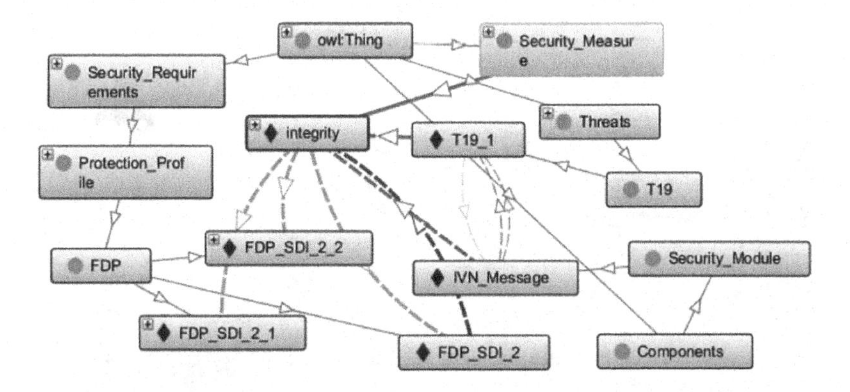

Fig. 4. Establishing relationships between different entities in the ontology model

entities in the ontology model to specify security requirements that address existing security issues. The rule-based approach defines and creates relationships among multiple entities in the ontology model to select a security requirement (e.g., FDP_SDI_2) according to common criteria for a particular threat (e.g., T19_1). The algorithm detects that the T19_1 exploits the data_integrity property, which is defined as protecting the asset (e.g., IVN_Message). Once a threat exploits this property, it is considered a vulnerable security point that needs more concern. Therefore, the proposed approach applies a set of rules to deduce new relationships that satisfy multiple properties to select the most appropriate security requirement (e.g., FDP_SDI_2) that could fit for addressing the threat (e.g., T19_1) and protecting the asset (e.g., IVN_Message). By automating the processes of understanding the internal relationships between the affected units, potential threats, and existing security vulnerabilities to specify an appropriate set of security requirements for mitigating potential cyber risks, we save a lot of time and effort of the expert.

6 Results

Here we present the concrete results of threat analysis and rule-based approach for mitigation inference and discuss its benefits.

6.1 Threats and Risks Discovered

ThreatGet discovers a considerable number of potential threats in our generalized threat model. Each unit, asset, and communication flow in our model has a set of security properties, which are used to mitigate risks. We have not initialized any applied security properties to any components, assets, and communication flow in this experiment. The tool detects 202 potential threats that impact this system design and evaluates each threat's risk to estimate the design's overall risk. The results of the analysis are presented in Fig. 5.

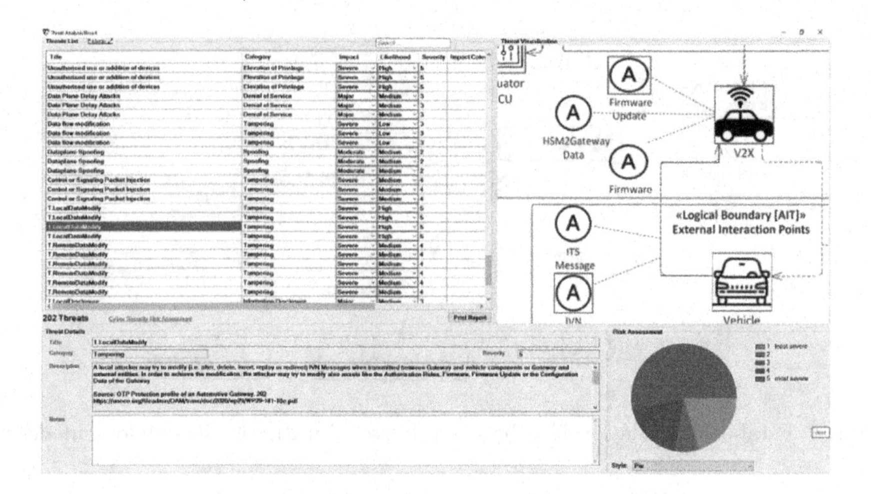

Fig. 5. The discovered threats and risks according to the security vulnerabilities

The figure shows that the selected threat "T.LocalDataModify" is identified by ThreatGet, which violates the integrity of the V2X and "IVN Message". Therefore, the threat is classified as a "Tampering" threat category. According to Microsoft's STRIDE model [4], the identified threats are classified, where each has a level of impact and likelihood parameter values. These parameters are used to estimate each threat's severity level to define the actual risk level. This process is essential to determine which risk could be mitigated or accepted. ThreatGet handles this process automatically. It evaluates these parameters to estimate an approximate severity risk level for each threat identified. Then the overall risk of the whole model is evaluated to determine the security of the model.

The integrity violation of the V2X unit and relevant critical assets is considered the critical security issue that needs to be considered in this system model. ThreatGet detects 42 threats that have a direct impact against the V2X and its assets. The risk severities of all V2X and assets threats are estimated, that ranging from severity level (Level-1) (i.e., low) to (Level-5) (i.e., critical). About 57% of these threats is estimated as a Level-5, where 12% of results is estimated as Level-1. In this work, we focus on the Level-5 threats, which are considered to have a high impact against the whole system model.

6.2 Traceability from Threats to Security Requirements

In order to address the current security issues, we need to specify the most appropriate security requirements able to address the malicious behavior of identified threats and perform risk mitigation accordingly. Therefore, we build an ontology representation model of all threats and relevant affected vehicular units (i.e., components and assets). We apply a set of inference rules to this model to deduce applicable security requirements. These rules automatically build a complete tracking path from threats to relevant security requirements, giving concise traceability from existing security issues to relevant solutions. Furthermore, we developed a Java algorithm based on SWRLAPI to automate the traceability process among multiple ontology entities and provide a graphical interface for demonstrating outcomes, as shown in Fig. 6. The threat "T.LocalDataModify" affects five assets (i.e., Firmware_Update, IVN_Message, Firmware, Authorisation_Rules, and Configuration_Data) are defined in the model, as illustrated in Fig. 6.

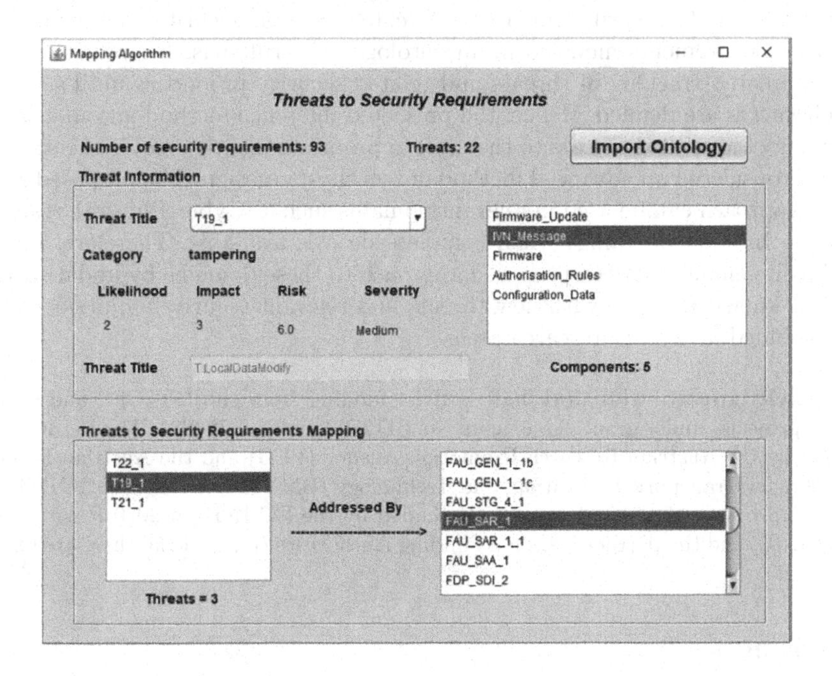

Fig. 6. Threats to security requirements mapping approach

In this example, the "IVN_Message" is selected to see how the proposed mapping approach could make the traceability from threats to security requirements for the affected unit (i.e., IVN_Message). The "**Threats to Security Requirements Mapping**" section gives more insight into all threats that impact the IVN_Message. According to the "T19_1" all inferred security requirements are listed on the right side, which defines a group of requirements selected by the proposed approach suggested to be a part of the affected asset to protect it from different potential threats. Outcomes of this process are illustrated in Fig. 4, which displays a security requirement "FDP_SDI_2" is inferred to address "T19_1" and protects the "IVN_Message".

In Fig. 4 we can observe four different elements of our ontology: *integrity* as a security property, *T19_1* as a threat resulting from our threat analysis, *FDP_SDI_2* a security requirement from Protection Profile and an *IVN_Message* which is the asset of our model.

7 Conclusion

In this work, we present an end-to-end approach for performing threat modeling and automating mitigation inference. To perform a thorough case study of our approach, we modeled a generalized AD architecture based on a mixed-criticality platform, typically used to develop autonomous driving applications. After obtaining the security threats, we apply the rule-based approach and automatically infer mitigation. We achieve this goal by leveraging the relationships between possible security threats and related assets, security measures, and security requirements modeled in an ontology. The rule-based approach allows for automated tracking of threats and related security properties until security requirements are defined. Hence, the presented inference method automates the mapping of identified threats to the most appropriate security requirements, and can be considered an advanced method of security-by-design. The proposed rule-based approach can be used in relevant domains such as Cyber-Physical Systems and the Internet-of-Things when complete data is available. Therefore, future work will demonstrate the proposed approach to these domains by updating the current knowledge base with new threats and relevant security requirements to be combined into other research areas.

Acknowledgment. This work has received funding from the IoT4CPS and AFarCloud projects, under grant agreements No. 6112792 and No. 783221. The IoT4CPS is funded by the Austrian Research Promotion Agency (FFG) and the Austrian Federal Ministry for Transport, Innovation, and Technology (BMVIT), within the "ICT of the Future" project. AFarCloud is partially funded by the EC Horizon 2020 Programme, ECSEL JU, and the partner National Funding Authorities (for Austria, these are bmvit and FFG).

References

1. Armengaud, E., Peischl, B., Priller, P., Veledar, O.: Automotive meets ICT—enabling the shift of value creation supported by European R&D. In: Langheim, J. (ed.) Electronic Components and Systems for Automotive Applications. LNM, pp. 45–55. Springer, Cham (2019). https://doi.org/10.1007/978-3-030-14156-1_4
2. Veledar, O.: New business models to realise benefits of the IoT technology within the automotive industry. Master's thesis, WU Executive Academy (2019)
3. SAE J3016: Taxonomy and definitions for terms related to driving automation systems for on-road motor vehicles. Technical Report, SAE International (2018)
4. Shostack, A.: Threat Modeling: Designing for Security. Wiley (2014). OCLC: ocn855043351
5. ThreatGet: Threat analysis and risk management. https://www.threatget.com/. Accessed 9 Feb 2021
6. Microsoft Threat Modeling Tool. https://www.microsoft.com/en-us/download/details.aspx?id=49168. Accessed 9 Feb 2021
7. ISA: The 62443 series of standards: industrial automation and control systems security (1–4) (2018)
8. ISO 15408: Information technology - security techniques - evaluation criteria for IT security Common Criteria - part 1: introduction and general model (2009)

9. Zhao, K., Ge, L.: A survey on the internet of things security. In: Ninth International Conference on Computational Intelligence and Security, pp. 663–667. IEEE (2013)
10. Veledar, O., et al.: Safety and security of IoT-based solutions for autonomous driving: architectural perspective (2019)
11. Burzio, G., Cordella, G.F., Colajanni, M., Marchetti, M., Stabili, D.: Cybersecurity of connected autonomous vehicles: a ranking based approach. In: 2018 International Conference of Electrical and Electronic Technologies for Automotive, pp. 1–6. IEEE (2018)
12. Shaaban, A.M., Schmittner, C., Gruber, T., Mohamed, A.B., Quirchmayr, G., Schikuta, E.: Ontology-based model for automotive security verification and validation. In: Proceedings of the 21st International Conference on Information Integration and Web-based Applications & Services, pp. 73–82 (2019)
13. Ruddle, A., Ward, D., Weyl, B., et al.: Security requirements for automotive onboard networks based on dark-side scenarios. EVITA Deliverable D **2**(3) (2009)
14. Islam, M., et al.: Deliverable d1.1 needs and requirements. HEAVENS Project (2016)
15. Ma, Z., Schmittner, C.: Threat modeling for automotive security analysis. Adv. Sci. Technol. Lett. **139**, 333–339 (2016)
16. Shaaban, A.M., Kristen, E., Schmittner, C.: Application of IEC 62443 for IoT components. In: Gallina, B., Skavhaug, A., Schoitsch, E., Bitsch, F. (eds.) SAFECOMP 2018. LNCS, vol. 11094, pp. 214–223. Springer, Cham (2018). https://doi.org/10.1007/978-3-319-99229-7_19
17. Macher, G., Schmittner, C., Veledar, O., Brenner, E.: ISO/SAE DIS 21434 automotive cybersecurity standard - in a nutshell. In: Casimiro, A., Ortmeier, F., Schoitsch, E., Bitsch, F., Ferreira, P. (eds.) SAFECOMP 2020. LNCS, vol. 12235, pp. 123–135. Springer, Cham (2020). https://doi.org/10.1007/978-3-030-55583-2_9
18. Shevchenko, N.: Threat modeling: 12 available methods 2018. https://insights.sei.cmu.edu/sei_blog/2018/12/threat-modeling-12-available-methods.html. Accessed 26 Apr 2021
19. Souag, A., Salinesi, C., Wattiau, I., Mouratidis, H.: Using security and domain ontologies for security requirements analysis. In: 2013 IEEE 37th Annual Computer Software and Applications Conference Workshops, pp. 101–107. IEEE (2013)
20. Magdy, A., Schmittner, C., Gruber, T., Baith Mohamed, A., Quirchmayr, G., Schikuta, R.: Ontology-based model for automotive security verification and validation. In: 21st International Conference on Information Integration and Web-Based Applications and Services, iiWAS 2019, December 2019
21. Bartsch, M., Bobel, A., Niehöfer, B., Wagner, M., Wahner, M.: OTP Protection profile of an automotive gateway (2020). https://unece.org/fileadmin/DAM/trans/doc/2020/wp29/WP29-181-10e.pdf
22. Enterprise Architect (2021). http://www.sparxsystems.com/. Accessed 4 Apr 2021
23. El Sadany, M., Schmittner, C., Kastner, W.: Assuring compliance with protection profiles with ThreatGet. In: Romanovsky, A., Troubitsyna, E., Gashi, I., Schoitsch, E., Bitsch, F. (eds.) SAFECOMP 2019. LNCS, vol. 11699, pp. 62–73. Springer, Cham (2019). https://doi.org/10.1007/978-3-030-26250-1_5
24. Shaaban, A.M., Schmittner, C.: ThreatGet: new approach towards automotive security-by-design. In: 28th Interdisciplinary Information Management Talks, pp. 413–419 (2020)
25. Ontotext: What are ontologies? (2018). https://ontotext.com/knowledgehub/fundamentals/what-are-ontologies/. Accessed 19 Feb 2021
26. W3C Member: SWRL: a semantic web rule language (2004). https://www.w3.org/Submission/SWRL/. Accessed 26 Apr 2021

27. O'Connor, M.J.: SWRLAPI (2019). https://github.com/protegeproject/swrlapi. Accessed 8 Feb 2021
28. O'Connor, M.J.: Reasoner not yet wired up (2020). https://github.com/protegeproject/swrlapi/issues/65. Accessed 8 Feb 2021
29. O'Connor, M.J., Das, A.K.: SQWRL: a query language for OWL. In: OWLED, vol. 529 (2009)
30. Bücs, R.L., Lakshman, P., Weinstock, J.H., Walbroel, F., Leupers, R., Ascheid, G.: Fully virtual rapid ADAS prototyping via a joined multi-domain co-simulation ecosystem. In: VEHITS, pp. 59–69 (2018)
31. RazorMotion (2021). https://www.tttech-auto.com/products/automated-driving/razormotion-tttech-auto/. Accessed 8 Feb 2021
32. D'Amato, A., Pianese, C., Arsie, I., Armeni, S., Nesci, W., Peciarolo, A.: Development and on-board testing of an ADAS-based methodology to enhance cruise control features towards CO_2 reduction. In: 2017 5th IEEE International Conference on Models and Technologies for Intelligent Transportation Systems (MT-ITS), pp. 503–508. IEEE (2017)
33. International Organization for Standardization: ISO 26262:2018 road vehicles - functional safety (2018)

Guideline for Architectural Safety, Security and Privacy Implementations Using Design Patterns: SECREDAS Approach

Nadja Marko[1], Joaquim Maria Castella Triginer[1](✉), Christoph Striecks[2], Tobias Braun[3], Reinhard Schwarz[3], Stefan Marksteiner[4], Alexandr Vasenev[5], Joerg Kemmerich[6], Hayk Hamazaryan[6], Lijun Shan[7], and Claire Loiseaux[7]

[1] Virtual Vehicle, Graz, Austria
{nadja.marko,joaquim.castellatriginer}@v2c2.at
[2] AIT Austrian Institute of Technology, Vienna, Austria
christoph.striecks@ait.ac.at
[3] Fraunhofer IESE, Kaiserslautern, Germany
{tobias.braun,Reinhard.Schwarz}@iese.fraunhofer.de
[4] AVL, Graz, Austria
stefan.marksteiner@avl.com
[5] Joint Innovation Centre ESI (TNO), Eindhoven, The Netherlands
alexandr.vasenev@tno.nl
[6] ZF Friedrichshafen AG, Friedrichshafen, Germany
{joerg.kemmerich,hayk.hamazaryan}@zf.com
[7] Internet of Trust, Paris, France
{lijun.shan,claire.loiseaux}@internetoftrust.com

Abstract. Vehicle systems engineering experiences new challenges with vehicle electrification, advanced driving systems, and connected vehicles. Modern architectural designs cope with an increasing number of functionalities integrated into complex Electric/Electronic (E/E) systems. Such complexity is extended, adding V2X (Vehicle-to-everything) communication systems, which provide remote communication services that collect, store, and manipulate confidential data. The impact on Safety, Security, and Privacy (SSP) of these new advanced technological systems requires the implementation of new processes during their development phase. Therefore, new product development strategies need to be implemented to integrate SSP mechanism across the entire product development lifecycle. The European H2020 ECSEL project SECREDAS proposes an innovative solution for Safety, Security and Privacy specifically for automated systems. The project outlines the shortcomings of existing SSP approaches and proposes its own approach to implementing SSP mechanism for the emerging technologies. This approach includes a reference architecture with SSP features implemented by a set of reusable Design Patterns (DPs) along with their associated technology elements. This guideline proposes rules for developing new architectural Safety, Security, and Privacy implementations in a product under development using Design Patterns.

Keywords: Safety · Security · Privacy · Design patterns · Systems engineering · Automated systems · Connected vehicles

I. Habli et al. (Eds.): SAFECOMP 2021 Workshops, LNCS 12853, pp. 39–51, 2021.
https://doi.org/10.1007/978-3-030-83906-2_3

1 Introduction

The automotive industry is constantly facing new challenges. Traditional mechanical components are being replaced by software-controlled ones, while vehicles functionalities are continuously enriched. To keep such functionalities up to date, manufacturers need to update their vehicular software frequently, prompting them to introduce over-the-air software updates [1]. Furthermore, new cars provide Internet services and communicate continually with smart devices and future connected cars will communicate with other cars and road infrastructures for improved road safety [2].

These new functionalities and associated technologies require attention for Safety, Security, and Privacy (SSP), which strongly affects systems engineering and development of vehicles. Over the last fifty years, the number of software-controlled components has increased significantly [3]; in fact, a modern vehicle contains more than 100 electronic control units with a total of about three million software functions [4]. To cope with the increasing complexity in the electrical and electronic (E/E) system design and the increasing demands on functionality safety, the automotive industry needs suitable tools. Besides safety, strongly interconnected systems such as driving assistance must also consider security threats from malicious agents. Fully autonomous driving will need to master V2X (vehicle-to-everything) communication to interact with intelligent transport infrastructures. Remote communication between these systems may encourage cyber-attacks with serious consequences in various areas such as security, mobility, and privacy [5]. Therefore, the collection, storage and communication of data (e.g. via V2X communication) must meet basic security and privacy requirements to ensure protection against data misuse and attacks [6]. To achieve such demanding levels of safety and cybersecurity, a range of analysis methods are required to ensure fail-operational E/E and trustworthy cyber-physical system architectures.

Reuse of components, subsystems, and partial solutions provides clear benefits: relying on best practices and already tested elements, developer can reduce errors and costs, simultaneously improving the time to market. Yet, the reuse task is far from trivial. Generic solutions have their scope and assumptions, often formulated implicitly. A question how to ensure adequate specialization of a generic element remains open, as standards provide little guidance on this aspect. One approach is to focus on assurance and information about a component to build safety arguments, e.g., with Goal Structured Notation [7]. This can assist in integration of safety-relevant components into individual systems. Another direction can be to properly classify different patterns to explicated assumptions and assist their reuse [8]. This paper complements standards and the mentioned directions by focusing on development processes. In particular, we pay attention to the need to tailor and iteratively assess design patterns according to a specific decision tree. In this way, we highlight the iterative nature of creating, adapting, and using design patterns, which is common in development companies.

The European H2020 ECSEL project SECREDAS[1] aims at innovative solutions for Safety, Security and Privacy specifically for automated systems. SECREDAS outlines

[1] https://secredas-project.eu/.

the shortcomings of existing SSP approaches and proposes its own approach to implementing SSP mechanisms for the emerging technologies. This approach includes a reference architecture with SSP features implemented by a set of proposed reusable Design Patterns (DPs) along with their associated technology elements. A DP is a best-practice template for a specific technical solution that allows developers to easily find solutions to common design problems. DPs are reusable, adaptable to specific needs within a given context, and engineering domain independent, reducing development time and effort. For example, a security DP may describe how technology elements, such as authentication, cryptographic libraries and secure elements can be employed to implement a secure V2X communication, or a security DP may be an authentication protocol for authenticating mobile devices in vehicle architectures. Technology elements can be considered the most basic technical building blocks for DPs and are domain-independent technologies at Technology-Readiness Level 7 or above. Technology elements are, for example, transport layer security (TLS), authentication and authorization mechanisms, or firewalls.

Developers who wish to use the SECREDAS approach may need assistance in understanding what steps to follow when developing a specific product. Therefore, this guideline provides rules for developing new Safety, Security and Privacy solutions using Design Patterns. The selection, collection and classification of the DPs should be done by the user, as described in the SECREDAS documentation [8].

The guideline supports the major steps during the development process, since DPs are applicable on multiple levels of abstraction (See Fig. 1). The levels of abstraction help identify the correct application within the development lifecycle. For example, the SECREDAS approach includes different groups of DPs based on technology elements that represent specific technology implementations at system level as well as at hardware or software level. It also includes other types of DPs based on technology elements that represent more abstract implementations spanning multiple levels of product development, such as the concept phase or the system level, including hardware or software layer (e.g., protocol specifications or mathematical models).

This guideline is based on the notion of development of [Safety] Elements out-of-context (SEooC), as proposed in ISO 26262 — Part 10: Guidelines on ISO 26262 — Chapter 9: "Safety Element out of Context" [9]. In SECREDAS, the SEooC concept has been generalized to Safety, Security, and Privacy elements. The ISO 26262 standard defines an SEooC as a safety-related element that is not developed for a specific item and not developed in the context of a particular vehicle. Therefore, SEooC development is based on assumptions and constraints regarding the operational context and intended use of the element, which translate into corresponding requirements during development and later on into obligations for the integration of the element into a particular system architecture. Similar to our DP approach, an SEooC can be a system, a combination of systems, a subsystem, a software component or a hardware component.

The ISO 21434 standard follows a similar approach [10]. ISO 21434 addresses the cybersecurity perspective in the engineering of E/E systems within road vehicles. In the context of this guideline, ISO 21434, Chapter 6.4.5: "Component Out of Context" is most relevant. According to the standard, items developed based on an assumed context represent generic components for different applications and for different customers. The supplier can make assumptions about the context and intended use of a component and derive requirements for the out of context development.

The Common Criteria (Common Criteria, Part 3: CAP (Composition Assurance Package) with Security Assurance Requirements class ACO (Assurance Composition)) [11] address the related problem of deriving security assurance for a composite based on assurance available for each of its constituents. CAP aims at mitigating the fundamental problem that security is not invariant under composition for well-defined special cases. However, the Common Criteria point out that the security assurance achievable with ACO is rather limited. Ultimately, the integrator must check that both functional and security interfaces are compatible and that the security properties of the assembled product are met.

This guideline follows a similar logic by adopting SEooC-like generic elements. Technology elements and designs are developed out of context based on assumptions and constraints, so they can be applied in different products (given, that the application context meets the assumptions and constraints). The SECREDAS approach extends SEooC-like generic elements for SSP and relates them to continuous improvement opportunities of DPs.

Note that the application of DPs by itself cannot guarantee the Safety, Security, and Privacy of the system under development. However, DPs are not intended to replace standardized SSP assessment activities that developers have to carry out throughout the entire product lifecycle. Therefore, this guideline provides rules to ensure that DPs and technology elements are developed with high quality and that they are used as intended and properly verified and validated after their application within a particular product.

2 The Guideline

This guideline proposes rules for developing new architectural Safety, Security, and Privacy implementations in a product under development using Design Patterns. This approach is developed in the context of the SECREDAS project. The guideline can be applied during product development of a system, but it does not replace any other activities that a developer must perform during the development of a product (see Fig. 1). At the same time, the product development should be adapted for the use of DPs and technology elements. For example, at the concept phase, it is important to early define functions and assets to find elements that potentially could link DPs to the product being developed. Furthermore, it is also important to know hazards/threats related to such assets/elements as well as further information such as requirements, designs, etc.

DPs and technology elements are defined for the fulfillment of specific requirements with an intention and motivation defined as a solution to a common problem. These requirements and their allocation to the architectural design are the bases for the guideline. Nevertheless, DPs and technology elements template contain a well-defined structure that accurately describe all the properties, context and constraints needed to use a DP. In the activity 1.1 "Design Patterns and Technology Elements selection" you can find the relevant attributes for the DPs and technology elements.

As an observation, remember that DPs and technology elements are created for the purpose of reuse. For this reason, all the DPs are structured with the same generic template and framed into a common reference architecture [8].

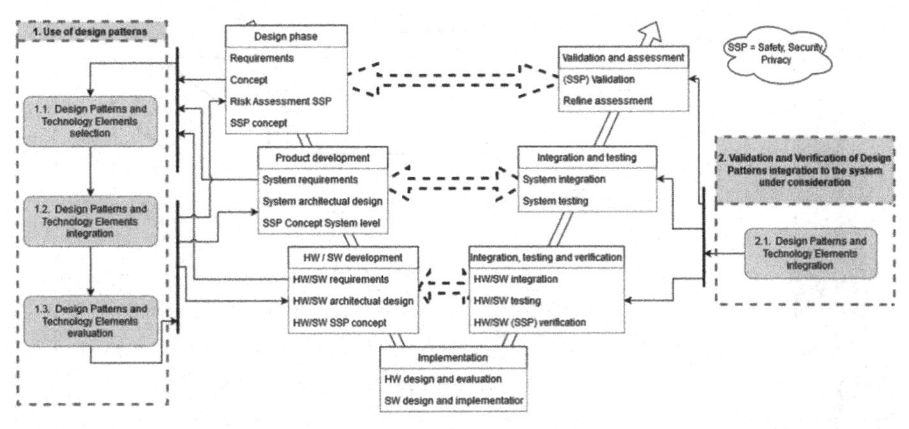

Fig. 1. Guideline for architectural implementations or extensions via new or existing design patterns.

Activity 1: Use of Design Patterns and Technology Elements

This activity provides the necessary steps to use DPs and technology elements for the product being developed divided into three main sub-activities (see Fig. 2). Activity 1.1 describes the required steps to select DPs and technology elements. Activity 1.2 analyzes the integrations of DPs and technological elements in the product being developed. Finally, Activity 1.3 evaluates the risk of such integrations and provides a final result.

As a preliminary action, the guideline requires the definition of functions and assets where DPs could potentially be used for the product being developed. Once these are defined, potential connected hazards/threats related to such assets must be defined. Note that this guideline supports multiple phases of the development process, which means that the guideline can be started at multiple levels of the product development, such as the concept phase or the system level, including hardware or software levels.

In addition, there are two other important steps before jumping into the selection of DPs and technology elements. The first one is the reference architecture. Such functions and assets must be clearly indicated where they are allocated in the reference architecture and which design specifications are required. The second relates to requirements. It is important to interconnect the requirements between the product being developed and the DPs and technology elements.

Activity 1.1: Design Patterns and Technology Elements Selection

Design Patterns and Technology Elements selection starts with finding DPs that could potentially be used to develop the new product or system. This selection interconnects product or system requirements and their allocation to the architectural design between the product being developed and the potential preliminary selections. To support such interconnection, DPs and technology elements contain a well-defined structure that accurately describes all the properties and context required for their (re)use in a generic product. Such relevant attributes are described in the deliverable "D3.6 Design Patterns description v2" in the SECREDAS project [12].

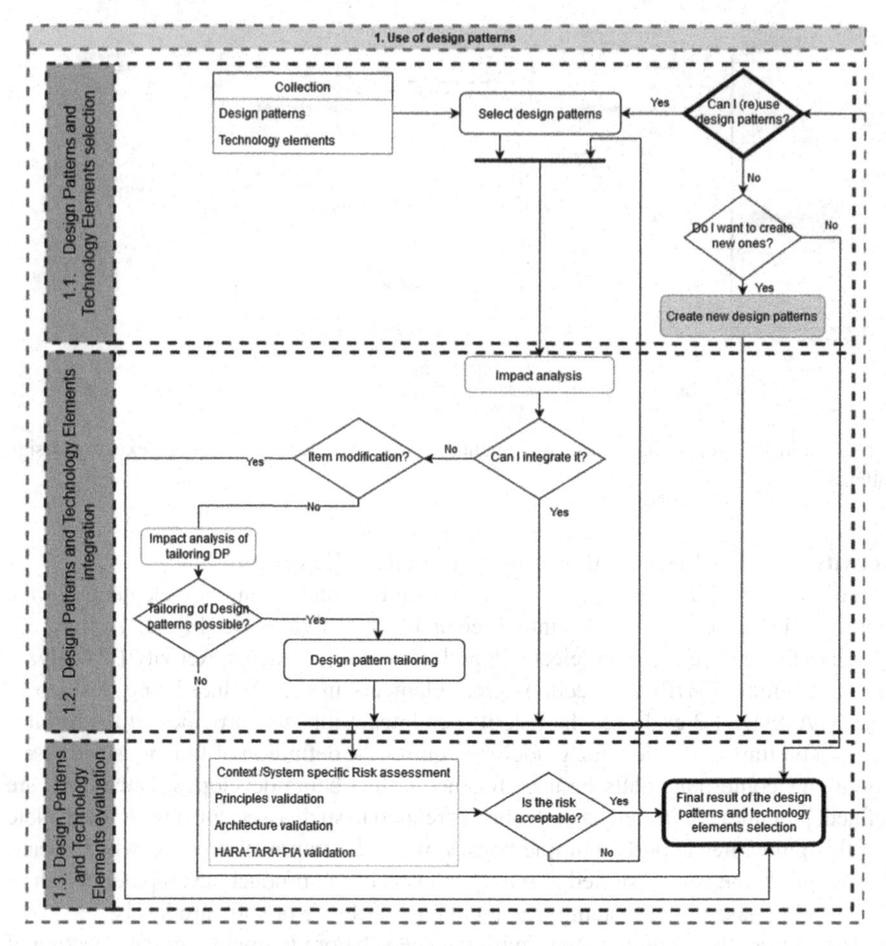

Fig. 2. Use of design patterns.

At this point, you take the list of DPs and for each element of the list make the decision if it can be integrated in your system design. Then, you need to compare the requirements and functionalities of your system with corresponded requirements and used technology elements in DP (see Fig. 2 'Select design patterns'). DPs provide an architectural concept on system level and include hardware and software designs which are used as guideline to solve a safety, security, or privacy problem. They already contain work products of the life cycle and their implementation specification in DP description.

In order to support this activity, the developers can use a list of 34 design patterns connected to multiple technology elements developed in the SECREDAS project [12]. Nevertheless, developers can also use other DPs collections or create new ones depending on their needs. To guaranty the correctness of such selection, the SECREDAS project also provides in the deliverable "D3.6 Design Patterns description v2" another guideline for developers to create their own new DPs or improve the Safety, Security or Privacy (SSP) of an existing one [12].

Activity 1.2: Design Patterns and Technology Elements Integration
The input for DPs and technology elements integration are the selection of potential DPs and technology elements for the product being developed from the previous activity. Such a list contains a preliminary selection that needs to be analyzed to evaluate the impact on Safety, Security and Privacy (SSP) in the context of the product under development.

Focus of this impact analysis is to examine new features, enhancements, or problems that DPs and technology elements address, and provides a rationale for their use, together with constraints and requirements. Thereby, the impact analysis shall identify and describe required modifications applied to the product being developed, including:

- Modification to the design: Considering the reference architecture, it identifies which design modification can result from requirements and how impact on the behavior of the system.
- Modifications of the implementation: Describes the degree of modifications intended to affect the specification or performance of the system, which may not affect, and which may be partially affected.
- Modifications related to the environment: Identify the new target environment (e.g. a different environment from the previous application of the selected DP), the changes to the operational situations or a different system location.
- Additional threats to be considered: The application of specific technology elements of DPs might open new problem fields, which have to be addressed and might require a modification of either the item or a tailoring of the DP/technology element to be applied (see below).

The impact analysis shall assess the implications of the modification with regards SSP, and identity and describe the SSP activities to be performed. All results of the impact analysis shall be collected and documented as rational for the integration response. Based on this analysis, the developers should decide if DPs and technology elements can be integrated or not to the product being developed, therefore:

- If the result of the impact analysis of the SSP is satisfactory, the guideline can proceed to integrate the DPs and technology elements without further changes or tailoring. Therefore, the user can directly proceed with the "Design Patterns and Technology Elements evaluation" activity 1.3.
- In case the results of the impact analysis of the SSP indicate major required modification or new threats, that need to be addressed when integrating the selected DPs respective technology elements, the user has two options: Either the item/product has to be modified or the DP/technology elements must be tailored with respect to the results of the impact analysis and the product context/environment:
  - Thus the first option sends the user back to the item/product definition to modify the product being developed in order to adapt it for the selected DPs and technology elements, and finally being able to integrate them.

- The second option introduces the possibility of "tailoring" one or multiple DPs and technology elements. The "tailoring" activity consist of customizing the DPs to the product being developed under consideration of the context and environment as well as combining multiple DPs to increase the SSP impact. To ensure, that by the tailoring (modification of the DPs/Technology Elements as well as combining them) no constraints are violated or new threats arise, a "tailoring" impact analysis must be performed. This tailoring and second SSP assessment reanalyzes the impact of modifying or combining different DPs and technology elements before rejecting completely the use of DPs and technology elements for the product development. If "tailoring" results is satisfactory with regard of SSP, the developers can proceed with the "Design Patterns and Technology Elements evaluation" activity 1.3, otherwise, the user returns to the chosen regular product development process and has to address the identified SSP risks without support of already proven DPs and respective technology elements.

Activity 1.3: Design Patterns and Technology Elements Evaluation
The last activity when using DPs and technology elements is the risk assessment for the specific context and system where DPs and technology elements are integrated (see Fig. 2). The purpose of this activity is evaluating whether the remaining risk after the integration of the DP is acceptable or not. The evaluation activity validates Safety, Security and Privacy of the integration of DPs and technology elements, ensuring that they are applied as intended and that they are correctly integrated into the reference architecture of the product being developed and its application context and environment.

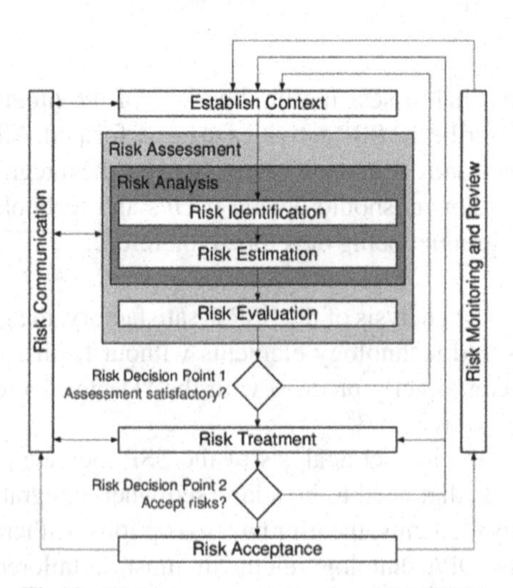

Fig. 3. Risk assessment, Source: ISO 27005.

The risk assessment, following Fig. 3, consists of a sequence of activities including risk identification, risk estimation and risk evaluation. Risk identification identifies risk sources, areas of impacts, events and causes, and their potential consequences/effects on the developed product when in use. Risk estimation involves the development of understanding of the risk, consideration of the causes and risk sources, their positive and negative consequences and the likelihood that those consequences can occur. The risk evaluation assists in decision making about risks are acceptable resp. which need treatment and their priority regarding treatment implementation.

The risk assessment is performed in the context of DPs and technology elements integration at the specific parts of the product being developed. For such specific context, there are three relevant areas which must be covered by validation activities:

- Principles validation: Security and privacy principles are condensed abstractions of best practices to provide guidance for system developers. The validation of the use of principles includes investigating relations, identifying conflicts, discussing trade-offs, and making as well as documenting decisions. As a result, the collected confirmation about security and privacy principles coverage is provided.
- Architecture validation: The reference architecture provides an allocation of DPs and their technology elements. Such allocation requires validation for the product being developed (e.g., compare context, intended use, etc.) to ensure the DPs are correctly applied into the analyzed architecture.
- HARA-TARA-PIA Validation: The SSP requirements that are covered by DPs and technology elements must be validated. Therefore, Hazard Analysis and Risk Assessment (HARA) for safety, Threat Analysis and Risk Assessment (TARA) and Privacy Impact Assessment (PIA) needs to be performed. To analyze and rate the potential dangers that can occur in case of incorrect implementation, malicious behavior, insufficient SSP coverage, conflicts, or unauthorized execution of a function, among others.

The risk assessment in the context of DPs and technology elements, can be complex to handle and difficult to assess. Such complexities and difficulties arise from the multiple levels of abstraction of the DPs and technology elements, the multiple application domains (Automotive, Rail, Health), and the interconnection of the expertise areas (Safety, Security and Privacy). To handle such variety, SECREDAS project designed an analysis framework to assess and categorize models used for SSP. The SECREDAS framework provides a structure that helps to identify and compare the various models used within SSP Risk Management, based on ISO 31000 [13].

The final activity collects all the results of the DPs and technology elements selection into a specific final document to demonstrate the correctness of the selection and to provide traceability for the SSP activities.

Activity 2: Validation and Verification of Design Pattern Integration to the System Under Consideration

The integration of Design Patterns (DPs) and their respective technology elements to the system under consideration must be verified and validated for Safety, Security and Privacy. This section describes different activities to be performed in order to evaluate the quality of such implementations (see Fig. 4). Activity 2.1 Design Patterns and technology elements validation and verification collects all the required activities to provide evidence of the adequacy of the SSP measures.

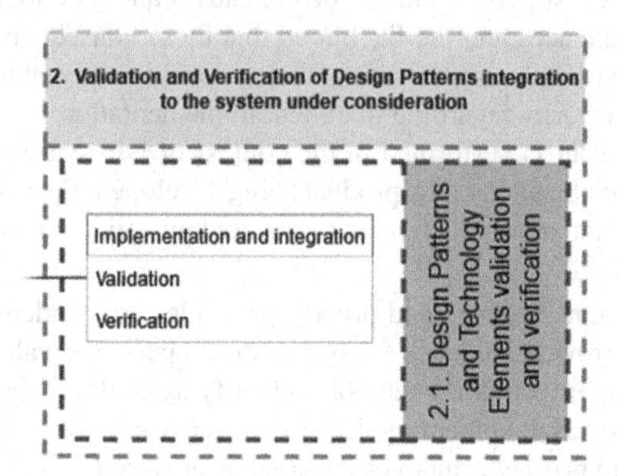

Fig. 4. Integration of design patterns and technology elements.

Activity 2.1: Design Patterns and Technology Elements Validation and Verification
The validation and verification activities provide arguments during different development phases to evaluate whether or not DPs and technology elements integration meet the specified SSP requirements and whether or not they are complete, consistent and valid.

First, the activity verifies that the selected DPs and their technology elements have been well applied in the architecture, and that they provide the expected SSP protection functions in the product (see Fig. 1). In addition, it also verifies that the SSP requirements are applied correctly once the integration has been made in the product being developed. After that, validation of the integrated DPs and technology elements provide evidence of appropriateness for the intended use and aims to confirm the adequacy of the SSP measures for the product being developed. SSP validation provides assurance that the SSP goals have been achieved, based on examination and test.

The required documentation for this step includes the identified SSP risks to the product/system connected to the selected DPs and technology elements and the SSP requirements to be satisfied, including the allocation in the reference architecture.

Verification of SSP Requirements and Architecture Design: Firstly, the SSP requirements must be verified to provide evidence for their correctness, completeness, and consistency within the context where they are used in order to create a sound basis for the further architecture/system verification. Subsequently, the reference architectural design must be verified to provide evidence that DPs and technology elements are consistent to achieve the required level of SSP according to the defined acceptance level of risk. Such an evaluation is concerned with checking specification-conform realization of the DPs and technology elements within the application context, as well as the consistency and compliance of the reference architectural design between the selected DPs and technology elements, and the system being developed.

Validation of Architecture Against Security Requirements: Validating the complete architecture, which integrates the selected DPs and technology elements back to the requirements, aims for demonstrating that all the SSP requirements have been covered. For example, to validate the architecture of a V2X gateway, the designer needs to show that all the security requirements specified in the V2X security standards have been met. Furthermore, the validation includes a security risk analysis conducted to demonstrate that the architecture which integrates the selected DPs mitigates the security risks to an acceptable level. The selection of the security analysis method is up to the system designers, as described in SAE J3061 [14]. In practice, any mature security risk analysis method, such as STRIDE [15], EBIOS [14], TARA [10] or STPA [16], is applicable.

Testing Safety, Security and Privacy (SSP): Verification and validation must be supported by testing activities. For example, security testing evaluates software system requirements related to security properties of assets that include confidentiality, integrity, availability, authentication, authorization and non-repudiation [14]. From the risk assessment, we can develop an overall test strategy, or simply develop a particular test (based on threats, vulnerabilities, requirements and assumptions), we can increase the test coverage or focus the test only on data inputs selected. The SECREDAS project introduces a comprehensive security-focused testing process based on ISO/SAE 21434 and a corresponding framework that facilitates automation, comparability, efficiency and scalability of security testing (particularly on the example of automotive systems) [17, 18].

3 Application Example

Design patterns use corresponding technology elements and assemble them in a distinct manner. They provide a description on how to apply them for certain usage scenarios. The patterns defined in SECREDAS are used in concrete implementations of the project's vehicle sensing, vehicle connectivity, in-vehicle networks, railway communications and healthcare IoT. All the technology required for these use cases integrate design patterns and technology elements during their development process, as proposed in this guideline.

To provide with an example, a couple of design patterns used in most of use cases are the *Sensor Data Analytics* pattern and the *Sensor Data Fusion Architecture* pattern. Both design patterns rely on the same technology element (named as well *Sensor Data Analytics*). However, the former design pattern uses a cloud architecture and includes driver monitoring, while the latter is dedicated to anomaly detection inside a vehicle. Depending on the setting (e.g., availability of cloud services), this allows for the re-use of the respective technology elements in different contexts. Those two patterns are to always be implemented the same way, which enables easy factorization and therefore lean and industrialized implementation of provenly safe, secure, and privacy-enabled building blocks.

4 Conclusion

In this paper, the authors presented a guideline for applying Design Patterns in a product under development for developing new architectural Safety, Security, and Privacy implementations using DPs. The guideline proposes rules and activities to carry out during product development using Design Patterns and technology elements. In addition, it supports developers or other potential users to develop SSP products using our SECREDAS approach.

This guideline enables complex systems to integrate SSP during their development process. The use of DPs and technology elements reduces the development effort by providing concrete, trustworthy solutions. They are cataloged, classified and reusable, adaptable to specific needs within a given context, and engineering domain-independent, reducing development time and effort.

Acknowledgements. This work has been partially funded by EU ECSEL Project SECREDAS. This project has received funding from the European Unions Horizon 2020 research and innovation programme under grant agreement No 783119. The publication was written at VIRTUAL VEHICLE in Graz and partially funded by the COMET K2 Competence Centers for Excellent Technologies Programme of the Federal Ministry for Transport, Innovation and Technology (bmvit), the Federal Ministry for Digital, Business and Enterprise (bmdw), the Austrian Research Promotion Agency (FFG), the Province of Styria and the Styrian Business Promotion Agency (SFG). We are also grateful to Netherlands Organization for Applied Scientic Research TNO for supporting this research.

References

1. Halder, S., Ghosal, A., Conti, M.: Secure over-the-air software updates in connected vehicles: a survey. Comput. Netw. **178**, 107343 (2020)
2. Coppola, R., Morisio, M.: Connected car: technologies, issues, future trends. ACM Comput. Surv. - Article **46**, 36 (2016)
3. Statista: Automotive electronics cost as a percentage of total car cost worldwide from 1970 to 2030. Statista, April 2019. https://www.statista.com/statistics/277931/automotive-electronics-cost-as-a-share-of-total-car-cost-worldwide/. Accessed 12 Apr 2021
4. Antinyan, V.: Revealing the complexity of automotive software. ResearchGate, July 2020
5. Alnasser, A., Sun, H., Jiang, J.: Cyber security challenges and solutions for V2XCommunications: a survey. Comput. Netw. **151**, 52–67 (2019)
6. Mujahid, M., Ghazanfar, A.S.: Survey on existing authentication issues for cellular-assisted V2X communication. Veh. Commun. **12**, 50–65 (2018)
7. Šljivoa, I., Juez Uriagereka, G., Puri, S., Gallina, B.: Guiding assurance of architectural design patterns for critical applications. J. Syst. Archit. **110**, 101765 (2020)
8. Marko, N., Vasenev, A., Striecks, C.: Collecting and classifying security and privacy design patterns for connected vehicles: SECREDAS approach. In: Casimiro, A., Ortmeier, F., Schoitsch, E., Bitsch, F., Ferreira, P. (eds.) SAFECOMP 2020. LNCS, vol. 12235, pp. 36–53. Springer, Cham (2020). https://doi.org/10.1007/978-3-030-55583-2_3
9. ISO 26262: Road vehicles – functional safety. International Organization for Standardization (2018)
10. ISO/SAE DIS 21434: Road vehicles – cybersecurity engineering. International Organization for Standardization (2020)
11. CC: Common criteria - part 3: security assurance requirements. Common Criteria (2017)
12. SECREDAS: D3.6 design patterns description v2, February 2021. https://secredas-project.eu/. Accessed June 2021
13. ISO/TC 262: ISO 31000, risk management. ISO (2018)

14. SAE: Cybersecurity guidebook for cyber-physical vehicle systems (J3061 ground vehicle standard). SAE International (2016)
15. Schostack, A.: Threat Modeling: Designing for Security. Wiley, Indianapolis (2014)
16. Young, W., Leveson, N.G.: An integrated approach to safety and security based on systems theory. Commun. ACM **57**(2), 31–35 (2014)
17. Marksteiner, S., et al.: A process to facilitate automated automotive cybersecurity testing. IEEE (2021)
18. Marksteiner, S., Ma, Z.: Approaching the automation of cyber security testing of connected vehicles. In: The Third Central European Cybersecurity, New York (2019)

Structured Traceability of Security and Privacy Principles for Designing Safe Automated Systems

Behnam Asadi Khashooei[1(✉)], Alexandr Vasenev[1], and Hasan Alper Kocademir[2]

[1] Embedded Systems Innovation (ESI), Netherlands Organisation for Applied Scientific Research (TNO), Eindhoven, The Netherlands
{behnam.asadikhashooei,alexandr.vasenev}@tno.nl

[2] Roche Diagnostics International AG, 6343 Rotkreuz, Switzerland
alper.kocademir@contractors.roche.com

Abstract. Creating modern safe automated systems like vehicles demands making them secure. With many diverse components addressing different needs, it is hard to trace and ensure the contributions of components to the overall security of systems. Principles, as high-level statements, can be used to reason how components contribute to security (and privacy) needs. This would help to design systems and products by aligning security and privacy concerns. The structure proposed in this positioning paper helps to make traceable links from stakeholders to specific technologies and system components. It aims at informing holistic discussions and reasoning on security approaches with stakeholders involved in the system development process. Ultimately, the traceable links can help to assist in aligning developers, create test cases, and provide certification claims - essential activities to ensure the final system is secure and safe.

Keywords: System security · Security analysis · Design traceability · System architecture

1 Introduction

The increasing complexity of modern high-tech systems, like vehicles or advanced manufacturing equipment, poses challenges to their creators [17]. Next to this is the ever-increasing demand to create connectivity between different systems and provide seamless solutions and services contributing to the concept of the internet of things (IoT). Since many of these high-tech systems are cyber-physical

This research is carried out as part of SECREDAS and INTERSECT projects. SECREDAS is funded by the ECSEL Joint Undertaking of the European Union under grant agreement number 783119. INTERSECT is a public private partnership funded by the Dutch National Research Council (Grant NWA. 1162.18.301).

I. Habli et al. (Eds.): SAFECOMP 2021 Workshops, LNCS 12853, pp. 52–62, 2021.
https://doi.org/10.1007/978-3-030-83906-2_4

systems, existing in both physical and cyber worlds, ensuring system safety is of high importance. The increased connectivity brings new challenges for system developers where the previously closed systems now become somewhat open. Therefore, there is a need to adapt systems design with considering security as a cruciality of the system.

Many consumer-based systems, e.g., healthcare systems that manage personal data obtained from diagnostic machines, deal with personal and sensitive information, where acquisition, processing, and storage of these information are subject to strict regulation like GDPR [12]. This adds a new layer of complexity in the design of such systems in which failure to comply with the regulation can cause financial and legal losses. This challenge is further magnified by providing connectivity to these systems and hence exposing them to outside world.

Security and privacy are system-wide properties, not only complex by themselves but also deeply intertwined with other system qualities like safety, performance, reliability, etc. To improve these properties, architects and designers shall maintain an overview of the system, as well as be able to dig deeply into details where needed. A common approach in security design is to apply best practices, instead of re-inventing or re-implementing the designs [8,18]. In such cases, well-developed components are combined into so-called design patterns which create solution templates that can be used in different applications.

To advance the state of the art and practice, a large EU research project called SECREDAS [15] pays significant attention to provide reusable solution designs in the form of design patterns. These security patterns employ existing technology elements (TE) that provide security functions. The main intent is to reuse well-known design solutions and existing tools and techniques rather than inventing solutions from scratch. The latter is too risky because of the likelihood of design and coding flaws which is specially important when the safety of the system is of paramount importance [9].

Design patterns shall be explicit what they aim to address and how, to facilitate their applicability and to be clear for users. Often, this can be expressed as aims or tenets, such as 'least privilege' for security or 'data minimization' for privacy. These expressions can be seen as principles that reflect best practices and are easy to communicate to stakeholders. Such principles in the domains of security [14] and privacy (listed e.g. in [13]) compress security and privacy domain-expertise and guide system developers. In general, the use of principles helps with investigating relations, identifying conflicts, discussing trade-offs, and making decisions.

Designing a system that addresses requirements of different stakeholders and complies with best practices is not always an exact science. Typically, there may be multiple refinement rounds needed until the system design complies in an optimal way with such needs. Even in that case, it may not be clear which design alternatives exist and which trade-offs have appeared as a result of design choices. Therefore, it is clearly beneficial to have a structure that organizes information, to provide a clear overview of the different aspects of a system design [5]. Such a structure can also help to see the tensions that naturally arise between various

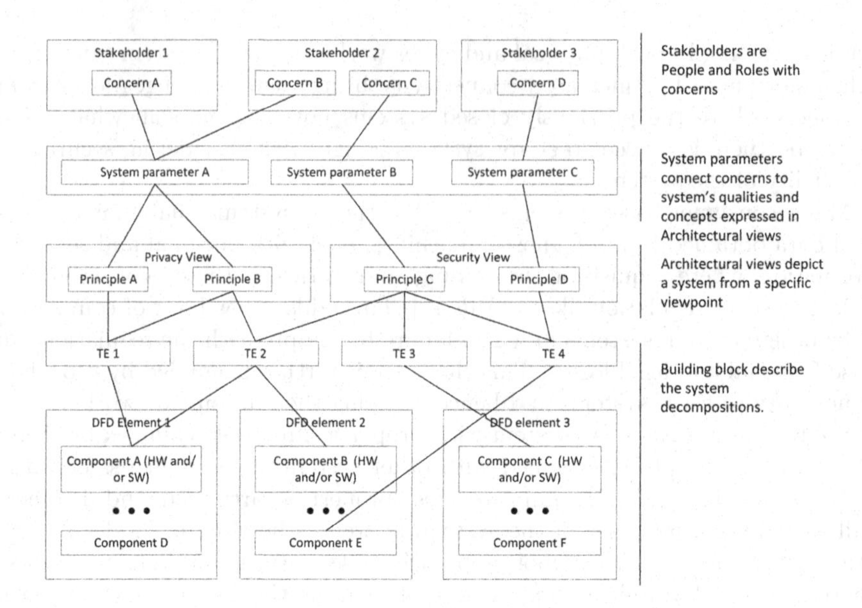

Fig. 1. Proposed structure to support traceability of links from stakeholders to system components based on security and privacy principles. Technology Elements (TEs) are existing industrial technologies that can be used to develop new solutions and Data Flow Diagrams (DFDs) graphically represent the flow of data within and across the systems boundaries.

concerns of stakeholders, e.g. the privacy and security concerns on one hand, and usability concerns on the other hand [6] and can facilitate the communication between experts. If available, this structure can enhance the traceablitiy of design decisions and rationale behind them which can further help in e.g. certification and test case developments.

In this paper, we argue that security and privacy principles can assist in the structural evaluation of a design and help solution/security architects in capturing and maintaining a holistic view of their design. Furthermore, we propose a method for facilitating traceability from stakeholders' concerns to elements of implementation, where principles act as a central connecting layer (see Fig. 1). This effort is in line with the architectural reasoning for system qualities (see, e.g., [11]), where we extend these lines of reasoning toward systems security. Here we illustrate the application of this method in the evaluation of an data transfer solution that utilizes several standalone systems with a wide range of stakeholders. The analyzed solution aims at providing healthcare data of a driver to an fitness assessment mechanism in vehicles which ensures healthiness of the driver and contributes to the overall safety of roads.

The paper proceeds as follows: Sect. 2 outlines the problem statement followed by our proposed method. Section 3 contains the illustrative use case example and, finally, Sect. 4 concludes the paper.

2 Problem Statement

Re-usable security and privacy solutions can be described based on the security and privacy principles to which they adhere. In this way, system architects, when thinking about existing or future systems, can benefit from the traceability that these principles offer. To enable such traceable links, it is desirable to introduce a structure that helps to discuss security and privacy in high-tech systems through principles.

2.1 Proposed Structure to Trace Principles to System Components

With security measures distributed across the system, it is beneficial to see how these measures are deployed and what they try to address. This e.g. can help in supporting traceability of requirements, assisting the future changes, and constructing certification claims.

By positioning these links in a larger picture of stakeholders and concerns, we aim to ensure that the structure in Fig. 1 will support traceability of links from stakeholders to system components. This structure is an adaptation of model-based system architecting (MBSA) layered approach (see [2,3,5]), which itself connects with the thread of reasoning in CAFCR method for embedded system architecting [11].

This adaptation provides views corresponding to the security and privacy viewpoints where principles act as a specific way to demonstrate the views. Each view will have a specific scope and will show how the system aspect is achieved. The views are then put in the context of the systems environment, with traceable connections to both the system stakeholders on the high-level as well as the building blocks on the low-level.

Specifically, we highlight the use of principles in the views and capture the connection of technology elements to the system components which are structured according to the data flow. The latter is particularly relevant for security and privacy concerns, as data-in-transit and data-at-rest aspects can be effectively depicted with data flow diagrams.

As depicted Fig. 1, the following layers are used to capture the information:

- Stakeholders: all persons or organizations that have rights, share, claims, or interests concerning the system or its properties (represented by key performance indicators, that quantify the desired system qualities) meeting their needs and expectations.
- System view: shows the key performance parameters of the system (see, e.g., [7]) and links those to the stakeholder needs and expectations.
- System aspects: privacy/security.
- Principles for privacy/security: condensed and codified abstractions of best experiences and lessons learned.
- Technology Elements (TEs): existing industrial proven technologies that can be used to develop new solutions.

- Data Flow Diagram (DFD) elements: elements of a DFD and their constituent components (HW, SW).

The relations between elements in and across layers enable traceability from stakeholder concerns to system parameters, principles, technology elements, and building blocks to achieve the desired functions in the system. Besides, these relations allow a structured way of thinking about available alternatives and resulting trade-offs [1]. The proposed structure in Fig. 1 makes relevant parts of the design explicit.

The traceability of constructs can indicate whether concepts overlap and help to spot, investigate, make, and highlight trade-offs explicitly. Particularly, such an approach can assist in:

- Indicating the level of compliance to existing principles. Arguably, the amount and the scope of used principles can indicate the maturity of the design. E.g., using MITRE cyber resiliency design principles [4] like concealment (reduce the visibility of a system from an external change agent) highlights that the developers should consider survivability as an advanced design concept.
- The structure suggested in Fig. 1 can help to organize information and support reasoning at different stages of a project.

It is worth to mention that applying this structure requires substantial knowledge and experience in security/privacy aspects and related technology elements as well as system architecting.

In the following section, we provide an example of how the proposed structure helps the holistic and systematic reasoning over the security and privacy concerns of a sample system. We illustrate this reasoning by analyzing the initial concept of a SECREDAS design pattern which concerns with Healthcare Data Exchange, [15].

3 Use Case Example

To illustrate application of the method, let us consider a design pattern where user's healthcare data are delivered from a healthcare cloud to a car through a user device. This data will be used in a third-party app within the car to validate the fitness of the driver. This pattern anticipates a case where car manufacturing companies need to assess the driver' health before the journey based on demands of official authorities. This helps to ensure safety of the vehicle and the surroundings, as a driver not fit for controlling the vehicle will not be allowed to start it.

The entities and connections involved in this design pattern are depicted in Fig. 2. The key systems involved in this design pattern include: a cloud system with corresponding identity management servers and policy Decision Point (XACML PDP), a traveler's device (e.g. smartphone), and a car. The diagram also depicts, the sequence of events leading to driver healthcare information being accessed from healthcare information cloud and being processed to deliver

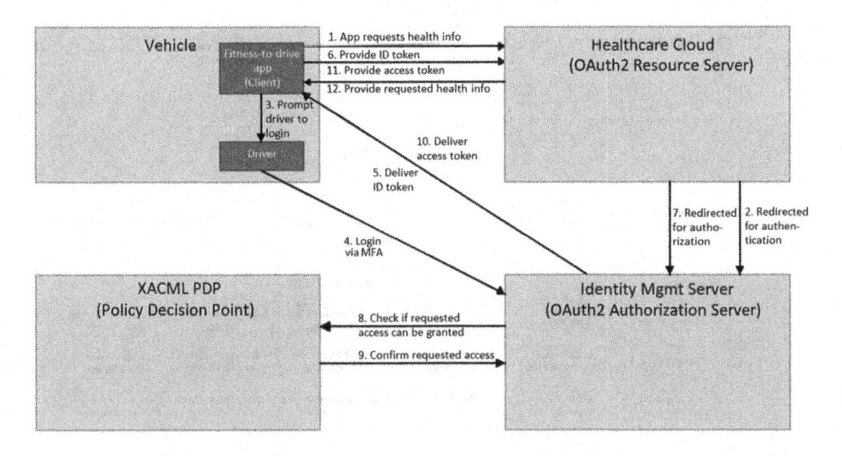

Fig. 2. Healthcare data exchange design pattern, initial data flow diagram [15].

fitness-to-drive assessment. It has a registration phase, an authentication phase, an authorization phase and a processing phase.

To analyze the design from security and privacy viewpoints, a series of discussions with the developer of the design pattern has been carried out in a step-wise fashion. At each step, the results of the analysis were provided to the developer of the design pattern, which helped him to further elaborate the design pattern. Finally, the proposed structure of Fig. 1 was used to create the overall view of the analysis as depicted in Fig. 3.

In the following section, we elaborate on different layers of Fig. 3 and the rationale behind them.

3.1 Stakeholders and System Parameters

From viewpoint of security and privacy, we identified the main stakeholders of this design pattern as depicted in Fig. 3-A. Since the driver's healthcare information is the subject of this design pattern, the driver's concerns on privacy and security are central to the case. Therefore, the manufacturing company of the vehicle needs to make sure that these data can be stored securely. Also, the driver can be concerned with a technician accessing the driver's information. This concern is captured by a line between the driver and the technician. The concern that the technician should have only a limited access to data is depicted by the link between Car manufacturing company and the technician.

Since product that use this design pattern target European Union market, GDPR compliance [12] is a must. Therefore, acquiring GDPR compliance entails the application of the privacy principles. On the other hand, since the data exchange happens over the network, ensuring a certain security level is crucial. This can be achieved by applying the best practices in terms of security principles.

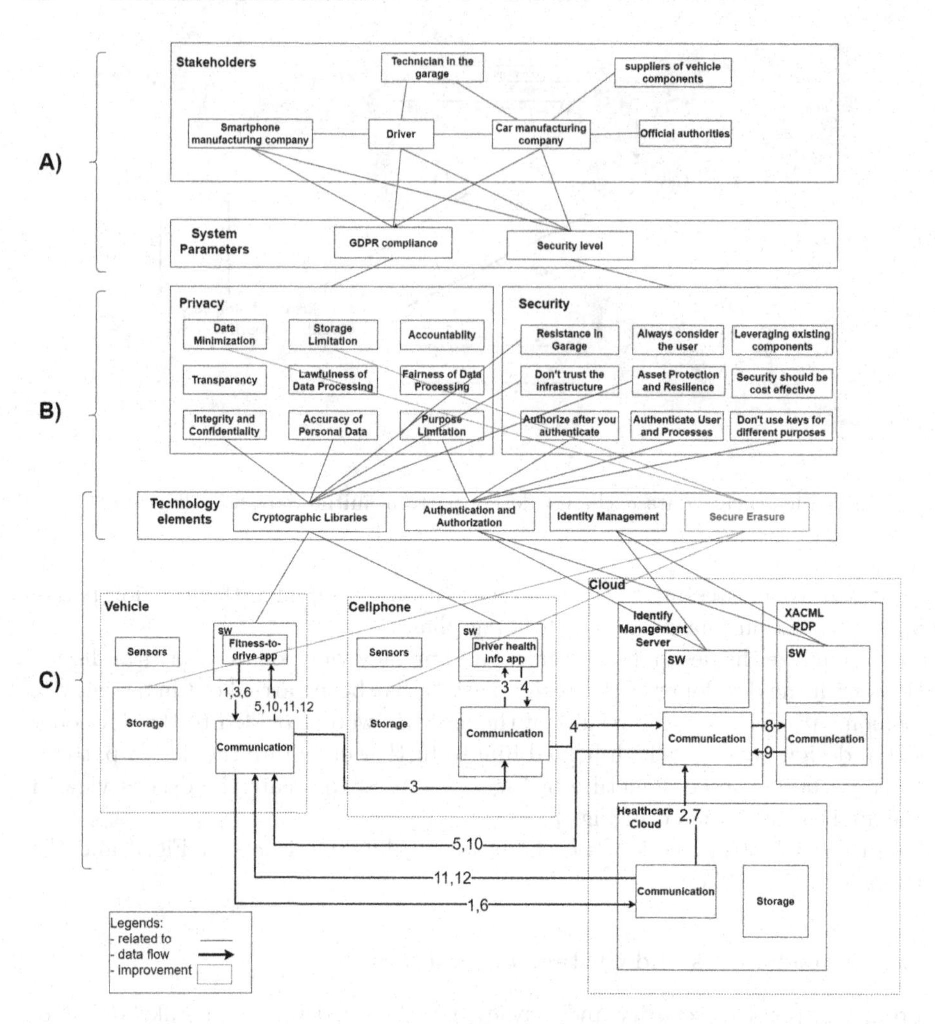

Fig. 3. Filled in structure to support traceability of links from stakeholders to system components based on security and privacy principles for illustrative example of Healthcare data exchange design pattern. The numbers on the data flow paths correspond to the sequence of actions indicated in Fig. 2.

3.2 Principles and Security and Privacy Technology Elements

To ensure traceabilility of stakeholders' concerns, based on the concerns of stakeholders, we identified the relevant principles (see Fig. 3-B). These privacy and security principles have been used by the architect in making the design decisions and selecting the technology elements for this design pattern. Table 1 provides the mapping from identified principles to the technology elements as well as the rationale behind the relevancy of the principles in the decision-making process. The left column of Table 1 indicates the principles discussed. Specifics of their embodiment in the design, if any, are indicated in the rest of the columns.

Table 1. Mapping of principles to technology elements in Healthcare data exchange design pattern.

#	Principles	Technology element	Rationale
S_1	Don't trust infrastructure	Cryptographic Libraries	To minimize the data leakage risks caused by trusting the security of underlining infrastructure e.g. phone and vehicle
S_2	Authenticate users and processes	Authentication and Authorization	To ensure proper access control within design pattern and on its boundary
S_3	Authorize after you authenticate	Authentication and Authorization	To explicitly check for authorization of the user e.g. driver after initial authentication
S_4	Always consider the users		Using a personal phone for authentication and authorization should be easy
S_5	Asset protection and resilience	Cryptographic Libraries	Data are stored in phone and car. Potentially, those assets can be stolen. The stolen phone use-case is in the scope
S_6	Security should be cost-effective		A balanced design between the security measures and their benefits is appropriate
S_7	Resistant in the garage	Cryptographic Libraries	The principle is relevant for crypto libraries because of cascade encryption (logged data are not even accessible in the garage)
S_8	Don't use keys for different purposes	Authentication and Authorization	Different keys for phone-to-car and phone-to-cloud connections. Regular update of access key for cloud
S_9	Leveraging existing components		
P_1	Data minimization	Secure Erasure	Only necessary personal data should be processed and when not needed, the data should be securely deleted
P_2	Accuracy of personal data	Cryptographic Libraries	Encryption add an extra layer of security to prevent alteration of data. On the other hand, the reliability of encryption solutions is important to prevent unwanted changes in data during encryption and decryption mechanisms
P_3	Accountability		The accountability of practice necessitates storage of audit logs
P_4	Purpose Limitation	Authentication and Authorization	Authorization (specifically xacml pdp) ensures that only the relevant data is transferred and nothing more
P_5	Fairness of Data Processing		
P_6	Transparency of data processing		
P_7	Storage Limitation	Secure Erasure	Only when necessary, the personal data can be stored for a limited period and should be deleted as soon as no more needed
P_8	Integrity and Confidentiality	Cryptographic Libraries	Encrypted driver healthcare data can only be decrypted in the car component which needs to work with this data. This way (except in the case of stolen encryption keys) as long as the car component can decrypt and parse the data, data integrity and provenance can be proven from the point that the sensitive data leaves the healthcare cloud until the point car component uses this sensitive data for processing fitness-to-drive assessment
P_9	Lawfulness of Data Processing		

To give an example, by enforcing the principle of "authenticate users and processes" (S_2 in Table 1), the architect can ensure that no healthcare data can be accessed without authorization. This high-level statement connects to the architect's design decision of selecting Authentication and Authorization technology elements in this design pattern.

As another example, the architect enforces the principle of "accuracy of personal data" (P_2 in Table 1) to protect the highly sensitive healthcare data. By this enforcement, he chooses to implement an encryption solution to add an extra layer of security to prevent alteration of data, on the other hand, he pays extra attention to the reliability of the encryption solution to prevent unwanted changes in data during encryption and decryption mechanisms.

3.3 Technology Elements to Design Elements

As the final layer, we identify the application of technology elements in components of different subsystems. To this end, we elaborated the DFD of Fig. 2 to account for components of the design pattern as shown in Fig. 3-C, and then we made the application of the technology elements in related components explicit. For example, the technology element of "Cryptographic Libraries" is implemented in both the car and the cellphone to ensure secure storage of data.

3.4 Holistic Analysis and Identifying the Points for Improvement

As can be noticed, the mere creation of the suggested structure in Fig. 1 has shed some light on the design structure and made the reasoning behind certain design decisions more explicit, which helps in the further analysis of this design pattern.

For illustrative purposes, a specific example may be helpful. The stakeholders "official authorities" and "suppliers of vehicle components" would like to have access to the generated data for example to understand the exact circumstances of a car accident. On the other hand, the driver's privacy concerns through principles of "storage limitation" and "data minimization" entail that only necessary personal, sensitive data about the driver should be used and should only be stored for the shortest necessary duration. Therefore, these data logs need to be appropriately protected against multiple potential adversaries (e.g. malevolent garage employees, vehicle or smartphone hackers) in different use case scenarios.

Therefore, one can see through the respective suggested structure, linking principles with technology elements and system components that the technology element of "Cryptographic Libraries" is applied to the storage in the car and ensures the appropriate protection of driver's healthcare data in the car. On the other hand, these stored data need to be securely erased which brings about a need for a new technology element "Secure Erasure" supporting the limited storage and irrevocable erasure of driver healthcare data. Furthermore, considering the security principle "Don't trust the infrastructure" and the links of the technology element "Cryptographic Libraries" to respective DFD elements, the need for additional encryption of the payload (driver health condition data) and respective logs on the smartphone and in the vehicle is identified.

Similarly, the design can further be analyzed by walking through different layers and creating a chain of reasoning that supports/shows the traceability of design decisions.

4 Discussion and Conclusions

The proposed method helps in identifying and resolving by discussing potential conflicts between stakeholder needs, best practices, and architecture. It also provides insights into potential gaps in solution architecture (e.g. the identification of a need for secure erasure functionality as described in Sect. 3.4).

The use case analysis, which followed the proposed method, helped to assess and improve the Healthcare Data Exchange solution in a constructive manner. Specifically, it assisted in obtaining a complete view of the main stakeholders, relevant principles, relevant technology elements, data flow path, and data storage. This holistic view improved the traceability by bringing out the relations between different aspects through the suggested structure. Furthermore, by identifying the constituent systems and their boundaries, the solution scope has been refined, and some conflicts were resolved. The use case owner noted that the application of the method helped to simplify the pattern and identify blind spots on data transfer.

The proposed structure can be complementary to the threat modeling approaches (e.g. [16]) to create traceability of security and privacy mechanisms by considering the attacker as an stakeholder and therefore obtain a holistic view of the design.

The graphical representation of the structure has its limitation as it is bounded by visual space, however, this limitation can be overcome by using digital tools (e.g. see [10] and DAARIUS tool [5]) that enable multi-layer representations.

In sum, this paper argued that application of principles, as high-level statements, to provide a structure for architectural reasoning from the security and privacy viewpoints. We illustrated the proposed structure with a use case that informed holistic discussions and reasoning on security approaches have been detailed. For future work, we aim to extend this approach to include other aspects of interest in system analysis, e.g., performance and cost, as well as to further adapt the method for seamless use in practical product creation processes.

References

1. Asadi Khashooei, B., Vasenev, A., Kocademir, H.A., Mathijssen, R.W.: Architecting system of systems solutions with security and data-protection principles. In: 2021 16th International Conference of System of Systems Engineering (SoSE) (SoSE 2021), June 2021

2. Bijlsma, T., van der Sanden, B., Li, Y., Janssen, R., Tinsel, R.: Decision support methodology for evolutionary embedded system design. In: 2019 International Symposium on Systems Engineering (ISSE), pp. 1–8. IEEE (2019)
3. Bijlsma, T., Suermondt, W.T., Doornbos, R.: A knowledge domain structure to enable system wide reasoning and decision making. Procedia Comput. Sci. **153**, 285–293 (2019)
4. Bodeau, D., Graubart, R.: Cyber resiliency design principles selective use throughout the lifecycle and in conjunction with related disciplines. Technical report, MITRE CORP BEDFORD MA BEDFORD United States (2017)
5. DAARIUS: DAARIUS methodology. Embedded Systems Innovation (ESI) (2019). https://esi.nl/research/output/methods/daarius-methodology. Accessed 18 Feb 2021
6. Halperin, D., Heydt-Benjamin, T.S., Fu, K., Kohno, T., Maisel, W.H.: Security and privacy for implantable medical devices. IEEE Pervasive Comput. **7**(1), 30–39 (2008)
7. Haponava, T., Al-Jibouri, S.H.: Identifying the KPIs for the design stage based on the main design sub-processes. In: Proceedings of joint CIB conference on Performance and Knowledge Management, 3–4 June, Helsinki, Finland. pp. 14–23. CIB (2008)
8. Laverdiere, M., Mourad, A., Hanna, A., Debbabi, M.: Security design patterns: Survey and evaluation. In: 2006 Canadian Conference on Electrical and Computer Engineering, pp. 1605–1608. IEEE (2006)
9. Marko, N., Vasenev, A., Striecks, C.: Collecting and classifying security and privacy design patterns for connected vehicles: SECREDAS approach. In: Casimiro, A., Ortmeier, F., Schoitsch, E., Bitsch, F., Ferreira, P. (eds.) SAFECOMP 2020. LNCS, vol. 12235, pp. 36–53. Springer, Cham (2020). https://doi.org/10.1007/978-3-030-55583-2_3
10. Moneva, H., Hamberg, R., Punter, T.: A design framework for model-based development of complex systems. In: 32nd IEEE Real-Time Systems Symposium 2nd Analytical Virtual Integration of Cyber-Physical Systems Workshop, Vienna (2011)
11. Muller, G.: CAFCR: A multi-view method for embedded systems architecting. Balancing Genericity and Specificity (2004)
12. Regulation, G.D.P.: Regulation eu 2016/679 of the European parliament and of the council of 27 April 2016. Official Journal of the European Union (2016). http://ec.europa.eu/justice/data-protection/reform/files/regulation_oj_en.pdf. Accessed 28 Jan 2021
13. Riva, G.M.: Privacy architecting of GDPR-compliant high-tech systems: the PAGHS methodology (2019). http://essay.utwente.nl/79359/
14. Saltzer, J.H., Schroeder, M.D.: The protection of information in computer systems. Proc. IEEE **63**(9), 1278–1308 (1975)
15. SECREDAS: SECREDAS project: an ECSEL joint undertaking (2021). https://secredas-project.eu/. Accessed 27 Jan 2021
16. Shostack, A.: Threat Modeling: Designing for Security. Wiley, Hoboken (2014)
17. Stolfo, S., Bellovin, S.M., Evans, D.: Measuring security. IEEE Secur. Priv. **9**(3), 60–65 (2011)
18. Yoshioka, N., Washizaki, H., Maruyama, K.: A survey on security patterns. Prog. Inf. **5**(5), 35–47 (2008)

Synchronisation of an Automotive Multi-concern Development Process

Martin Skoglund[1(✉)], Fredrik Warg[1], Hans Hansson[2], and Sasikumar Punnekkat[2]

[1] RISE Research Institutes of Sweden, Borås, Sweden
martin.skoglund@ri.se
[2] MRTC, Mälardalen University, Västerås, Sweden

Abstract. Standardisation has a primary role in establishing common ground and providing technical guidance on best practices. However, as the methods for Autonomous Driving Systems design, validation and assurance are still in their initial stages, and several of the standards are under development or have been recently published, an established practice for how to work with several complementary standards simultaneously is still lacking. To bridge this gap, we present a unified chart describing the processes, artefacts, and activities for three road vehicle standards addressing different concerns: ISO 26262 - functional safety, ISO 21448 - safety of the intended functionality, and ISO 21434 - cybersecurity engineering. In particular, the need to ensure alignment between the concerns is addressed with a synchronisation structure regarding content and timing.

Keywords: Functional safety · Cybersecurity · Multi-concern · SOTIF · Automotive · ISO 26262 · ISO 21448 · ISO 21434

1 Introduction

The complexity of embedded systems that are developed and integrated by manufacturers into modern cars is increasing. Advanced driver assistance systems (ADAS) progress by leaps and bounds, and soon the advent of vehicles with automated driving systems (ADS) is upon us. The chief design principle is no longer that of a fail-safe system, i.e., which has the option to become unavailable when a problem occurs. ADSs with level 3 and level 4 features according to SAE J3016:2018 [17], which intend to remove active supervision by a human driver, will need to replace the driver with fail-operational systems. There is also an increased reliance on connectivity and sensors to attain this new level of automation [21]. However, external communication and environmental sensors make the vehicles susceptible to security threats that may incapacitate or fool the ADS. Therefore, in engineering an ADS, both safety and security and their interplay must be addressed.

I. Habli et al. (Eds.): SAFECOMP 2021 Workshops, LNCS 12853, pp. 63–75, 2021.
https://doi.org/10.1007/978-3-030-83906-2_5

In many domains, including automotive, standards are often used to ensure quality concerns are appropriately treated. There is, however, a lack of experience in dealing with security in safety engineering and vice versa. Even though fundamental requirements for cooperation between concerns exist - e.g., ISO 26262:2018 expresses the need to take interdependence with cybersecurity into account - there is little guidance on transforming these requirements into a practical process that makes adherence to several standards with different concerns possible. Such details are also missing in the recently released technical report ISO/TR 4804, dealing with safety and cybersecurity specifically for ADSs; however, it does point to the three road vehicle standards ISO 26262 (functional safety), ISO 21448 (safety of the intended functionality), and ISO 21434 (cybersecurity engineering) for dealing with safety and security. In addition, there is a need for procedures to assess the conformity of an identified minimum set of standards for a dependable and secure system. ISO/TR 4804 will be expanded into a technical specification, ISO 5083, but the completion of this initiative is a long way off, and it is not yet known what it will contain. In the meantime, there is an urgent need for hands-on guidelines on best practices for multi-concern development, using the already available (or soon to be available) standards.

The contributions of this paper are intended to support the implementation of a multi-concern development process that could operate within the current standardisation landscape. It contains a unified chart, organised around a generic V-model, describing relevant processes, artefacts, activities, and a mechanism for synchronisation regarding content and timing between concerns at each step in the process.

2 Methodology and Scope

The paper builds upon work within the SECREDAS project [16]. The goal of the corresponding task in the project is to provide a hands-on guideline for continuous multi-concern qualification/certification, focusing on safety and security concerns. Multi-concern and continuous development are in this investigation considered as different aspects. This paper concentrates on refining the guideline regarding multi-concern development in the automotive domain for phases before the release to market. Phases after the first release are more relevant to address in relation to continuous development and successive releases.

The methodology entails surveying the applicable standards for relevant findings affecting continuous and multi-concern certification as a first step. A hypothesis regarding how to order and characterise the information is put forward in a unified development V-model, generalised to suit all standards. It was considered essential to have a starting point and iterate rather than analyse details and then generalise. The findings were inventoried and aggregated into this V-model. Analysis of the finding then resulted in assembling a chart and guideline for a multi-concern development process, encompassing safety and security concerns for the automotive domain. The complete survey of the standardisation landscape in the SECREDAS project was more vast in scope than the results presented here, encompassing general system engineering, functional safety and information security, and other domains in addition to automotive.

3 Guideline for a Multi-concern Development Process

The process outline for multi-concern development resulting from our analysis of standards can be seen in Fig. 1. The figure is illustrating the life-cycle for nominal functions with added activities for functional safety and cybersecurity. The interplay between concerns in the coloured phases is described in this chapter. Phases dealing with software and hardware development had to be omitted due to space constraints, and the development of components is expected to have a lesser need for alignment if inconsistencies are appropriately dealt with in previous phases. Synchronisation and coordination will be necessary as the system is integrated. However, the pattern is similar to the system development phases.

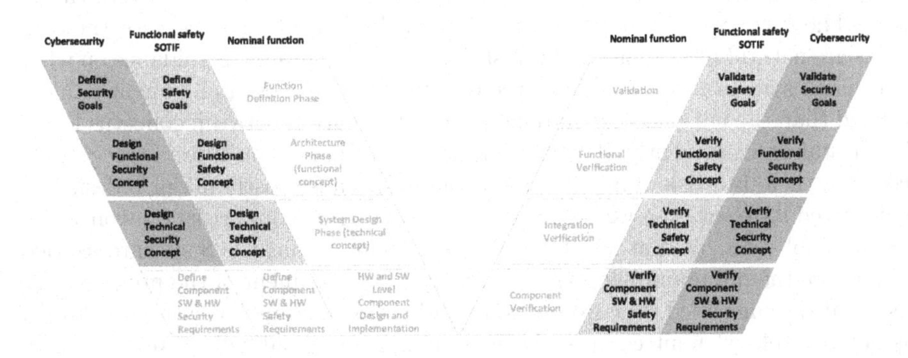

Fig. 1. Multi-concern development lifecycle, with current scope highlighted.

The used set of applicable standards can be narrowed down to a non-dispensable minimum relevant for the specific implementation and domain. In this paper, the chosen domain is automotive, with an envisioned implementation of an ADS with advanced environmental sensors susceptible to manipulation. The relevant standards for a non-dispensable minimum are, ISO 26262 [7] (addressing functional safety), PAS ISO 21488 [8] (addressing the safety of the intended functionality, SOTIF) and final draft of the ISO/SAE 21434 [10] (addressing cybersecurity). ISO/TR 4804 [9] can also be considered; however, as it gives some guidance but does not contain any normative requirements, it will not be directly considered in the multi-concern life-cycle described in this chapter.

Development of the nominal function can be regarded as the backbone of the process structure following a V-model. The V process model is an extension of the waterfall model in which each phase of development resulting in a successively refined design (shown on the left leg of the V) has a corresponding verification phase[1] (shown in the right leg). Additional concerns are expanding the process scope with additional activities for each concern in each phase. In our figures and tables, functional safety and SOTIF are considered to intertwine into a general safety concern, while cybersecurity is maintained separately. The separation is due to scope, process, and treatment; it and is not intended to suggest an orthogonal outcome in the implementation. The suggested synchronisation is to address inconsistencies when concerns overlap. For example, cybersecurity threats and exploits that affect safety must be identified and resolved, and correspondently, safety assured integrity guarantees need to be put on the cybersecurity mitigations. The structure of all three aforementioned standards follows or can be adapted to the V-model development life-cycle. The multi-concern activities in phases on the left side of the V is referred to as co-design and discussed in Sect. 3.2, while on the right side, we talk about co-verification (including validation and assessment) in Sect. 3.3.

For the identified process phase scope (Fig. 1) there are 8 synchronisation points, 4 for the co-design part of the V model, and 4 for the co-verification part. The suggestion is that these 8 points in the combined process for the concerns are natural places for synchronisation that gives practical guidance on what and when synchronisation should occur. Thus, the proposed alignment is more refined than the normative requirements for cooperation between concerns found in the standards currently. If synchronisation is infrequent or imprecise regarding content, there is a risk of missing issues that may cause costly major redesigns if discovered late—or even worse, resulting in too high residual risks if inconsistencies are not discovered at all. Therefore, the claim is that the 8 synchronisation points are the minimum necessary for an effective alignment of the processes. To assess if the suggested synchronisation is sufficient to align concerns successfully is not possible without comprehensive evaluation and validation, to be investigated in the future. A possible evaluation scheme could compare loosely aligned processes, the suggested synchronisation approach, and a seamless process approach. A loosely aligned process would constitute the normative requirements in the standards where hazard and threat analysis and resulting risks are unified, and a seamless process approach would be a tailored total alignment of concerns with no distinction. Parameters of evaluation would be throughput and total effort in design and verification.

As mentioned in Sect. 2, the investigation is limited to activities from the initial design of the function up to validation thereof. This is the initial step for the future endeavour to present recommendations for a complete multi-concern *continuous* development, which also require guidelines encompassing the deployment and assessment activities during and after a release to market (e.g., maintenance, audits, incident detection and handling, over the air updates).

[1] It should be noted that it is primarily a model expressing dependencies in the refinement of design and verification phases for traceability, and does not necessarily mean the entire development project is performed in this sequence.

3.1 Elaboration on Concerns

Aspects of each concern under consideration can be allocated to the categories of means to attain dependability and security put forward by Avizienis et al. [1]. The categories are fault prevention (preventing the introduction of faults, e.g., with good engineering practices), fault tolerance (avoiding failures with a system design that can tolerate the existence of faults, e.g., redundancy), fault removal (removing faults from the system with, e.g., rigorous testing), and fault forecasting (evaluation of the system to forecast likely incidence and consequences of faults). Figure 2 shows the sources and interrelation of faults treated by each of the three standards and whether the sources stem from the operational context, i.e., the environment in which the vehicle will operate or from system development.

The functional safety concern addresses malfunctioning behaviour that stems from random hardware faults, foreseeable non-malicious misuse, and systematic faults. As shown in Fig. 2, random hardware faults and non-malicious misuse stem from the operational context. The risks introduced by these aspects are mitigated by fault forecasting and fault removal and implemented in the system development. There might be a need for further mitigation by fault tolerance mechanisms. Unavoidable fault modes need to have verified diagnostic coverage, e.g., diagnostics, monitoring and exception-handling mechanisms. The function is analysed for risks, specified, implemented, and tested in the system development process, potentially introducing systematic faults. The rigour of the process provides the primary prevention mechanism against systematic faults, and testing rigour provides additional fault removal.

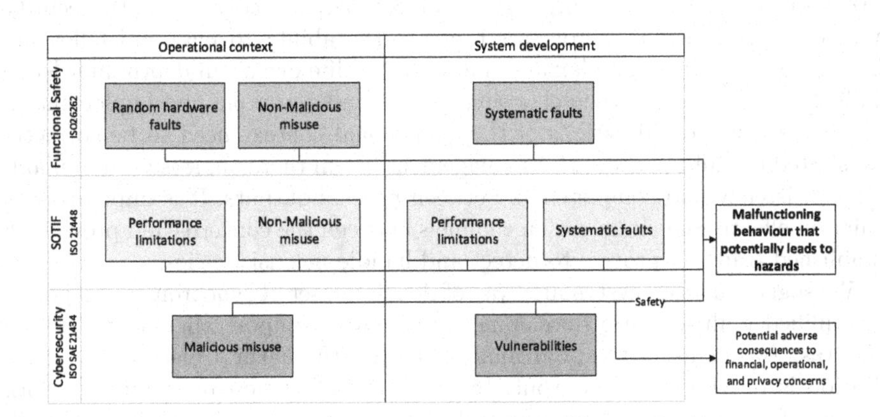

Fig. 2. Detailed multi-concern scope

An analogous logic applies to the safety of the intended functionality and cybersecurity, where risks of malfunctioning behaviour due to the operational context need to be mitigated and correctly addressed in the system development. During the system development, measures are needed against the introduction of systematic faults and vulnerabilities. Cybersecurity, in general, addresses the broader scope of adverse consequences to financial, operational, safety, and privacy concerns brought on by malicious misuse exploiting vulnerabilities, but here the focus is on safety issues.

In a multi-concern development life-cycle, there is a need to align fault forecasting, fault prevention, and fault tolerance between the concerns in the co-design of the system. Additionally, there is an advantage in coordinating and aligning the fault removal activities in the co-verification, and validation [23]. Figure 1 illustrates this combined life-cycle for a nominal function with added functional safety, SOTIF, and cybersecurity activities.

For each synchronisation point, one work product is identified from each concern for the reasons mentioned above. Thus, the synchronisation is incremental both in time and in content. The timing of synchronisation is dependent on the development phase, and the subject is dictated by the content of the identified work products. The selection of work products is the most seminal one for each refinement of the system. The mapping of each concern into a unified process revealed a natural set of triplets of work products to be synchronised both in time and content. A larger process scope would contain more synchronisation points. The goal of the work presented here is to enhance and formalise the alignment between concerns by introducing natural synchronisation points in a development life-cycle in the following Sections.

3.2 Co-design

A multi-concern approach has many prerequisites on the capabilities of the development organisation—organisational competence covering both security and safety areas and the processes thereof—the ability to plan and follow up the progress of a very complex and multi-faceted life-cycle—in-depth knowledge of all the consideration when choosing sensors and actuators and implementing controllers. The considerations of the operational context need to be correctly transferred to multi-concern risk management. From there on, it is vital to maintain consistency and completeness between process outputs. It is imperative to maintain effective communication channels between the concerns and provide an established trade-off process to detect and handle inconsistencies.

We suggest using a systematic way of defining a set of synchronization points in a unified multi-concern development process to support alignment between concerns. A conclusion from our analysis is that the natural place for synchronisation is on the process output. In Fig. 3 the allocation to significant work products from the three standards is illustrated. The synchronisation points in the co-design are expected to support a common understanding of the operational context. In the system design, the synchronisation points align analysis methods, countermeasures, and requirements. In general, for co-design, the main

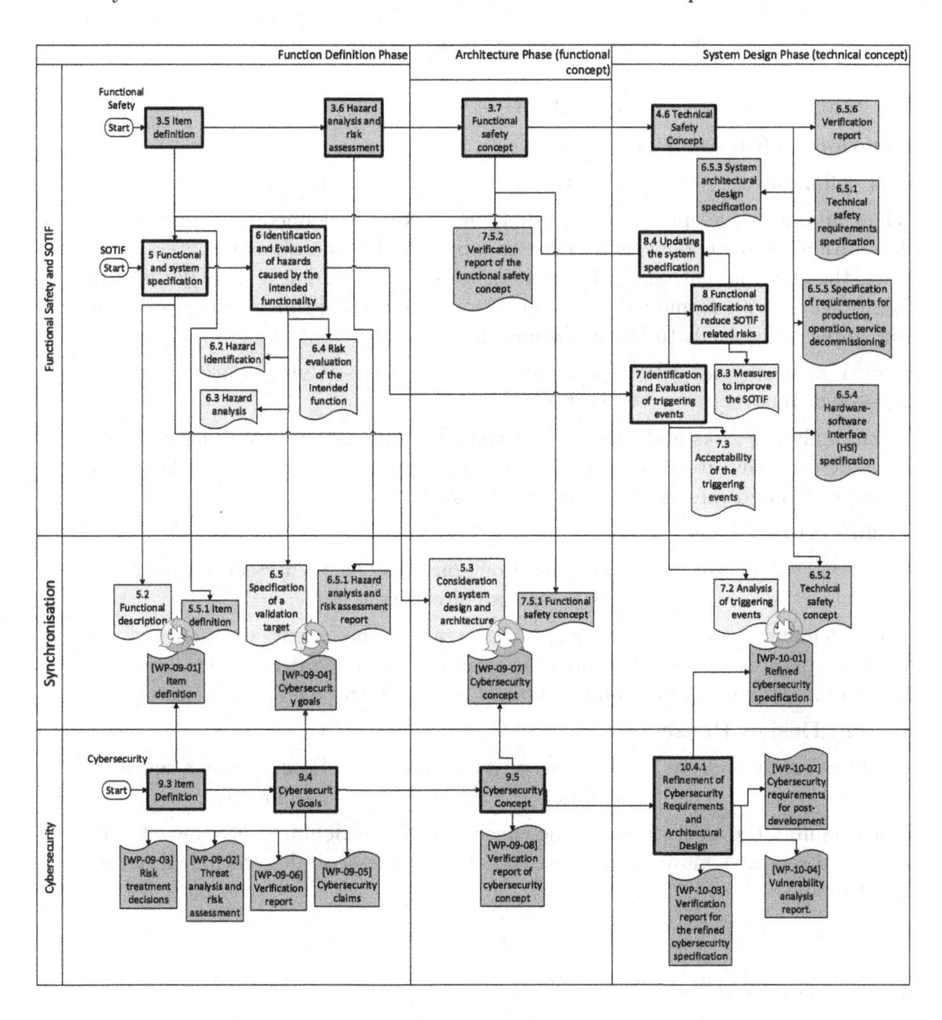

Fig. 3. Multi-concern process for co-design

advantage of treating all concerns in parallel is completeness, i.e., that all concerns and their interplay are considered throughout all phases, reducing the risk of missing inconsistencies that may cause malfunctioning behaviour of the system. Common for all synchronisation points in Table 1 is that the rigour of the analysis and process provides an added fault prevention mechanism, and review stringency delivers additional fault removal of inconsistencies.

Table 1. Co-design synchronisation points

ISO 26262	ISO 21488	ISO 21434
Function Definition Phase		
Item definition	Functional description	Item definition
When product development is initiated, the system is defined and described—elaboration on dependencies and interactions in the operational context, e.g., users, the environment, and other systems. Each concern has a slightly different scope (see Fig. 2), formalising the operational context, primary the overlap in fault forecasting needs to be synchronised.		
Hazard analysis and risk assessment report	Specification of a validation target	Cybersecurity goals
When analysing risks and establishing goals for the system stemming from the different concerns, it is essential to align assumptions of exposure, likelihood, consequences and interplay of faults in the work products.		
Architecture Phase		
Functional safety concept	Consideration on system design and architecture	Cybersecurity concept
While defining a functional concept that realises the intent and goals of the function, there is a need to remove inconsistencies in the degradation strategy, architectural requirements, fault tolerance measures and validation criteria.		
System Design Phase		
Technical safety concept	Analysis of triggering events	Refined cybersecurity specification
When refining the concept, the same considerations as defining the concept need to be addressed. A more comprehensive analysis is essential as technical details are available.		

3.3 Co-verification

The V-model's right side covers checking whether the system adheres to given properties formalised as requirements. Co-verification aims to act as fault removal activity for the particular concerns and presents an opportunity to investigate the interplay, implementing a well-thought-out integration test strategy coupled with verification and validation activities. An integration test strategy coordinates the use of test environments and test techniques for the concerns—considers how the techniques need to interact with the system—how close to production intent the system needs to be for the results to be relevant.

We suggest introducing systematic synchronisation points in the multi-concern co-verification process to maximise synergies in the use of test environment, and methods [23]. Therefore, synchronisation is allocated to the verification specification outputs. In Fig. 4 identified work products is shown. Test strategies are a subject of synchronisation, test reports not, since it assumed that any inconsistencies discovered in the testing results would be fed back into an appropriate co-design phase for analysis before planning the subsequent inte-

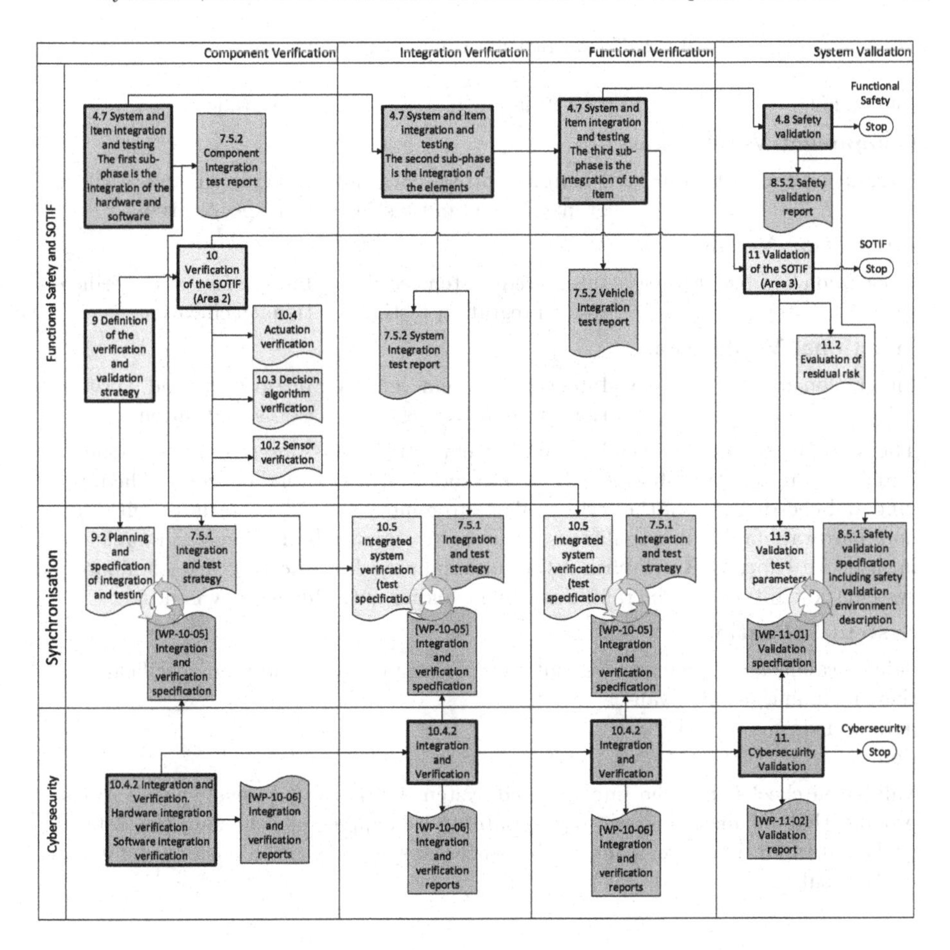

Fig. 4. Multi-concern process co-verification and validation.

gration stage. The bulk of testing required is not safety- or cybersecurity specific per se, but rather aimed at ensuring correct nominal function and good product quality in general and would require minimal alignment attention. Common to all synchronisation points in Table 2 is that testing rigour delivers fault removal.

Co-assessment. The aggregated evidence from all preceding safety and security-related activities proving that the system is safe and secure will need to be independently assessed. ISO 26262:2018 expresses the need to take interdependence with cybersecurity into account, but there is no support in transforming these requirements into a practical assessment process that checks adherence to ISO 21434. The same is true for ISO 21448, where it is stated that the examination of the results of the SOTIF activities can be considered in ISO 26262-2:2018 functional safety assessment, but not further explained how. ISO 21434 has a separate section that addresses cybersecurity assessment; however, the exact

Table 2. Co-verification synchronisation points

ISO 26262	ISO 21488	ISO 21434
Component Verification		
Integration and test strategy	Planning and specification of integration and testing	Integration and verification specification
Integration Verification		
Integration and test strategy	Integrated system verification (integration testing)	Integration and verification specification
Functional Verification		
Integration and test strategy	Integrated system verification (system testing)	Integration and verification specification
The integration steps are dealing with system components, integrating components into a system, and the integrated system, respectively. A general concern is the planning and coordinating of the testing efforts to remove the risks associated with random hardware faults, non-malicious misuse, performance limitations, and malicious misuse. An effort that establishes the underpinning for detecting and eliminating systematic faults and vulnerabilities while avoiding introducing new issues.		
System Validation		
Safety validation specification including safety validation environment description	Validation test parameters	Validation specification
Validation checks that the implemented system fulfils the top-level specifications, proving that all measures are appropriate and adequate in achieving the defined goals and validation criteria for all the concerns; synchronisation ensures alignment of the results.		

connection to ISO 26262 assessments is a non-normative annex. For example, one link to ISO 26262 states that the safety risks are best defined within the ISO 26262 scope and collected. All standards, however, suggest the use of the ISO 26262 scheme of independence for assessment.

The conclusion is that assessing if the interplay between concerns has been adequately addressed is not possible if the assessments are conducted independently for each concern. However, there is little practical guidance for conducting co-assessment and what it would entail, a topic that needs to be addressed soon. The assumption is that the suggested synchronised development process (Sect. 3) supports a joint incremental assessment/certification process with a similar approach, which might be enhanced by an assurance case template approach [3].

4 Related Work

Safety and security work has primarily been performed as independent activities, although there are several recent efforts towards unification or harmonising these concerns [12]. However, the endeavour to combining safety and security aspects during the development processes has been identified as non-trivial due to the high interference between these aspects and their respective treatment by Huber et al. [6]. One of the challenges identified, and one that is targeted by this paper, is the lack of experience, standards, and guidelines concerning the combination of the safety and security domains. Huber et al. also conclude that utilising a conceptual model unifying relevant documentation artefacts from requirements engineering, system modelling, risk assessment, and evidence documentation is the way forward, which aligns well with the suggested synchronised process in Sect. 3.

An alternative, more ad-hoc approach to synchronisation is suggested by Martinez et al., where interference analyses trigger co-engineering meetings and trade-off analyses [14]. The bottom-up triggering mechanism with a loose connection to governing processes could be a disadvantage for assessment, process control, and ease of adoption but might be preferable in agile development.

Supporting investigations into commonalities and cross-fertilization can give guidance on optimising and streamlining planning and implementing synchronisation concerns [11,15]. There are substantial efforts in investigating the safety and security interplay [4,5] that are useful in addressing them simultaneously. There is a need for special attention to bridge the gap between standards for automotive security engineering and hands-on, actual-system testing for verifying and validating automotive cybersecurity [13,22], that could be integrated into the process structure suggested in this paper.

There are several frameworks put forward that addresses safety and security. In general, our approach can, on a high level, be seen as a specialisation of [18]; set in a different standardisation landscape. However, we suggest keeping the life-cycle domain-specific to retain risk calibration to automotive. To handle the complexity for co-engineering safety and security concerns, it essential for tool support, which might be remedied by the explicit systematisation and management of commonalities and variabilities [2]. However, the scalability of the approach needs further investigation.

As a starting point to identify the scope of the applicable standard for the work presented here, the standards were inventoried [20] and enhanced [19].

5 Conclusions

This paper focuses on refining synchronisation guidance targeting multi-concern in several crucial development phases within the automotive domain—providing the basis for implementing a multi-concern process that operates within the envisioned relevant standardisation landscape. The result put forward builds on a unified process chart, organised around a generic V-model, describing relevant processes, artefacts, and activities, and a mechanism for synchronisation regarding content and timing between concerns at each step in the process. In addition, each concern is further refined and allocated to the categories of means to attain dependability and security.

The synchronisation suggested to be inserted in the process output on a non-dispensable standard set, ISO 26262 [7] (addressing functional safety), PAS ISO 21488 [8] (addressing the safety of the intended functionality, SOTIF) and final draft of the ISO/SAE 21434 [10] (addressing cybersecurity). The synchronisation of concerns when developing ADS aids the elimination of inconsistencies between concerns as early as possible due to the support of a common understanding of the operational context. For co-design, the main advantage of treating all concerns in parallel is completeness, i.e., that all concerns and their interplay are considered throughout all phases. In addition, the introduction of synchronisation points in the multi-concern co-verification process maximises synergies in using the test environment and methods. The evaluation of the effectiveness of the proposed synchronisation has not been investigated. As alignment does not alter the process or the outputs directly, a future evaluation could compare the added effort of alignment to the effort saved by the enumerated benefits.

The current investigation is a work in progress and limited to the initial design of the function up to validation thereof and presented here as the initial step for the future endeavour to support a complete multi-concern life-cycle—support covering the deployment and assessment activities addressing *continuous* development where co-assessment of interdependent concerns would be of particular interest.

Acknowledgement. This work was supported by the SECREDAS project with the JU Grant Agreement number 783119, and the partners national funding authorities.

References

1. Avizienis, A., Laprie, J.C., Randell, B., Landwehr, C.: Basic concepts and taxonomy of dependable and secure computing. IEEE Trans. dependable Secure Comput. **1**(1), 11–33 (2004)
2. Bramberger, R., Martin, H., Gallina, B., Schmittner, C.: Co-engineering of safety and security life cycles for engineering of automotive systems. ACM SIGAda Ada Letters **39**(2), 41–48 (2020)
3. Chowdhury, T., et al.: Safe and secure automotive over-the-air updates. In: Gallina, B., Skavhaug, A., Bitsch, F. (eds.) SAFECOMP 2018. LNCS, vol. 11093, pp. 172–187. Springer, Cham (2018). https://doi.org/10.1007/978-3-319-99130-6_12
4. Favaro, J.: AQUAS d1.3: Report on the evolution of co-engineering standards
5. Folkesson, P., Svenningsson, R., Söderberg, A., Wallerström, M., Montan, S.: HEAVENS d4 - interplay between safety and security
6. Huber, M., Brunner, M., Sauerwein, C., Carlan, C., Breu, R.: Roadblocks on the highway to secure cars: an exploratory survey on the current safety and security practice of the automotive industry. In: Gallina, B., Skavhaug, A., Bitsch, F. (eds.) SAFECOMP 2018. LNCS, vol. 11093, pp. 157–171. Springer, Cham (2018). https://doi.org/10.1007/978-3-319-99130-6_11
7. ISO: ISO 26262:2018 Road vehicles - Functional safety (2018)
8. ISO: ISO/PAS 21448:2019 Road vehicles - Safety of the intended functionality (2019)

9. ISO: ISO/TR 4804:2020 Road vehicles—Safety and cybersecurity for automated driving systems (2020)
10. (ISO SAE): ISO SAE DIS 21434 (e) road vehicles - cybersecurity engineering
11. Lautieri, S., Cooper, D., Jackson, D.: SafSec: commonalities between safety and security assurance. In: Redmill, F., Anderson, T. (eds.) Constituents of Modern System-safety Thinking, pp. 65–75. Springer, London (2005). https://doi.org/10.1007/1-84628-130-X_5
12. Lisova, E., Šljivo, I., Čaušević, A.: Safety and security co-analyses: a systematic literature review. IEEE Syst. J. **13**(3), 2189–2200 (2019)
13. Marksteiner, S., et al.: A process to facilitate automated automotive cybersecurity testing. arXiv preprint arXiv:2101.10048 (2021)
14. Martinez, J., Godot, J., Ruiz, A., Balbis, A., Ruiz Nolasco, R.: Safety and security interference analysis in the design stage. In: Casimiro, A., Ortmeier, F., Schoitsch, E., Bitsch, F., Ferreira, P. (eds.) SAFECOMP 2020. LNCS, vol. 12235, pp. 54–68. Springer, Cham (2020). https://doi.org/10.1007/978-3-030-55583-2_4
15. Piètre-Cambacédès, L., Bouissou, M.: Cross-fertilization between safety and security engineering. Reliab. Eng. Syst. Saf. **110**, 110–126 (2013). https://doi.org/10.1016/j.ress.2012.09.011
16. Pype, P.: SECREDAS project – SECREDAS will increase consumer trust in connected and automated transportation and medical industries. https://secredas-project.eu/
17. SAE: SAE J3016 - Taxonomy and Definitions for Terms Related to Driving Automation Systems for On-Road Motor Vehicles (2018)
18. Schmittner, C., Ma, Z., Schoitsch, E.: Combined safety and security development lifecylce. In: 2015 IEEE 13th International Conference on Industrial Informatics (INDIN), pp. 1408–1415. IEEE (2015). http://ieeexplore.ieee.org/document/7281940/
19. Schoitsch, E., Schmittner, C.: Ongoing cybersecurity and safety standardization activities related to highly automated/autonomous vehicles. In: Zachäus, C., Meyer, G. (eds.) AMAA 2020. LNM, pp. 72–86. Springer, Cham (2021). https://doi.org/10.1007/978-3-030-65871-7_6
20. Shan, L.: SECREDAS project deliverable d10.2 state-of-the-art analysis and applicability of standards (2019)
21. Skoglund, M., Thorsén, A., Arrue, A., Coget, J.B., Plestan, C.: Technical and functional requirements for V2X communication, positioning and cyber-security in the HEADSTART project. In: Proceedings of ITS World Congress 2021 (2021)
22. Skoglund, M., Warg, F., Hansson, H., Punnekkat, S.: Black-box testing for security-informed safety of automated driving systems. In: 2021 IEEE 93rd Vehicular Technology Conference (VTC2021-Spring), pp. 1–7 (2021). https://doi.org/10.1109/VTC2021-Spring51267.2021.9448691
23. Skoglund, M., Warg, F., Sangchoolie, B.: In search of synergies in a multi-concern development lifecycle: safety and cybersecurity. In: Gallina, B., Skavhaug, A., Schoitsch, E., Bitsch, F. (eds.) SAFECOMP 2018. LNCS, vol. 11094, pp. 302–313. Springer, Cham (2018). https://doi.org/10.1007/978-3-319-99229-7_26

Offline Access to a Vehicle via PKI-Based Authentication

Jakub Arm[1](✉), Petr Fiedler[2], and Ondrej Bastan[1]

[1] FEEC, Brno University of Technology, Brno, Czechia
Jakub.Arm@vut.cz
[2] CEITEC, Brno University of Technology, Brno, Czechia

Abstract. Using modern methods to control vehicle access is becoming more common. At present, various approaches and ideas are emerging on how to ensure the access in use cases reflecting car rental services, car sharing, and fleet management, where the process of assigning car access to individual users is dynamic and yet must be secure. In this paper, we show that this challenge can be resolved by a combination of the PKI technology and an access management system. We implemented a vehicle key validation process into an embedded platform (ESP32) and measured the real-time parameters of this process to evaluate the user experience. Utilizing the SHA256-RSA cipher suite with the key length of 3072 bits, we measured the validation time of 46.6 ms. The results indicate that the user experience is not worsened by the entry delays arising from the limited computing power of embedded platforms, even when using key lengths that meet the 2020 NIST recommendations for systems to be deployed until 2030 and beyond.

Keywords: PKI · X.509 · Certificate · Access management · Key management

1 Introduction

Controlled vehicle access is ensured via various technical solutions. In the consumer domain, wireless technologies based on RF (Radio Frequency) and NFC (Near Field Communication) have prevailed. Currently, cars allow access (unlocking the door and starting the engine) only in the presence of the correct code transmitted from the user′s unique mechanical key with an embedded transponder; thus, the user must have such a key. Although this widely used approach is considered secure enough, threats such as replay-attack and relay-attack still need to be addressed. Moreover, the solution exploiting a physical key is not convenient for scenarios like granting access to the company fleet vehicles or car sharing, i.e., scenarios where the physical key needs to be passed from one person to another.

These scenarios require a rather quick transfer of the access, ideally without the need to manipulate with the physical key. In general, solutions based on traditional car keys with an integrated RF transponder in the car sharing sector tend to cause many problems. Multiple challenges need to be solved to ensure a reliable and secure operation of key control management, including that presented [1]:

I. Habli et al. (Eds.): SAFECOMP 2021 Workshops, LNCS 12853, pp. 76–88, 2021.
https://doi.org/10.1007/978-3-030-83906-2_6

- Combating unauthorized vehicle usage;
- resolving lost and misplaced car keys;
- keeping track of the keys when the user is an outside contractor;
- reducing the risk if a third party is involved.

1.1 Standard Car Access Methods

A rental/car share (RCS) and other management systems can employ barcodes, QR (Quick Response) codes (or NFC/RFID), GPS, and mobile apps coupled with a wireless network to enable customers to bypass the reservation desk. The backend servers and mobile applications communicate with a lock module integrated in the vehicle to maintain the valid codes (keys) and control the access. Presently, such an approach is patented by [2].

Another patent [3] describes the use of Smart card systems in connection with transportation services; the smart card acts as a user ID that not only controls the access to the vehicle but also manages the driver-specific settings.

The above patents describe a car sharing and fleet management system ensuring support for symmetric cipher technologies (mostly provided by the AES-CBC cipher algorithm), which is prone to various attacks, namely, the replay attack, cipher algorithm attack, brute force, password guessing, and cipher key compromising; such situations occur especially if insufficient key lengths are used (below 128 bits). Current recommendations for future-proof security solutions (2030 and beyond) advise using symmetric encryption algorithms with a key size of at least 128 bits [17]; however, to allow truly long-term protection against mathematical attacks and quantum computing, a key size of 256 bits is sometimes recommended [18].

1.2 Using PKI in the Automotive Domain

The intention to use PKI (Public Key Infrastructure) in the automotive domain is not new. In this context, for example, let us point to reference [4], which already in 2006 outlined the possibilities of incorporating more advanced security methods, including the application of PKI, in car sharing, fleet management, and various other scenarios. However, as regards car sharing or fleet management, no detailed solution involving an architecture diagram and/or a detailed description of the functionality and performance has been published to date. The PKI-based technology can be advantageously employed in the car access scenario to facilitate access rights validation including offline validation through validating the digital certificate. In general, the most typical scenarios for the use of PKI in the automotive field are:

- V2X communication;
- car or car component authentication;
- secured car access.

The current role of the PKI technology in the automotive environment rests mainly in aiding message security (message authentication) within the V2X communication scenario, where communication between cars or cars and RSUs (Road-Side Units) must be digitally signed to prevent misuse of the system by an attacker or another malicious entity [5].

A generalized scenario for the application of PKI in motoring relies on authenticating the car to the outside world (third parties), in terms of authentication exploiting a unique Vehicle ID assigned to each vehicle by the vehicle manufacturer. Such a functionality is beneficial in the following use cases [6]:

- identifying the vehicle throughout its lifecycle (in repair shops, for example);
- upgrading the software or firmware in the control units of the car;
- ensure secure communication in telematic applications, including vehicle tracking and fleet management;
- securely manage or replace car keys if these are, for example, lost or broken.

Additionally, vehicle authentication is a prerequisite for V2X certificate enrollment. Another scenario employing PKI in the automotive domain is the security of networked devices (ECUs - Electronic Control Units) inside the car [7]. The goal is to prevent potential external and internal attackers from modifying the devices or spoofing the firmware during the firmware upgrade. These tasks require reliable and trusted key management at the level of the in-vehicle communication networks.

1.3 PKI Problems and Limitations in Car Access Management

The implementation of the PKI technology in vehicle access management also introduces some technical challenges and limitations. The major issues include insufficient computing resources in automotive equipment. Further, we have to consider the challenging automotive qualification process as regards long-term component reliability, and, importantly the fact that the PKI concept itself brings some specific challenges. The distributed authentication can be exploited advantageously in standard situations; however, problems may occur in non-standard situations, e.g., if the key was issued correctly but returned unexpectedly before expiring (e.g., late cancelation of a vehicle reservation). A solution utilizing the standard revocation process is inadequate in some of the scenarios.

The revocation process can be performed via various methods or instruments: the CRL - Certificate Revocation List, CRT - Certificate Revocation Tree, OCSP - Online Certificate Status Protocol, Novomodo, or short-lived certificates. Each of these options has its advantages and drawbacks, such as computational complexity, connectivity dependence, and available memory [8]. Although the OCSP technique appears to be the best choice for embedded devices, it needs permanent Internet connection to remain functional. Regrettably, such connectivity may not be feasible, and offline car access key validation has to be relied on. This lack of connectivity is typical in the underground garage environment or during a DoS (Denial of Service) attack. For the scenario (use case car sharing and fleet management) concerning access to a vehicle, it is, therefore, most advantageous to use the short-lived certificate method, possibly in combination with the OCSP.

By evaluating the computational complexity of the security algorithms providing PKI operations (pairing, request creation, certificate signing, authentication), it has been found [9] that the longest operation rests in the validation of a certificate against a CA certificate or even a certificate chain. In [9], the author states that the best verification times on the Raspberry PI platform are 2.542 s in the RSA algorithm with the key length of 1024 bits and 8.905 s in the ECC algorithm with the key length of 1024 bits. Verification exceeding 1 s is hardly acceptable for the average user; it is accepted that 1 s corresponds approximately to the limit of the user's flow of thought [19] and thus also the limit of the broadly acceptable time delay while interacting with a vehicle. In bad weather conditions, above all, any noticeable delay is highly unwelcome.

Even when using computing platforms with limited computing power compared to standard platforms, we have to reduce the time required for these operations, or, more concretely, the key validation task.

1.4 Alternative Approaches to Car Access

Importantly, a set of other, more sophisticated, approaches towards key management security for cars are available to resolve the problems above; these options involve, above all, principles on a blockchain basis. According to our survey, however, such approaches are only theoretical or have remained in the prototype phase at best.

The instruments and tools include, for example, the P3KI-based system, or Decentralized Offline Authorization for IoT (Internet of Things) in general. This system is based on the Web-of-Trust concept, using rescinding instead of explicit centralized certificate revocation. In this setup, the nodes "change their minds" about who they trust and to what degree they do so [20].

Applying Smart Contracts (based on the blockchain theory) to secure car sharing systems allows these schemes to operate without a trusted intermediary. The use of smart contracts deployed on the Ethereum blockchain ensures that full-fledged car sharing functionalities along with various countermeasures to tackle malicious behavior are provided [10]. While this approach is theoretical and has not been employed yet, it embodies one of the possible ways the car sharing solution in the future.

SePCAR is a privacy-enhancing protocol to facilitate car access provision based on the public ledger. It delivers generation, update, revocation, and distribution mechanisms for access tokens to shared cars, together with procedures to solve disputes and to deal with law enforcement requests in, for instance, car accidents. The proof-of-concept implementation shows that SePCAR takes 1.55 s to provide car access [11]. We find this concept promising and appreciate the mathematical proof; however, incorporation into a real system necessitates a significantly higher amount of work, mostly in the form of standardization.

Through an extension and continuation of SePCAR, the HERMES tool [12] was designed and evaluated. HERMES securely outsources vehicle access token generation operations to a set of untrusted servers by concealing the secret keys of vehicles and the transaction details from the servers, including the vehicle booking details, access token information, and user and vehicle identities. The tool is built on the multipart computation protocols HtMAC-MiMC and CBC-MAC-AES. The authors suggest that the generation of a token is 42 times faster compared to the previous SePCAR and access token operations performed on a prototype vehicle on-board unit, which take approximately 62.1 ms. We assume this is a perfectly acceptable value because, from the user's perspective, delays close to 100 ms are considered negligible and perceived as instantaneous.

In access control, standard ICT techniques already exist, including but not limited to Radius (described in RFC 3579) and Diameter (described in RFC 3588). However, these are intended for advanced access control in intangible objects (service, software). The techniques also define a specific approach to securing the communication, i.e., physical permission or denial of the communication, the actual way of negotiating with a participant, and permanent connection to the server. We, therefore, consider such solutions unsuitable for the vehicle access scenarios.

1.5 The Security of the PKI System

PKI is based on X.509 certificates, which exploit one of the following asymmetric cipher algorithms:

- RSA (utilizing the factoring of large numbers, and the RSA problem);
- DSA, El Gamal (based on non-effective solving of the discrete logarithm);
- ECC (elliptic curves).

From the perspective of security, without side-channel attacks, PKI exploits the complexity of the applied cipher algorithm.

In source [13], the authors use a testbed evaluating the performance and energy consumption of Transport Layer Security (TLS) 1.2 ECC (Elliptic Curve) Cryptography and RSA (Rivest-Shamir-Adleman) cipher suites (that comply with the TLS 1.3 standard requirements, whose introduction is currently at the preparatory stage). The results show that ECC outperforms RSA in both energy consumption and data throughput for all of the tested security levels. Moreover, the importance of selecting a proper ECC curve is highlighted.

Certificates according to X.509 version 3 (RFC 3280 and its descendant, the RFC 5280 [21]) standard, are mostly utilized for the authentication and securing of communication. The standard defines the structure of the certificate and processes concerning certificate formation and validation. As the cipher suite, the SHA256-RSA or the SHA256-ECDSA is most often used followed by Base64 encoding. In addition to implementation and architecture issues, this technology also suffers from security flaws relating to the encryption mechanism itself; among the most feared attacks on the cipher algorithm is the creation of a hash collision, where two valid certificates are involved, one of which is forged and potentially dangerous. In addition, if an attacker has a valid CA certificate with the CA attribute, he or she can generate additional certificates based on this spoofed certificate, which will be valid during the authentication process [14].

An analysis of the new standard for secured communication [15] shows that some encryption algorithms (e.g., AES-CBC or MD-5) are no longer supported. Regrettably, to date, the TLS 1.3 standard has not yet been fully implemented and deployed in the embedded world. Therefore, technologies according to the TLS 1.2 standard, which is already known to contain security issues, are still being employed in new devices. The TLS1.3 standard also brings support for the embedded world, i.e., devices with less computing power, and IoT devices, i.e., devices with limited connectivity and higher security requirements. The authors also claim that some vulnerabilities remain present despite the state-of-the-art status of the TLS 1.3.

The validation process of a certificate which complies with the X.509 standard is performed against the ancestor certificate, which itself is validated against its ancestor certificate. This hierarchic queue of certificates is called a chain of certificates [16]. Such mechanism implies that a leaf (client) certificate must be validated against its ancestor certificate (a car sharing company or a fleet management service in our case).

2 The Proposed Concept

The essence of the proposed approach lies in using the PKI (Public Key Infrastructure) technology based on X.509 certificates to assure the security of the in-vehicle access control. In the car sharing and fleet management scenario, the access is managed by a backend server (Fig. 1). In our solution, the key empowers the user only for a short time (short-lived certificates to assure off-line certificate revocation), and the key is linked to the individual vehicle reservation. Furthermore, the reservation cancelation function based on OCSP (Online Certificate Status Protocol) is also implemented in our solution to manage changes in the status of the individual keys (reservations); the backend server uses the OCSP-based revocation; this function, obviously, requires an active connection. When the vehicle is off-line, the short-lived certificate technology is used as a backup to increase the resilience of the solution against a DoS attack at the connectivity.

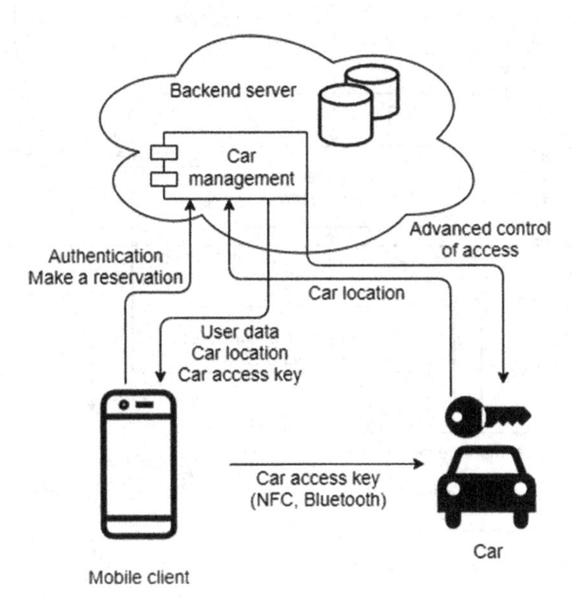

Fig. 1. The block diagram of the vehicle access system (car sharing or fleet management).

2.1 Architecture

By adopting PKI for the purposes of the access control system in physical objects, we, map services normally provided by the standard components of the PKI system (RA, VA, CA) to the individual components of the car sharing system. The access device (an in-vehicle ECU controlling the door lock via a CAN communication network) is integrated into the physical object (a vehicle), thus acting as a validation authority (VA). As such, it verifies the validity of the certificates that are presented to it when the user attempts to open the vehicle. Within the certificate, an active reservation with valid parameters is embedded. A mobile user application (installed in the client's phone), i.e., the reservation service, acts as a registration authority (RA) because it ensures the creation of new keys (reservations) with certain parameters (the start and end of the reservation; company name; and, optionally, vehicle designation) by forwarding these requests to the certification authority (CA).

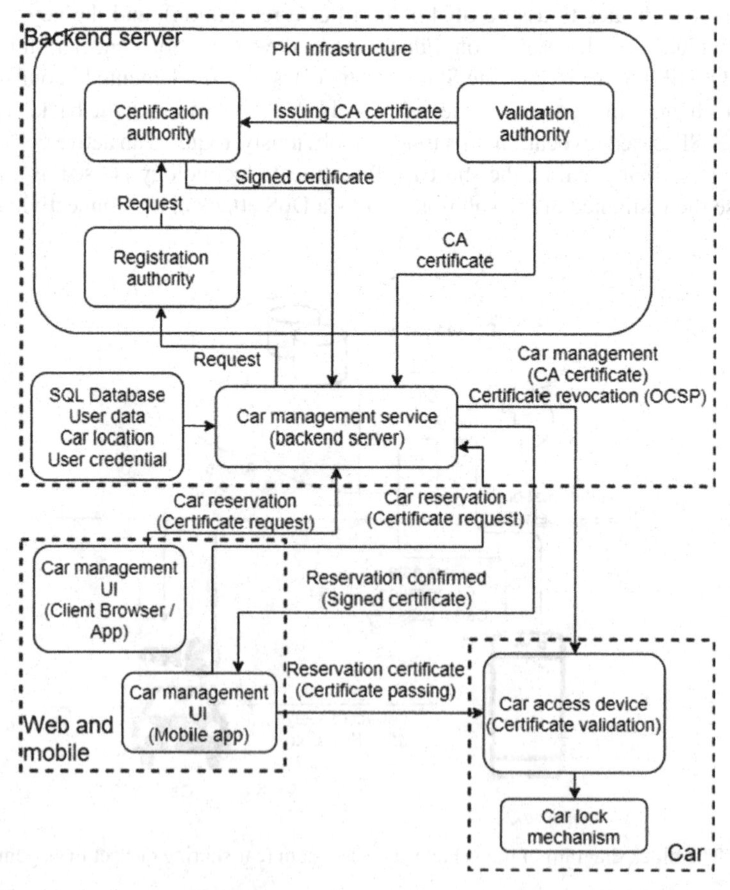

Fig. 2. The block diagram of the proposed car access system, where the PKI technology is incorporated to ensure car key creation and validation.

The advantage of this solution is that the protected physical asset (a vehicle) does not have to be actively connected to the Internet during the key validation process (reservation). This corresponds to situations where, for example, the car is in a garage or a place without Internet connectivity. On the other hand, the business model of the company providing the service must be set appropriately to prevent misuse of the possibility of making reservation just before accessing the object (vehicle) while the object itself is without active connectivity. These situations are outlined at the end of the paper.

Figure 2 describes a diagram of the proposed solution, where the PKI infrastructure ensures the dynamic access control to the vehicle. The system is designed as a server database application in conjunction with a mobile application providing an interface for the customers. The individual components of this system, therefore, take over the individual functions of the PKI technology.

This architecture thus represents a case where the entire PKI-based infrastructure runs and is provided by the operator company (car sharing or fleet management). The second, and perhaps the more frequent case will be the architecture, where the PKI infrastructure, including the root certificates, will be outsourced to a specialized company providing the PKI services. In such cases, the service providers (car sharing or fleet management) will only be intermediaries, i.e., they will maintain intermediate certificates (Fig. 3).

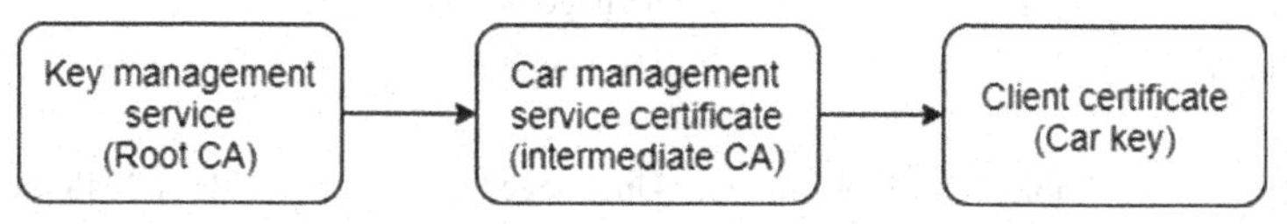

Fig. 3. A certificate chain reflecting the mode in which the key management service can be autonomous, instead of being a part of the car sharing or fleet management system.

2.2 X.509 Adaptation

The virtual reservation key is represented by a valid certificate according to the X.509v3 standard (described in RFC 3280). As mentioned earlier, this standard defines not only the process of handling certificates but also their structures. Thus, the certificate (acting as a key to the vehicle) encapsulates parameters such as the beginning and end of the reservation, or, optionally, the object designation in the case of using one root key for multiple devices (vehicles). The user to whom a valid certificate is issued (e.g., a paid reservation) produces this certificate (sent utilizing the short-range technology) at the car access device of the given object, whose access is controlled. To allow usage in the vehicle access control scenario, i.e., in the use case for car sharing and fleet management, the attributes of the certificate are filled in according to Table 1.

Table 1. The attributes modified against the X.509 standard.

Attribute	Description	Meaning in our scenario	Sample data
Version	Certificate version	Unchanged	1
Serial number	Certificate serial number	Certificate serial number of car-sharing company PKI	2
Signature algorithm	Cipher suite description	Unchanged	SHA256-RSA
Validity	Certificate validity	The time span which defines the exercisability of the virtual key	
Not Before			May 9 13:30:00 2021 GMT
Not after			May 10 15:00:00 2021 GMT
Subject	Subject of the certificate	Customer (car sharing) or employee (fleet management)	Customer name
Subject public key Info		Unchanged	
Public key algorithm	Public key algorithm identification	Unchanged	RSA 2048 bit
Public key		Unchanged	Data
Issuer unique identifier	(optional)	Not used	
Subject unique Identifier	(optional)	Not used	
X509v3 extensions	(optional)	Not used	

2.3 Special Situations

When using the PKI technology to represent and manage virtual keys for use cases like car sharing and/or fleet management, various situations may arise due to the combination of business models and security technologies that do not meet the limitations of the applied technologies.

Such situations include late cancellation of the reservation of a vehicle, which itself is off-line. In such a situation, the canceled reservation will be considered valid when validated because the car access device has no response from the OCSP server. The possible solutions are:

- Do not accept the reservation cancelation request when the vehicle is parked in a location without connectivity. This solution should be reflected in the business model and service conditions.
- Do not deliver the certificate (create certificate) until the reservation is non-cancelable. Another situation leading to potential trouble is one where, for some reason, two or more certificates (virtual keys) are valid at the same time. This includes the scenario in which there is a group of people traveling together and thus needing to access the same vehicle. Another scenario that might require concurrent access to the vehicle includes various emergencies, including the need to access the vehicle for unplanned maintenance/service. Therefore, the basic solution described above still has to be enhanced to accommodate for such non-standard uses, e.g,. by using additional attributes that would in more detail define the rights associated with the given certificate (unlock the vehicle, switch on the vehicle, drive the vehicle, service the vehicle, etc.).

2.4 Evaluating the Validation Time for a Certificate

We measured (Table 2) the validation time of the end certificate (virtual vehicle key) against the root certificate on an Espressif ESP32 platform, employing an Extensa 32-bit core @ 240 MHz, 530 KiB SRAM, with a cryptographic HW Accelerator to support SHA-256, AES, RSA, and RNG. We used Mbed TLS libraries with TLS 1.2 version as the internal API for the cryptographic operations (TLS 1.3 is not fully implemented at present). For all measurements, the root and client certificates were already in the ESP32 memory, and the number of levels of the certification chain equals one, i.e., we validated the client certificate against one parent certificate (issued by the car sharing service or fleet management service). During the measurements, all necessary services had already been initialized (performed in the MCU start-up phase).

The executed measurements show the results of the verification of a certificate based on RSA (1024 bits, 2048 bits, 3072 bits, and 4096 bits) and ECDSA exploiting the X9.62 curve (256 bits). Each result was calculated from 1,000 measurements, and therefore the average, maximum, and variance values are provided. For each algorithm, we separately distinguished the length of the encryption key.

Table 2. The measurement of the validation time.

Algorithm	Key length [bits]	Average [ms]	Max [ms]	Variance [ms]
RSA	1024	6.15	10.62	0.04
RSA	2048	21.63	31.88	0.21
RSA	3072	46.61	64.88	0.67
RSA	4096	81.22	108.43	1.48
ECDSA	256	635.15	636.05	0.05

2.5 Limitations of the Study

While the incorporation of the PKI technology into the vehicle access system (car sharing or fleet management) was found successful, several limitations arose simultaneously, due in particular to the limitations of PKI technology and associated standards. For example, to ensure the non-repudiation and traceability of the vehicle "ownership" a car key management system (the backend part) cannot be implemented in such a way that the business operator can freely create and pass car keys. This is surely an advantage from the security perspective but also a drawback because there is no full privileged master in the system, capable of solving every special situation that was unaccounted for during the business process design.

In addition, the system is vulnerable to X.509 attacks, i.e., the following issues:

- The private root could be compromised.
- Revocating the root certificate is a challenge.
- The car access rights are aggregated in a single file, implying problems with the car key management (mostly in a mobile app.).
- The root CA cannot prohibit intermediated CAs (car sharing companies) from issuing certificates (car keys) for vehicles they do not own.
- The car key management company (car sharing or fleet management) has to form a long-term certificate to create the public authority imported into every car access device. Before the root certificate expires, the new root certificate must be uploaded in every device. Such a situation can be solved by a secured OTA update.

3 Conclusion

We proposed and defined a car management system architecture, exploiting the PKI technology for the car access scenario. Using the X.509, we defined the attributes of a car access key/certificate and the process of creating, signing, and validating it as incorporated into the car access system. On the ESP32 platform, we performed multiple measurements of the certificate validation process duration. The results show that the validation of the RSA certificate with the key length of 3,072 bits takes only 46.61 ms on the average and 64.88 ms at the maximum. This key length meets the protection level recommendations relating to the period after 2030, as provided by NIST [17] and NSA [18]. Moreover, the results indicate that a solution requiring the validation of multiple certificates to grant access to a vehicle can provide satisfactory short entry delays.

Thus, low computing power does not constitute a problem if we utilize key lengths that provide a high level of security for both the present time and the near future.

Acknowledgment. The research was supported by Brno University of Technology. A portion of the work was carried out under the support of the core facilities of CEITEC – the Central European Institute of Technology. The investigation procedures received funding from the following projects: FV30037 Research and development of new control systems for purchasing platforms, FV40196 Research and development of the monitoring of immobile persons who are tethered to the bed in terms of risk of suffocation – decubitus, FV40247 Cooperative robotic platforms for automotive and industrial applications, TF04000074 Digital representation of Assets as a configurable AAS for OT and IT production systems, TH02030921 The sophisticated wireless network with elements of IoT for plant protection and water management, FEKT-S-20–6205 Industry 4.0 in automation and cybernetics, and 783119–1 SECREDAS Product Security for Cross-Domain Reliable Dependable Automated System (H2020-ECSEL, EU). All of these grants and institutions facilitated the efficient performance of the research and associated tasks.

References

1. Morse Watchmans: Optimizing Fleet Management with Key Control. https://www.morsewatchmans.com/hubfs/Content%20offers/Fleet-Management-Whitepaper.pdf. Accessed May 2021
2. Enterprise Holdings Inc.: Rental/Car-Share Vehicle Access and Management System and Method. US20140309842A1. Inventors: Jefferies, J.E., Rod, W., DeMay, Gurgen, L., Lachinyan. US patent
3. Smart card systems in connection with transportation services. US20060157563A1. Inventor: David Marshall. US patent
4. Adelsbach A., Huber U., Sadeghi AR.: Secure software delivery and installation in embedded systems. In: Lemke, K., Paar, C., Wolf, M. (eds) Embedded Security in Cars. Springer, Berlin, Heidelberg (2006). https://doi.org/10.1007/3-540-28428-1_3
5. Thales eSecurity. Securing the connected vehicle. http://go.thalesesecurity.com/rs/480-LWA-970/images/ThalesEsecurity_Connected_Vehicle_sb.pdf. Accessed May 2021
6. Nexus: PKI for vehicle ID. https://doc.nexusgroup.com/display/PUB/PKI+for+vehicle+ID. Accessed May 2021
7. Alfred, J.: Automotive security and trust management: The case for PKI. BlackBerry Technology Solutions, Certicom. https://www.certicom.com/content/dam/certicom/images/pdfs/wp-automotive%20security%20and%20trust%20management-jul2015.pdf. Accessed May 2021
8. van Oorschot P.C.: Public-key certificate management and use cases. In: Computer Security and the Internet. Information Security and Cryptography. Springer, Cham (2020). https://doi.org/10.1007/978-3-030-33649-3_8
9. Dahlmann, F.: PKI For Automotive Applications. Bachelor thesis. Supervisor: Prof. Dr. Thomas Eisenbarth. Universitat zu Luebeck (2018). Accessed May 2021
10. Madhusudan, A.: Applying Smart Contracts to Secure CarSharing Systems. Master thesis. Supervisor: Prof. dr. ir. Bart Preneel. KU Leuven (2018). Accessed May 2021
11. Symeonidis, I., Aly, A., Mustafa, M.A., Mennink, B., Dhooghe, S., Preneel, B.: SePCAR: a secure and privacy-enhancing protocol for car access provision. In: Foley, S.N., Gollmann, D., Snekkenes, E. (eds.) ESORICS 2017. LNCS, vol. 10493, pp. 475–493. Springer, Cham (2017). https://doi.org/10.1007/978-3-319-66399-9_26
12. Symeonidis, I., Rotaru, D., Mustafa, M.A., Mennink, B., Preneel, B. Papadimitratos, P.: HERMES: Scalable, Secure, and Privacy-Enhancing Vehicle Access System (2021)
13. Suárez-Albela, M., Fraga-Lamas, P., Fernández-Caramés, T.M.: A practical evaluation on RSA and ECC-based cipher suites for IoT High-security energy-efficient fog and mist computing devices. Sensors. **18**(11), 3868 (2018). https://doi.org/10.3390/s18113868
14. Kaminsky, D., Patterson, M.L., Sassaman, L.: PKI layer cake: new collision attacks against the global X.509 infrastructure. In: Sion, R. (ed.) FC 2010. LNCS, vol. 6052, pp. 289–303. Springer, Heidelberg (2010). https://doi.org/10.1007/978-3-642-14577-3_22
15. Levillain, O.: Implementation flaws in TLS stacks: lessons learned and study of TLS 1.3 benefits. In: Garcia-Alfaro, J., Leneutre, J., Cuppens, N., Yaich, R. (eds.) CRiSIS 2020. LNCS, vol. 12528, pp. 87–104. Springer, Cham (2021). https://doi.org/10.1007/978-3-030-68887-5_5

16. van Oorschot P.C.: Public-key certificate management and use cases. In: Computer Security and the Internet, pp. 217–221. Information Security and Cryptography. Springer, Cham (2020). https://doi.org/10.1007/978-3-030-33649-3_8
17. Barker, E.: NIST special publication 800–57 part 1 revision 5: . NIST (2020). https://doi.org/10.6028/NIST.Spp.800-57pt1r
18. Commercial National Security Algorithm (CNSA) Suite Factsheet, MFS U/OO/814670–15, NSA, January 2016. https://apps.nsa.gov/iaarchive/library/ia-guidance/ia-solutions-for-classified/algorithm-guidance/commercial-national-security-algorithm-suite-factsheet.cfm. Accessed May 2021.
19. Nielsen, J.: Usability Engineering. Academic Press, San Diego, CA, USA. (1994)
20. Jehle, G.: P3KI Explained: Decentralized Offline Authorization for IoT, version 13, October 2019. P3KI_Explained__Decentralized_Offline_Authorization_for_IoT__1.3.pdf. Accessed May 2021
21. Cooper, D. et al.: Internet X.509 Public Key Infrastructure Certificate and Certificate Revocation List (CRL) Profile, May 2008. https://datatracker.ietf.org/doc/html/rfc5280. Accessed June 2021

HEIFU - Hexa Exterior Intelligent Flying Unit

Dário Pedro[1,2,3](✉), Pedro Lousã[1], Álvaro Ramos[1], J. P. Matos-Carvalho[4], Fábio Azevedo[5,6], and Luís Campos[1]

[1] PDMFC, Lisbon, Portugal
{dario.pedro,pedro.lousa,alvaro.ramos,luis.campos}@pdmfc.com
[2] Centre of Technology and Systems, UNINOVA, Lisbon, Portugal
[3] Electrical Engineering Department, FCT, NOVA University of Lisbon, Lisbon, Portugal
[4] Cognitive and People-Centric Computing Labs (COPELABS), Universidade Lusófona de Humanidades e Tecnologias, Campo Grande, Lisbon, Portugal
joao.matos.carvalho@ulusofona.pt
[5] Beyond Vision, Ílhavo, Portugal
fabio.azevedo@beyond-vision.pt
[6] Electrical and Computing Engineering Department, FEUP, U.Porto, Porto, Portugal

Abstract. The number of applications for which UAVs can be used is growing rapidly, either because they can perform more efficiently than traditional methods or because they can be a good alternative when there are risks involved. Indeed, as a result of some incidents that could have resulted in disastrous accidents, the European Union is tightening regulations regarding the use of drones and requiring formal training as well as logged missions from those who want to use UAVs above a certain MTOW for whatever reason, whether domestic or professional. If the application requires BVLOS flights, the limitations become much more stringent. In this article HEIFU is presented, a class 3 hexacopter UAV that can carry up to an 8 kg payload (having a MTOW of 15 kg) and a wingspan of 1.5 m, targeting applications that could profit much from having fully automated missions. Inside, an AI engine was installed so that the UAV could be trained to fly, following a pre-determined mission, but also detect obstacles in real-time so that it can accomplish its task without incidents. A sample use case of HEIFU is also presented, facilitating the temporal replication of an autonomous mission for an agricultural application.

Keywords: HEIFU · UAV · Drones · Collision avoidance · Resilience · Artificial Intelligence · Deep learning

1 Introduction

Although regarded as toys by many people, the range of applications that can use Unmanned Aerial Vehicles (UAVs) is growing fast, either because they can perform more efficiently than conventional methods or because they can be a good

I. Habli et al. (Eds.): SAFECOMP 2021 Workshops, LNCS 12853, pp. 89–104, 2021.
https://doi.org/10.1007/978-3-030-83906-2_7

replacement whenever there are risks involved [1,2]. While using UAVs however, many of these applications run behind an ideal efficiency, because they usually require an experienced pilot to take good advantage of their characteristics, as it is still deemed too risky to leave them operating autonomously. As a matter of fact, due to some incidents that could have caused catastrophic accidents [3–8], the European Union is tightening the regulations regarding the usage of drones (e.g. Reg. 945/2019/EU and Reg. 947/2019/EU) and demanding formal training as well as logged missions to those who want to use UAVs above a certain Maximum Take-Off Weight (MTOW) for whatever purpose, be it recreational or professional [9]. The restrictions are even higher if the application may require Beyond Visual Line-Of-Sight (BVLOS) flights [10]. Targeting applications that could profit much from having fully automated missions [11–13], PDMFC in collaboration with Beyond Vision has decided to develop the Hexacopter Exterior Intelligent Flying Unit (HEIFU), a class 3 hexacopter UAV that can carry up to 8 kg payload (having a MTOW of 15 kg) and has a wingspan of 1.5 m. This aircraft is illustrated in Fig. 1.

Fig. 1. HEIFU in flight.

Inside, an Artificial Intelligence (AI) engine was installed so that the UAV could be trained to fly, following a pre-determined mission, but also detecting obstacles (either static or dynamic) in real-time so that it can accomplish its task without incidents. Its geographic awareness engine also comprises dynamic no-fly zones imposed by the country's authorities in order to ensure that it does not fly over forbidden zones even if instructed to. Also, both PDMFC and Beyond Vision participate in the European SECREDAS project, which aims to ensure security, safety and privacy on automated vehicles. This will enable to further increase both the security and safety of the HEIFU flights on the road to fully automated missions, which can run from the cartographic assessment of a large field to the urgent transportation of critical goods (like human organs) without the need of an experienced pilot. In the following chapters, we will first describe the hardware on-board (Sect. 2) and the background software (Sect. 3). The collision avoidance subject is discussed on Sect. 4 and Sect. 5 approaches the connectivity issues. Section 6 shows some field results achieved while Sect. 7 targets future work.

2 Hardware

In order to be able to carry out fully automated missions, the UAV's hardware must be very robust, implying not only high-quality materials, but also tight specifications assembly to ensure a mechanically solid device. It is also important to ensure redundancy so that even in case of sensor failure, the UAV is still able to continue following the pre-determined mission, using other sensors. The HEIFU UAV was developed based on the architecture depicted in Fig. 2.

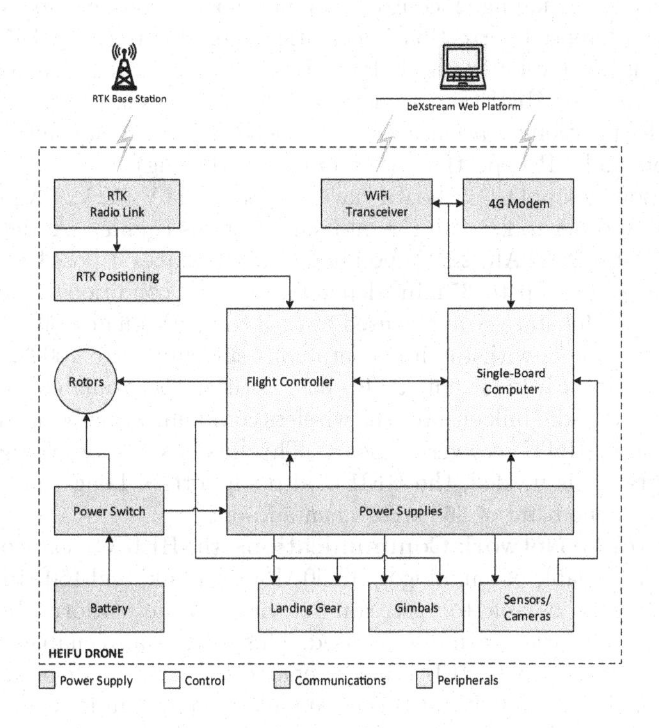

Fig. 2. Hardware reference architecture.

Being the main components the following:

1. **Single-Board Computer (SBC)**: the "brains" of the system, which ensures remote control and real-time status from and to the beXStream platform,

respectively. HEIFU counts on a Jetson Nano [14] System-on-Module (SoM) which comprises a 128-core Nvidia Maxwell architecture-based Graphics Processing Unit (GPU), together with a quad-core ARM57 Central Processing Unit (CPU), and 4 GB of 64-bit LPDDR4 memory. This architecture was chosen to facilitate the implementation of AI algorithms within the core of HEIFU applications portfolio;

2. **Flight Controller**: responsible for ensuring a stable flight throughout the whole mission, following the pre-defined path from the take-off to the landing. For this function, the ProfiCNC Cube Black flight controller [15] was chosen, as it counts on triple sensor redundancy, can control up to 8-rotor UAVs in multi-rotor configuration and can also control up to 6 auxiliary devices on-board (like gimbals, landing gear and others). The flight controller is permanently connected to the on-board SBC in order to receive commands and transmit status data, which is then relayed to the beXStream platform;
3. **RTK Positioning**: based on a u-blox ZED F9P Real Time Kinematics (RTK) [16] positioning receiver, it can achieve much better accuracy than conventional positioning since it counts on a base station capable of delivering RTK differential correction information. For this purpose, PDMFC and Beyond Vision developed the beRTK Base Station, which can be supplied together with the HEIFU to achieve a positioning accuracy in the range of 10 cm. The positioning accuracy is extremely important in some core applications of the HEIFU portfolio (like precision farming);
4. **Power**: power supplies on-board have to deliver 24 V, 12 V, 5 V, 4 V, 3.3 V and 1.8 V in order to keep all the on-board systems running. A high-capacity battery (Li-po 6S 22 Ah, 22.2 V nominal voltage) makes it possible to achieve a flight length of up to 35 min, depending on the conditions (wind, speed, payload). A solid-state switch is used to turn the UAV on and off. This switch must be capable of withstanding continuous currents up to 200 A;
5. **RTK Radio Link**: if an RTK base station is available, a 2.4 GHz IEEE802.15.4 radio link ensures the wireless communication between the base station and the RTK receiver on-board. This link ensures a range up to 3 km. If longer range is needed, the HEIFU also supports a Long Range (LoRA) radio link in the band of 868 MHz as an add-on;
6. **Wi-Fi/Mobile Network Communications**: the HEIFU counts on a native 4G modem, capable of handling up to 50 Mbps (uplink) and 150 Mbps (downlink) in order to be able to upstream 8 K video to the platform. If available, Wi-Fi communications can also be used. This module also enables Bluetooth communications. The HEIFU is also prepared to carry a 5G modem to profit from the higher bandwidth of this communication system if needed;
7. **Rotors**: the six UAV motors are 24VDC brushless rotors with integrated Electronic Speed Control (ESC) powered directly from the battery. Each of the rotors has a peak power of 1 kW, capable of generating up to 4.6 kg of thrust at 6393 rpm. The rotation speed is controlled by the Flight Controller via a Pulse Width Modulation (PWM) signal. The rotors come together with 18" (18 * 6.1) carbon-fiber propellers, ensuring a light and robust solution;

8. **Payload Accessories**: HEIFU can carry up to 8 kg payload (adding to the 7 kg native weight) which can be placed using a slider tray below the battery. Depending on the mission, the payload can be used to give the beXstream platform user full awareness of what is happening on-board, namely by the usage of cameras, sensors like Light Detection And Ranging (LiDAR), on-board sensors and other useful devices. The HEIFU counts on a power board on this tray to provide power to the payload devices (24 V, 12 V, 5 V and 3.3 V available). These devices can use the several interfaces available at the SBC (UART, SPI, I2C, USB 3.0, Ethernet) to exchange data with the platform. Several types of cameras were already tested and integrated into the system, namely RGB, multispectral, thermal and 360° cameras. The HEIFU also uses a depth camera as it is used by the collision avoidance algorithms to have depth perception;
9. **Frame Structure**: The HEIFU frame is based on a carbon-fiber skeleton, due to the lightweight and mechanical robustness of this material, encased in a ABS plastic enclosure, giving it environmental protection, aesthetics and character. Due to the big wingspan, the HEIFU arms are foldable, relying on a specially conceived arm-lock that guarantees that the arms do not fold while flying. The structure is tightly assembled together using torque-controlled hexalobular Inox A2 screws guaranteeing the mechanical robustness of the assembled kit.

All this components were integrated in the Printed Circuit Board (PCB) illustrated in Fig. 3

Fig. 3. HEIFU electronics.

When targeting autonomous missions, the most important aspect that we have to assure is safety, namely if we think on the potential damage that could cause a 15 kg object falling from the sky. The Reg. 945/2019/EU divides the drones into categories, depending on their MTOW, to impose bigger restrictions on the largest drones to enhance safety. The HEIFU was designed to be compliant with these regulations and as such, it follows all the recommendations and rules imposed by them. Further, to be able to target fully autonomous flights, the HEIFU was designed to overcome potential threats while flying, whether predictable or unpredictable, so to minimize the risk of the flight. Below, we list the major threats HEIFU faces and how we handled them to ensure, as much as possible, that it lands safely no matter what happens:

- **Weather conditions**: rain and wind are the biggest threats. But the HEIFU can withstand up to 70 km/h winds (already tested) and although the current version is not supposed to be used while raining, we are preparing a water resistant version which will withstand rain as well;
- **Static/dynamic obstacles**: collision avoidance processing on-board allows the HEIFU to divert from obstacles in its way, so to avoid collision and fall (see Sect. 4);
- **Battery failure**: the HEIFU main battery is continuously monitored. In case of failure or if it reaches a pre-defined low-battery voltage threshold, the HEIFU flight controller tries to land immediately in the most controlled manner possible;
- **Rotor failure**: Although a very unusual situation, being an hexacopter, the HEIFU is able to maintain its flight even if a rotor fails and/or stops working. In that case, it will try to land immediately;
- **Sensor failure**: the HEIFU flight controller has triple redundancy on its navigation sensors allowing it to identify and discard erroneous input data from a damaged or failing sensor;

3 Software

HEIFU's software was developed with the architecture depicted in Fig. 4. The core elements are built on top of the Robotic Operating System (ROS) [17] architecture, using its abstractions and message passing structures to implement all necessary features. The publisher/subscriber paradigm is used to communicate across blocks through ROS topics. The beXStream is a cloud platform that mimics a UAV Control Station with the added benefit of allowing the pilot to control a UAV remotely through the internet. The only requirement for the pilot and the UAV is an internet link to the site. A complete package was made available in open-source as a ROS package[1].

A desirable (i.e. secure and reliable) UAV system requires the creation of several software components. The ROS libraries were used for the modules running in the UAV because they allow for greater flexibility and easier integration of future features. Furthermore, it includes a number of off-the-shelf nodes that can be tweaked or reused to fit new needs. The remainder of this section explains each part represented in Fig. 4 using some of the terminology inherent to ROS.

[1] You can download HEIFU ROS package at http://wiki.ros.org/heifu.

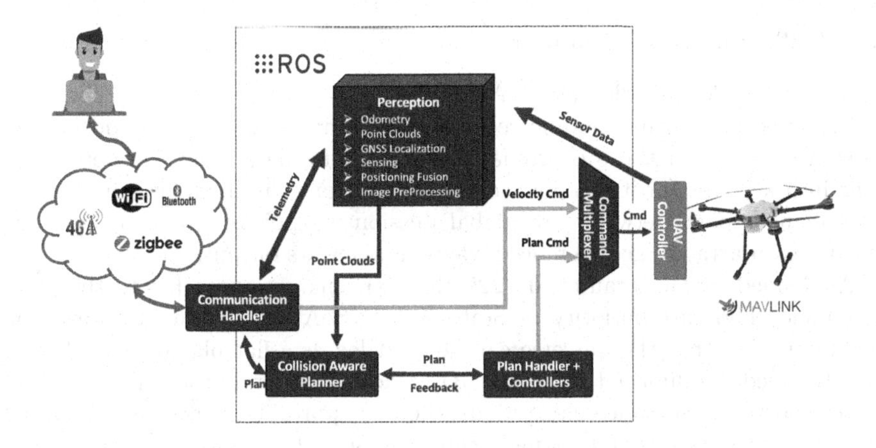

Fig. 4. HEIFU's ROS architecture.

3.1 Perception

The presented architecture's central component is perception block. It encapsulates a series of functionalities that can be divided into multiple nodes (typically one for each sensor), with higher-level nodes receiving input data from those that are actually managing the physical sensors.

The following are the nodes that deal with sensors directly:

- **Image Node**: Receives camera data and publishes it in sensor Image message format. This seems to be a straightforward node, but it can be incredibly complex. ROS sends images in its own message format, but developers can also use image nodes that come with various image processing libraries, such as OpenCV [18].
- **Odometry Node**: This node will approximate the location of the UAV in relation to its starting point. This can be accomplished by using the data from the UAV motion sensors, conducting visual odometry estimations, or employing some kind of fusion algorithm that combines methods. This node's data is published on a message format known as navigation odometry messages.
- **Inertial Measurement Unit (IMU) Node**: In charge of managing the IMU sensors (accelerometer, gyroscope, magnetometer, and barometer) and posting sensor IMU messages to the ROS network on a regular basis.
- **GNSS Node**: Obtains data from the Global Navigation Satellite System (GNSS) and periodically publishes navigation messages on the ROS network.

Until publishing data to the ROS network, each node will apply filtering algorithms. It is important to flush out incorrect sensor readings to prevent polluting the top-level nodes with noise. As a result, in addition to the first set of perception nodes, additional nodes that use the filtered data may be integrated. Positioning nodes [19], 3D nodes [20] (using form from motion), and depth images to point cloud nodes [21] are a few examples.

3.2 Collision Aware Planner

The Collision Aware Planner (CAP) module is in charge of determining a collision-free path. This module receives one or more coordinates and creates a direction between those coordinates when taking data from the perception layer into account (obstacle point-clouds). In a long autonomous mission, the CAP plans a path between two global mission waypoints. These trajectories, which are an array of georeferenced waypoints, are its output.

As defined by Galceran et al. [22], there are many approaches to the planning issue. The vast majority of State-of-Art (SoA) algorithms are based on the assumption that the universe can be modeled as a flat planar disk. Hert et al. [23] added 2D information to the 3D world and several other approaches to address obstacle avoidance navigation. Another approach involves interpolating the user's target points and setting escape callbacks that are enabled if an obstacle is identified [24]. Section 4 avoidance will include more information on this topic. The CAP generates a trajectory, which is a series of points separated by distance tolerances, and updates the schedule handler.

3.3 Plan Handler

The plan handler interpolates the CAP trajectory points with movement and behavior restrictions. The joint state data from the UAV actuator's encoders and an input set point are fed into the plan handler node. To control the output, it employs a generic control feedback loop system, usually a PID controller. Since the number of joints on a UAV is normally small, a regular action controller can be created to convert the trajectory provided by this CAP into commands for the UAV controller.

3.4 Command Multiplexer

Prioritizing safety and control issues is mandatory in today's UAVs. Safety restrictions imply that switching from an autonomous action to manual operation must be done almost instantly. As a result, all of the input points must be multiplexed into a single convergence point capable of communicating with the hardware controller.

The Command Multiplexer (CM) subscribes to a list of topics that publish commands and multiplexes them based on a set of priorities. The UAV is controlled by the feedback with the highest priority, which becomes the active controller. Timeout (no answer from an input) or subject locking may be used to adjust the active controller (some inputs might be locked, being discarded). In fact, the node will take various input topics from various issuers and emit the messages of the highest priority issuer (blocking the others). This is particularly helpful when the UAV is flying autonomously and the pilot wishes to take over. If the pilot has the highest priority, any order he issues will make him the active controller.

3.5 Communication Handler

A connectivity module converts ROS publisher/subscriber into websockets in order to interface with this system. This allows the beXStream platform to operate a fleet of UAVs from any distance [2]. The kit also manages WiFi, 4G, and 5G handover. This is accomplished by building a stream in all possible communication networks and choosing the one with the greatest bandwidth at all times. When the UAV is authenticated, it is given a dedicated channel from which it can stream video and audio data using the Real-time Transport Protocol (RTP), according to [25].

4 Collision Avoidance

Creating a route plan from point A to point B while avoiding barriers and responding to environmental changes is an easy task for humans but not so simple for an autonomous UAV. These activities raise obstacles that each UAV must resolve in order to achieve maximum autonomy. A UAV uses sensors to sense its surroundings (to a certain degree of certainty) and to create or change its map. Different decision and planning algorithms may be used to evaluate motion behavior that lead to the target position. The kinematic and dynamic constraints of the UAV should be addressed during route planning.

The exploration of the shortest routes is used to solve problems ranging from basic spatial route planning to the choosing of an optimal action sequence needed to achieve a specific objective. Since the world is not necessarily understood in advance, this form of preparation is often confined to areas that have been planned in advance and correctly defined prior to the planning phase.

Shortest paths discovery is used to solve problems in different fields, from simple spatial route planning to the selection of an appropriate action sequence that is required to reach a certain goal. Since the environment is not always known in advance, this type of planning is often limited to the environments designed in advance and described accurately enough before the planning process. Routes may be calculated in fully understood or partially known environments, as well as in completely unknown environments where sensed data is used to determine the optimal UAV motion.

Planning in existing settings is an ongoing testing field that serves as a basis for more complex situations in which the world is unknown *a priori*.

As proposed in [26], the problem of Collision Avoidance is divided into two categories:

- Static collision s_c - Collisions between the UAV and any obstacle that moves much slower than the UAV are represented. Using the universe as a reference, it is assumed that an object can cause a static collision if it moves slower than 5% v_{max} of the maximum speed of the UAV.

[2] The developed beXStream platform can be accessed by the link-https://bexstream.beyond-vision.pt.

- Dynamic collision d_c - Collisions between the UAV and any obstacle that moves quicker than the computation of point clouds, allowing the route planner to plot a protected path that avoids the collision. Using the universe as a reference, it is assumed that an object will cause a dynamic collision if it moves faster than 5% v_{max} of the UAV.

4.1 Static Collision Avoidance

As stated above, the Static collision avoidance module on HEIFU is responsible for ensuring its safety and integrity when encountering static or nearly static objects. An Intel Realsense D435i is used to perceive the surroundings by means of a depth point cloud [27]. At each timestamp, the point cloud is ego-compensated into a global reference frame. Highly dynamic obstacles result in a blurred representation that might mislead the analysis and generate strange avoidance paths. That's the reason why there is a separate module to deal with them.

The representation of the world around the drone is memory-hungry, as there is a vast amount of points generated at each data frame. To overcome this, data is often voxelized into Bkd-trees (Fig. 5). Octomap [28] is a known and successful framework that concatenates the data into an octree encoding 3D cubical voxels of different resolutions. The leafs represent the cubic spaces as *free*, *occupied*, or *unknown* (not seen yet), based on a probabilistic approach dependent on the number of points that lie inside. Taking advantage on the GPU unit of its SBC Jetson Nano, HEIFU uses GPU-Voxels [29], which is very similar to the Octomap framework. It outperforms Octomaps by exploring the highly-parallelized capabilities of a GPU, instead of being only CPU-based.

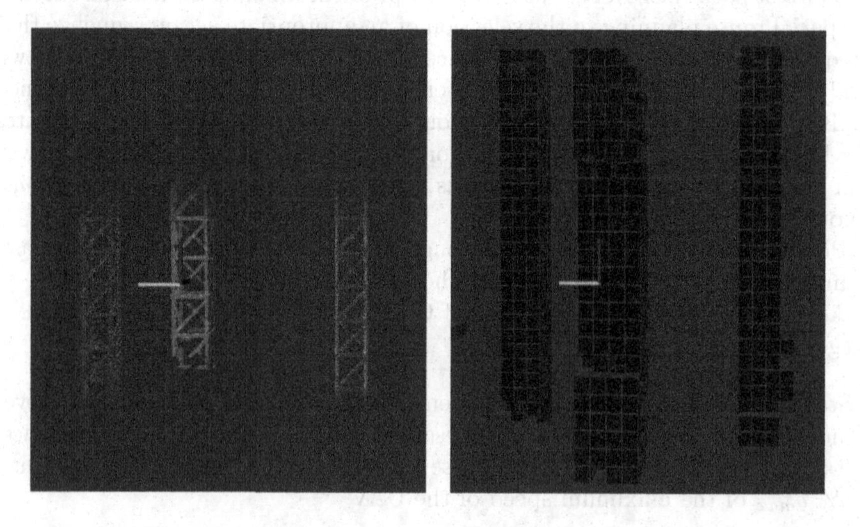

Fig. 5. Point cloud of obstacles (left) and its voxelized representation (right).

The algorithm for obstacle detection on HEIFU is further optimized when comparing to SoA solutions. Instead of checking if any *occupied* voxel is a threat for the current planned path, it works inversely: checks if there is any occupied voxel near a requested list of 3D positions. Taking advantage on the distance map from *VoxelMap* of GPU-Voxels, the algorithm follows a strategy similar to [24] and checks for obstacles lying inside a cylindrical safety volume approximated by a set of spheres. Using a near-to-far checking strategy with a list of sphere centers, potential threats are evaluated from the drone to the waypoint. This way, if an obstacle is found, the checking process can be immediately terminated and requested a new avoidance path to the planner (Fig. 6). In [30] is made a deeper analysis on the advantages of these strategies by implementing a simple reactive obstacle avoidance solution.

With the new information of the surroundings, a new avoidance path is calculated based on an adapted version of Rapidly-exploring Random Tree (RRT) [31] that favours the direction of the desired final point. In order to avoid wasting too much time in a constantly changing or unknown environment, the planner is requested to generate a solution only within a predefined workspace around the UAV, providing intermediate goal positions. Whenever the workspace limits are reached, new paths are requested to the planner until the final waypoint is successfully achieved.

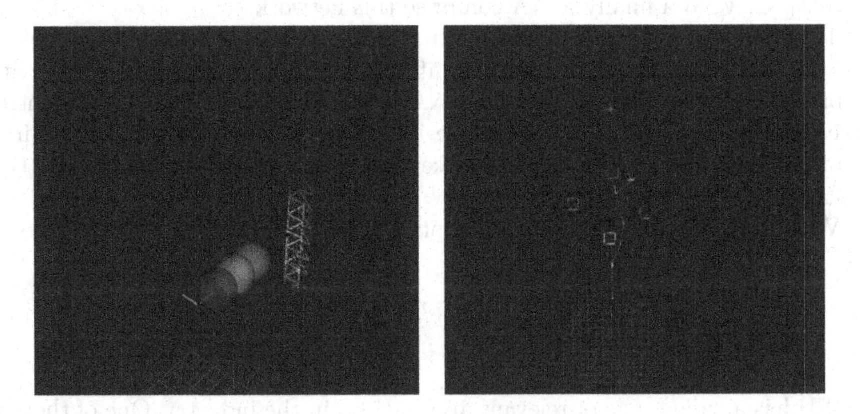

Fig. 6. Safety volume checking for threats (left) and avoidance path generated by RRT (right) to reach the waypoint (red circle). (Color figure online)

The Dynamic collision avoidance is an harder task, which involve fast processing and quick reaction. HEIFU collision avoidance algorithm uses a pipeline of Neural Works, which were trained in the ColANet dataset [32]. Further details can be found in [26].

By implementing a priority manager for the drone's commands, HEIFU is able to navigate safely in challenging environments. The commands for avoiding dynamic objects moving towards the drone have greater priority due to their urgency.

5 Online Connectivity

The beXStream platform was created to incorporate all UAVs into a single platform [33,34]. It is a cloud-based platform that enables remote control of UAVs. This platform has five primary functions:

1. Collect and analyze data produced by multiple UAVs during each flight, with the option of replaying a specific flight/mission.
2. Real-time communication with UAVs, including receiving telemetry, video data, and transmitting commands.
3. Parallel execution of computationally intensive AI algorithms that aid in consumer decision-making.
4. Take command of these UAVs by uploading missions with predefined way-points.
5. Plan swarm missions using several UAVs.

The platform is the final component of the overall architecture depicted in 4, which envisions stable long-distance control of several autonomous UAVs. A BE, a relational database for non-temporal data (such as UAV identities and users with permissions for that UAV), a time-series database for high rate temporal data (such as UAV flight telemetry), a Media-Gateway server for incoming video from the UAV, and finally a WA comprise this network.

The Backend (BE) was built with NestJS, the relational database with MySQL, the time series database with InfluxDB, the Media-Gateway server with WebRTC, and the Web Application (WA) with Angular 8. The BE also introduces service layers to facilitate future integrations with external APIs. Since all of the modules are created as Docker images that run in Kubernetes, this solution is simple to scale.

When a UAV is turned on, it attempts to link to the BE via websockets over 4G, 5G, or WiFi.

6 Field Tests Results

HEIFU has a wide range of relevant applications in the industry. One of them is the potential to agricultural survey mapping with high accuracy, when connected to an RTK base.

Thus, in the precision agriculture sector, multispectral maps of the same region can be produced on different days while maintaining high mapping precision. Because of this aspect, any temporal difference in the health condition of a plant can be identified and acted upon immediately in order to fix the crop's health problems.

HEIFU receives a series of points for a given task and runs them automatically to build georeferenced maps without the need for human interference. Figure 7 builder depicts an example of a mission created by the beXStream platform, with each stage of the mission being sent to the UAV.

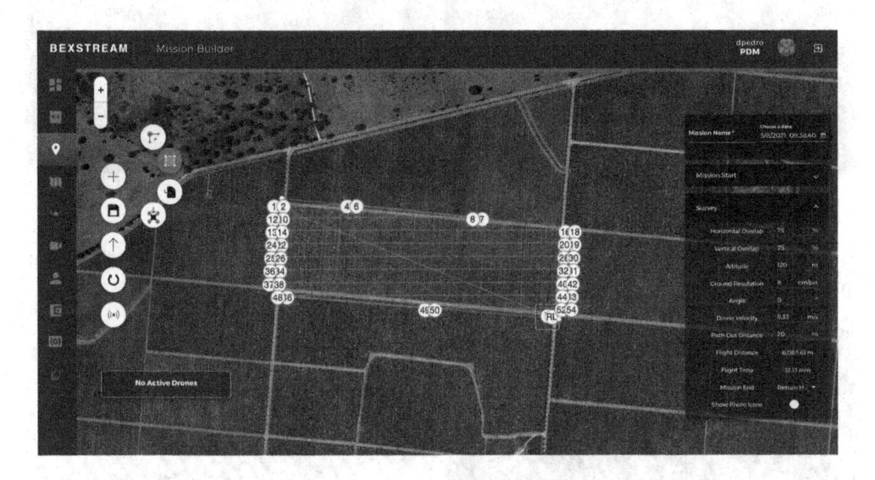

Fig. 7. Mission builder – example of a Survey Mission.

Figure 8 shows the result of 2 missions in the same area with a difference of 3 h. The first figure (Fig. 9a) was carried out at 9 am while the second mission (Fig. 9b) was carried out at 12 pm.

From Figs. 9a and 9b it is possible to conclude that due to the high precision replication of autonomous missions, multispectral maps can be created knowing the position of each pixel of the map will coincide for different readings (of the same study area) at different times of the day. Without HEIFU's high position accuracy, the map positioning error would increase from 4 cm to 1 m, which is a lot when it comes to precision agriculture topic. This enables the study of the mapped area and helps agricultural research.

Considering an isolated map (in time and area), Color Infrared (CIR) (Fig. 9) and Normalized Difference Vegetation (NDVI) (Fig. 9) maps can be used as input for a fault detection algorithm, also based in Artificial Intelligence, that detects vineyard canopy gaps. Detecting those gaps is of extreme importance, especially in large farms. It avoids the time-consuming human labour of going through all the corridors, and can point out beforehand, key points of potential interest for closer analysis. Gaps in the canopy might indicate some plague and/or disease on the vineyard that, if detected early, may avoid potential great losses in production. If the potential gaps are confirmed as threats to the culture, a corrective action may be immediately triggered, saving time and potentially the crop.

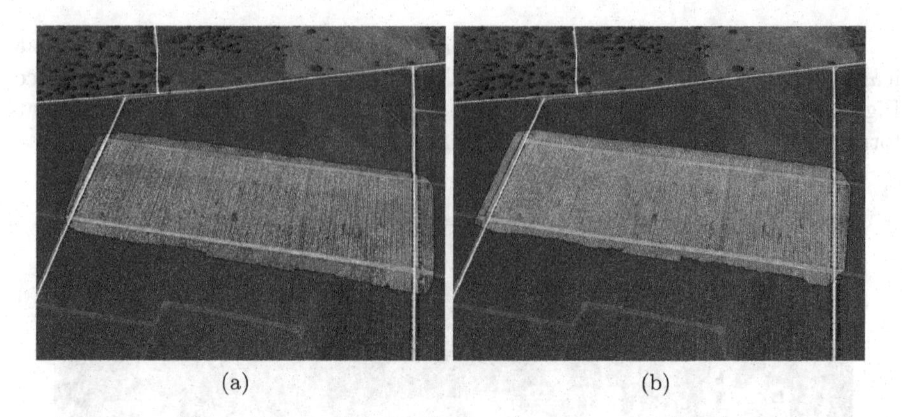

Fig. 8. Thermal mapping results with HEIFU's RTK base: a) Mapping at 9 am; b) Mapping at 12 pm.

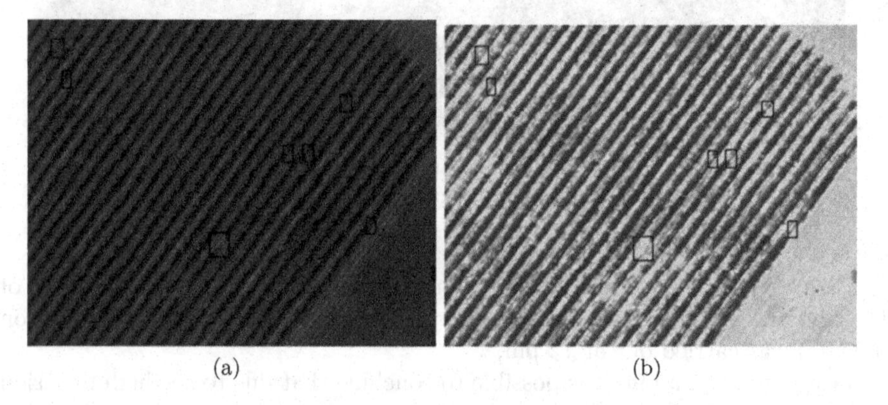

Fig. 9. Multispectral indexes with AI applied in these multispectral maps to detect vineyard canopy gaps: a) CIR map; b) NDVI map.

7 Conclusions and Future Work

In this article, an autonomous vehicle has been presented, entitled HEIFU. It can operate in the harshest conditions because it was designed to be easily adapted to multiple industries. Some of the problems identified as critical in literature were taken into consideration when formulating HEIFU electronics, mechanics, and software architecture framework. The UAV development was segmented and described in those areas. Furthermore, an example of application was outlined, which was the high precision temporal replication of an autonomous mission, allowing an analysis of the evolution of crops. The data gathered from the drone in these flights help farmers to make decisions and anticipate some possible problems.

HEIFU is a door for multiple areas in research, that can be explored by thc community. In particular, the authors wish to explore the following areas in the near future:

- Precision farming: Autonomous surveys using Multispectral and Thermal cameras to predict diseases in crops.
- Security: Constant Monitoring with detection and report algorithms.
- Transportation: Fast, Reliable, and secure transportation of critical items, such as an organ between hospitals.
- Tourism: Remote sightseeing with 360° cameras mounted on UAVs.

References

1. Zhu, X., Pasch, T.J.: Understanding the structure of risk belief systems concerning drone delivery and Aaron Bergstrom. a network analysis. Technol. Soc. **62**, 101262 (2020)
2. Enemark, C.: Drones, risk, and moral injury. Crit. Military Stud. **5**(2), 150–167 (2019)
3. King, D.W., Bertapelle, A., Moses, C.: UAV failure rate criteria for equivalent level of safety. In: International Helicopter Safety Symposium (2005)
4. BBC. Drone' hits british airways plane approaching heathrow airport (2016). https://www.bbc.com/news/uk-36067591. Accessed 19 May 2019
5. CBC Canada. Drone that struck plane near quebec city airport was breaking the rules — cbc news (2017). http://www.cbc.ca/news/canada/montreal/garneau-airport-drone-quebec-1.4355792. Accessed 19 May 2019
6. BBC. Drone collides with commercial aeroplane in canada (2017). https://www.bbc.com/news/technology-41635518. Accessed 19 May 2019
7. Goglia, J.: Ntsb finds drone pilot at fault for midair collision with army helicopter (2017). https://www.forbes.com/sites/johngoglia/2017/12/14/ntsb-finds-drone-pilot-at-fault-for-midair-collision-with-army-helicopter/. Accessed 19 May 2019
8. Rawlinson, K.: Drone hits plane at heathrow airport, says pilot (2016). https://www.theguardian.com/uk-news/2016/apr/17/drone-plane-heathrow-airport-british-airways. Accessed 19 May 2019
9. Bassi, E.: From here to 2023: civil drones operations and the setting of new legal rules for the European single sky. J. Intell. Robot. Syst. **100**(2), 493–503 (2020). https://doi.org/10.1007/s10846-020-01185-1
10. Fang, S.X., O'Young, S., Rolland, L.: Development of small UAS beyond-visual-line-of-sight (bvlos) flight operations: system requirements and procedures. Drones **2**(2), 13 (2018)
11. Matos-Carvalho, J.P., et al.: Static and dynamic algorithms for Terrain classification in UAV aerial imagery. Remote Sens. **11**(21), 2051 (2019)
12. Salvado, A.B., et al.: Semantic navigation mapping from aerial multispectral imagery. In: 2019 IEEE 28th International Symposium on Industrial Electronics (ISIE), pp. 1192–1197 (2019)
13. Matos-Carvalho, J.P., Fonseca, J.M., André, M.: UAV downwash dynamic texture features for terrain classification on autonomous navigation. In: 2018 Federated Conference on Computer Science and Information Systems (FedCSIS), pp. 1079–1083 (2018)

14. NVIDIA. NVIDIA Jetson Nano Developer Kit — NVIDIA Developer (2019)
15. PX4. Pixhawk Autopilot (2017)
16. U-blox. ZED-F9P module u-blox F9 high precision GNSS module (2019)
17. ROS. Powering the world's robots (2007). https://www.ros.org/. Accessed 19 May 2019
18. Bowman, J., Mihelich, P.: Camera Calibration - ROS Wiki (2014)
19. Kalman, R.E.: A new approach to linear filtering and prediction problems. Trans. ASME-J. Basic Eng. **82**(Series D), 35–45 (1960)
20. Turner, D., Lucieer, A., Watson, C.: An automated technique for generating georectified mosaics from ultra-high resolution unmanned aerial vehicle (UAV) imagery, based on structure from motion (SFM) point clouds. Remote Sens. **4**(5), 1392–1410 (2012)
21. Keselman, L., Iselin Woodfill, J., Grunnet-Jepsen, A., Bhowmik, A.: Intel(R) realsense(TM) stereoscopic depth cameras. In: IEEE Computer Society Conference on Computer Vision and Pattern Recognition Workshops (2017)
22. Galceran, E., Carreras, M.: A survey on coverage path planning for robotics. Robot. Auton. Syst. **61**(12), 1258–1276 (2013)
23. Hert, S., Tiwari, S., Lumelsky, V.: A terrain-covering algorithm for an AUV. Autonomous Robots, pp. 17–45 (1996)
24. Azevedo, F., et al.: Collision avoidance for safe structure inspection with multirotor UAV. In: 2017 European Conference on Mobile Robots, ECMR 2017 (2017)
25. Paul, S., Paul, S.: Real-time transport protocol (RTP). In: Multicasting on the Internet and its Applications. Springer, US (1998)
26. Pedro, D., et al.: Ffau-framework for fully autonomous UAVS. Remote Sens. **12**(21) (2020)
27. Matos-Carvalho, J.P., Dário, P., Miguel Campos, L., Fonseca, J.M., Mora, A.: Terrain classification using w-k filter and 3D navigation with static collision avoidance. In: Advances in Intelligent Systems and Computing (2020)
28. Hornung, A., Wurm, K.M., Bennewitz, M., Stachniss, C., Burgard, W.: OctoMap: an efficient probabilistic 3D mapping framework based on octrees. Auton. Robots, **34**(3), 189–206 (2013)
29. Hermann, A., Drews, F., Bauer, J., Klemm, S., Roennau, A., Dillmann, R.: Unified GPU voxel collision detection for mobile manipulation planning. In: 2014 IEEE/RSJ International Conference on Intelligent Robots and Systems. IEEE (2014)
30. Azevedo, F., Cardoso, J.S., Ferreira, A., Fernandes, T., Moreira, M., Campos, L.: Efficient reactive obstacle avoidance using spirals for escape. Drones **5**(2), 51 (2021)
31. Lavalle, Steven M.: Rapidly-exploring random trees: A new tool for path planning. Computer Science Dept., Iowa State University, Technical report (1998)
32. Pedro, D., Mora, A., Carvalho, J., Azevedo, F., Fonseca, J.: ColANet: a UAV collision avoidance dataset. In: IFIP Advances in Information and Communication Technology (2020)
33. Pino, M., Matos-Carvalho, J.P., Pedro, D., Campos, L.M., Seco, J.C.: Cloud Platform, U.A.V., for precision farming. In: 12th International Symposium on Communication Systems, Networks and Digital Signal Processing. CSNDSP (2020)
34. Nakama, J., Parada, R., Matos-Carvalho, J.P., Azevedo, F., Pedro, D., Campos, L.: Autonomous environment generator for UAV-based simulation. Appl. Sci. (Switzerland) (2021)

Testing for IT Security: A Guided Search Pattern for Exploitable Vulnerability Classes

Andreas Neubaum(✉), Loui Al Sardy(✉), Marc Spisländer(✉), Francesca Saglietti(✉), and Yves Biener

Software Engineering (Informatik 11), University of Erlangen-Nuremberg, Martensstr. 3, 91058 Erlangen, Germany
{andreas.neubaum,loui.alsardy,marc.spislaender, francesca.saglietti,yves.biener}@fau.de

Abstract. This article presents a generic structured approach supporting the detection of exploitable software vulnerabilities of given type. Its applicability is illustrated for two weakness types: buffer overflowing and race conditions.

Keywords: Software vulnerability · Buffer overflow · Race condition · Symbolic execution · SMT-solver · Intelligent testing

1 Introduction

Complex software systems with safety demands evidently require the application of accurate verification techniques before being released for operation. Inappropriate software behaviour may be due to logical faults originating from flawed, incomplete or ambiguous requirement specifications or from incorrect refinement and/or implementation activities.

Apart from their temporal origin, the faults potentially resulting at code level usually differ also with respect to their behavioural effects, as well as in terms of the number of instructions and of inputs affected. Other than for hardwired circuits, where typical physical faults affecting logical gates (e.g. stuck-at-0-faults or stuck-at-1-faults) may be classified and localized, in general due to their problem-specific nature software fault types cannot be satisfactorily enumerated.

On the other hand, in addition to one-of-a-kind faults individually arising during development, software may also be affected by typical vulnerabilities, i.e. by known patterns of weak design which may be more easily identified and maliciously exploited; in fact, insider know-how concerning the presence of a vulnerability of given type may enable attackers to provoke operational scenarios with critical effects.

Typical vulnerability patterns whose relevance was confirmed by recent security incident analyses [8, 15] include the following ones:

- *buffer overflowing* which may be exploited to overwrite memory space in an illegal way, in particular such as to enable the unplanned execution of code portions;
- *race conditions* which may be exploited to provoke an anomalous behaviour by enforcing an unintended interleaved execution of parallel threads.

I. Habli et al. (Eds.): SAFECOMP 2021 Workshops, LNCS 12853, pp. 105–116, 2021.
https://doi.org/10.1007/978-3-030-83906-2_8

Due to the relevance of such vulnerability classes, it is felt that verification should include, in addition to functional and structural coverage testing, also activities devoted to the systematic search for the presence of given exploitable weakness types. For this purpose, an approach targeting the detection of exploitable buffer overflows was defined, implemented and successfully evaluated within the project SMARTEST. Within the successor project SMARTEST2 the underlying concept was generalized and is currently being transferred to the systematic detection of race conditions. The insight and experience gained hereby are reported in the following sections.

2 Related Work

Generic runtime errors (not necessarily related to specific vulnerability types) are targeted by several fuzzing approaches, among them AFL [24] and Radamsa [18]. Both are based on the execution of randomly varying inputs; in addition, AFL captures testing progress (without optimizing it) by a code coverage metric. Such generic approaches are likely to miss the detection of specific vulnerabilities (e.g. buffer overflows or race conditions) in case their exploitation poses complex logical constraints (s. [1]).

Alternative approaches, a. o. KLEE [5] and QSYM [23], aim at maximizing code coverage by applying symbolic execution to generate predicates (so-called path constraints) characterizing all inputs traversing given control flow paths. By means of constraint solvers like STP [11] or Z3 [9], corresponding test cases can then be automatically generated, executed and checked for runtime errors. Although more systematic than the fuzzers mentioned above, these approaches also suffer from inherent limitations, as the achievement of high code coverage does not necessarily ensure the detection of flawed program behaviour.

While the techniques mentioned so far address any arbitrary flawed behaviour, further approaches resp. tools focus on the detection of vulnerabilities of given type.

For example, in order to detect race conditions, both CHESS [16] and KISS [19] check different interleaved executions w.r.t. the validity of predefined assertions, based on user-defined input data and use cases. Both of them limit their analysis to a subset of potential interleaved executions, CHESS by bounding the number of thread switches, KISS by excluding switch sequences which cannot be generated via a stack-based algorithm. Such assertion-based approaches may miss the occurrence of race conditions, as the use of assertions capturing complete program correctness is likely to be unrealistic.

MEMICS [17], on the other hand, limits the consideration of race conditions to the occurrence of predefined reading/writing patterns in interleaved executions of parallel threads. By means of bounded model checking it aims at proving or disproving the possibility of interleaved executions meeting these patterns. Also this pattern-oriented approach may miss critical race conditions whenever these are not captured by appropriate patterns.

Further approaches, e.g. [4], do not primarily aim at exposing race conditions, but rather at filtering out correct interleaved executions which do not require further analysis. In this sense, they do not solve the problem addressed, but can provide helpful support in restricting the search for race conditions.

The technique proposed in this article differs from those mentioned above, as it determines the occurrence of a race condition by comparing different program executions rather than by analysing executions individually.

The generic search pattern presented in this article aims at overcoming all the limitations mentioned above: it allows to focus on different vulnerability types, e.g. buffer overflows or race conditions. Rather than relying on limited evidence (as provided by incomplete assertions, non-exhaustive anomalous patterns or code coverage) it supports the systematic search for exploitable vulnerabilities by making use, whenever possible, of targeted and complete analytical information as provided by symbolic execution. Where logical complexity or undecidability prevents from carrying out a systematic search, the combined use of heuristics may be considered.

3 Guided Search Pattern for Arbitrary Vulnerability Classes

In this section a strategy for guiding the search of exploitable vulnerabilities is introduced in generic terms, thus giving rise to a pattern which may be reused to address different weakness classes. It consists of the following 3 steps:

Step 1: Restriction of the Search Space
This step is devoted to restricting the search space (typically, the input space) to those scenarios for which an exploitation of the vulnerability class considered cannot be excluded a priori. A restriction of the input set may be implicitly induced by filtering out part of the logic (typically, control flow graphs) whose execution can be proven to be immune to the attack types considered.

In case no part of the logic presents constructs which may be targeted by attacks of the type considered, the analysis can stop here, as this vulnerability type could be successfully excluded from being exploitable even by well-informed attackers.

Step 2: Analysis and Determination of the Search Target Constraint
This step is devoted to the analysis and determination of a logical constraint whose satisfiability can be used to imply the presence of exploitable vulnerabilities of given type. Ideally, the constraint to be identified represents a necessary and sufficient condition for input and execution mode (e.g. interleaved instructions of parallel threads) resulting in misbehaviour. Typically, such constraints can be constructed if the program considered is fully statically analysable, in particular, if the number of loop iterations can be bound a priori; symbolic execution [3] can then be used to provide at any execution step i symbolic expressions (E_i, A_i) where E_i provides a necessary and sufficient constraint to be satisfied by the input in order to reach position i and A_i provides for any (intermediate or final) variable an expression reflecting its symbolic evaluation up to the conclusion of step i.

Step 3: Search for Scenarios Satisfying the Target Constraint
This step is devoted to a search for inputs and execution modes satisfying the constraint derived in step 2. If satisfiability is fully analysable with affordable effort (e.g. by an SMT-solver), the outcome of this analysis may be sufficient to exclude critical attacks or to demonstrate their potential via concrete scenarios.

In general, however, constraint satisfiability may not be decidable or require prohibitive effort to be analysed. Search for vulnerabilities must then be supported by heuristics (e.g. intelligent testing strategies based on evolutionary techniques) which, though not guaranteeing termination and success, may be combined with statically gained insight to optimize the chances of a guided search.

4 Instantiation of the Guided Search Pattern for Buffer Overflow

4.1 Introduction of the Vulnerability Class *Buffer Overflow*

A buffer overflow (resp. underflow) occurs by writing data into memory space beyond (resp. beneath) the boundaries limiting the memory allocated for any buffer. This vulnerability may be exploited by an attacker to enable the execution of unplanned code portions (e.g. by redirecting the instruction pointer [7, 10]). Without restricting generality, the following considerations will address buffer overflowing which is considered to be among the most relevant software security weaknesses, as confirmed by statistical reports [8, 14] and recent exploits [15, 20].

4.2 Guided Search Structure

The guided search approach SMARTEST-SWE developed within the project SMARTEST to detect buffer overflows is summarised in the following; for details, the reader is kindly referred to [2]. It results by instantiation of the generic pattern described in Sect. 3. According to its structure, the 3 steps are illustrated below.

Step 1: Restriction of the Search Space
Each instruction B with a direct or indirect access to a buffer variable requires an individual search for buffer overflows. For each such B the corresponding search space can be restricted by removing control flow paths not affecting the buffer access considered. In the following a search will address one arbitrarily selected B.

Step 2: Analysis and Determination of the Search Target Constraint
The analysis and determination of the search target constraint for buffer overflows can be derived by considering for a given path prefix p ending in B (determined by a finite sequence of instructions) the following predicates:

- $C_E(p)$ denotes the necessary and sufficient constraint on inputs for executing p; it is determined by the condition E_i as provided by symbolic execution (s. Sect. 3);
- $C_O(p)$ denotes a necessary and sufficient condition on inputs to indicate an *overflow* (i.e. an access beyond predefined buffer boundaries) at the conclusion of p; it can be determined by making use of the expressions A_i as provided by symbolic execution (s. Sect. 3).

The occurrence of a buffer overflow at the end of the execution of path prefix p requires and is implied by the satisfiability of the following constraint on p:

$$C(p) := C_E(p) \wedge C_O(p)$$

All constraints concerning path prefixes ending in B can be integrated into one prefix-independent constraint

$$C(B) := \bigvee_{p \text{ ends in B}} C(p)$$

Assuming an upper bound of the number of loop iterations to be known, *C(B)* provides a constraint of finite length on the search target. Otherwise, in order to detect a buffer overflow at least one constraint *C(p)* concerning a path prefix *p* out of a potentially infinite set of such path prefixes must be fulfilled.

Step 3: Search for Scenarios Satisfying the Target Constraint
If the number of loop iterations can be bounded a priori, the analysis of the necessary and sufficient constraint *C(B)* can be addressed by an SMT-solving tool like Z3; this does not guarantee, however, that the satisfiability of *C(B)* can be proven or disproven within an affordable time limit. In such cases as well as in cases where – due to lack of knowledge about the maximum number of loop iterations – the constraint *C(B)* may involve the disjunction of an infinite number of path prefix constraints *C(p)*, heuristics-based techniques can be addressed.

This can be achieved, for example, by testing a functionally equivalent code including an additional control flow branch distinguishing whether or not a buffer overflow occurs after executing instruction *B* [2]. In other words, traversal of the *true*-arc of the branch would result in an execution of *B* overflowing the buffer. In order to maximize the chances to do so, local optimization based on genetic algorithms was applied (s. [12, 21, 22]). It consists of guiding the search for appropriate inputs by a fitness function which captures both the *sub-path distance* (i.e. the number of branching nodes separating the closest node already reached from the branching node) and the *branch distance* (i.e. the degree of fulfilment of the branching condition), where the latter is normalized to a range [0; 1] in order to allow for the domination of the former.

4.3 Evaluation

In [1] the guided search approach SMARTEST-SWE was compared with the fuzzing tools AFL [24] and Radamsa [18] on the basis of 5 code examples (s. Table 1): the 2 simplest ones (columns 2, 3) allow for being analysed also by appropriate SMT-solvers; the remaining 3 (columns 4, 5, 6) include loops not allowing for a complete static analysis, thus requiring the application of heuristics.

The original comparison was extended to include the fuzzer QSYM [23] under identical conditions concerning the 5 codes and the experimental setup. QSYM is a public domain hybrid fuzzer aiming at the maximization of code coverage by concolic execution, i.e. by combining random testing with symbolic execution to derive constraints to be evaluated by the SMT-solver Z3 [9]. The comparison (s. Table 1) confirms the results presented in [1], i.e. the superiority of the guided search approach at the price of the latter being restricted to a particular vulnerability type.

Table 1. Summary of fuzzing results (including results from [1]): successful runs out of 5 runs, average execution time of successful runs, timeout after 5 h

Fuzzer / Example	Skey_Challenge		Socket Printf		Fb_realpath		Read & write		Turing machine	
AFL	5	<4 s	5	<11 m	0	Timeout	0	Timeout	0	Timeout
Radamsa	5	<1 s	5	39 s	1	<2 h	0	Timeout	0	Timeout
QSYM	5	<6 s	5	<11 m	0	Timeout	0	Timeout	0	Timeout
SMARTEST-SWE	5	<1 s	5	<1 s	5	<1 s	5	<8 s	5	<2 h

5 Instantiation of the Guided Search Pattern for Race Condition

5.1 Introduction of the Vulnerability Class *Race Condition*

It is well-known that in case of more threads being executed in parallel the relative order in which their instructions are executed plays a fundamental role for the correctness of the outcome. A good design requires a preliminary analysis of the logical dependency of the threads including considerations on reading/writing access to common variables. In particular cases, however, operational conditions may differ from the usage assumed during the planning and design phase: for example, if only one thread execution was originally considered, or if the original assumptions about concurrent behaviour were successively violated via cyber-security attacks. The purpose of the following approach is to identify the chances of potential attackers to trigger incorrect concurrent behaviour such as to disable such attacks by appropriate corrective actions before operation.

In more detail, attackers may succeed in manipulating the order in which instructions belonging to different threads are alternately executed, hereby enforcing a different sequence of instructions than originally intended.

In the following, an *interleaving* will denote any sequence of alternating thread instructions; a pair of such interleavings reveals the occurrence of a *race condition* whenever, executing a common input, the two interleavings result in unacceptably deviating behaviours. Race conditions have revealed as particularly relevant, as confirmed by [14, 15]. The approach to detect race conditions described in the subsequent sections considers pairs of interleavings and analyses whether they may result in different outputs.

5.2 Guided Search Structure

In this sub-section the generic guided search pattern introduced in Sect. 3 will be applied to the detection of race conditions. Hereby, it is assumed that:

- for each thread the number of control flow paths is finite and known a priori;
- the number of threads and their codes are statically determined before operation;
- the execution of any basic instruction (i.e. assignments and predicate evaluations) cannot be interrupted by other threads;
- predicate evaluations do no result in side effects.

In general, such assumptions impose restrictions on the programs to be analysed; in the particular case of safety-critical applications within industrial contexts, however, they may be considered as acceptable, also in view of existing coding standards (s. [6, 13]).

Step 1: Restriction of the Search Space
Whenever the execution of an input involves only accesses to thread-dependent local variables, the occurrence of race conditions can be excluded. Therefore, the search space can be reduced by filtering out the set of all such inputs which can be identified by a preliminary static analysis.

Step 2: Analysis and Determination of the Search Target Constraint
In case of race conditions the search target must characterise inputs which may result in different behaviour depending on how the corresponding thread executions are interleaved. Therefore, the search target must address inputs as well as different schedulings resulting in different outputs.

Assuming n concurrent threads $T_1, \ldots, T_n$, an interleaving I is characterised by an n-tuple $(p_{T_1}, \ldots, p_{T_n})$ consisting of thread-specific control flow paths (p_{T_i} path of T_i) as well as a scheduling S defining how the execution of $T_1, \ldots, T_n$ is interleaved:

$$I = \left((p_{T_1}, \ldots, p_{T_n}), S\right)$$

Therefore, the search constraint involves different interleavings, e.g.

$$I_j = \left((p'_{T_1}, \ldots, p'_{T_n}), S_j\right) \text{ and } I_k = \left((p''_{T_1}, \ldots, p''_{T_n}), S_k\right)$$

which are executable by the same input and result in different outputs. In other words, in order to trigger a race condition by executing an input x under two schedulings S_j and S_k, the following conditions are necessary and sufficient:

- input x executes paths $\left(p'_{T_1}, \ldots, p'_{T_n}\right)$ under scheduling S_j;
- input x executes paths $\left(p''_{T_1}, \ldots, p''_{T_n}\right)$ under scheduling S_k;
- the two executions result in different output values.

In the particular case that a specific scheduling strategy was specified as providing acceptable behaviour, the above conditions can be applied to compare reference interleavings (i.e. interleavings based on the reference scheduling strategy) with alternative ones. Otherwise, the comparison must involve any possible pair of potential interleavings.

Symbolic execution [3] supports the formalisation of the above three conditions; for each interleaving I it determines a logical predicate E_I characterising those inputs which traverse it as well as expressions $(A_I(z_1), \ldots, A_I(z_m))$ indicating the values assigned to the output variables $(z_1, \ldots, z_m)$ depending on the value of the corresponding input. The output expressions will be abbreviated by:

$$A_I := (A_I(z_1), \ldots, A_I(z_m))$$

Applying symbolic execution to the pair $\left(I_j, I_k\right)$ of interleavings to be compared yields the following constraint denoted as $RCC(I_j, I_k)$:

$$E_{I_j} \wedge \mathrm{E}_{I_k} \wedge \left(\mathrm{A}_{I_j} \neq \mathrm{A}_{I_k}\right)$$

On the whole, considering all relevant pairs (I_j, I_k) of interleavings, the overall race condition constraint *RCC* on inputs reads:

$$\exists I_j, I_k : E_{I_j} \wedge \mathrm{E}_{I_k} \wedge \left(\mathrm{A}_{I_j} \neq \mathrm{A}_{I_k}\right)$$

Step 3: Search for Scenarios Satisfying the Target Constraint
As already mentioned, the constraint identified is necessary and sufficient for the occurrence of race conditions. Therefore, the presence of an exploitable vulnerability of this type is guaranteed by the satisfiability of *RCC* and excluded by its unsatisfiability. The constraint analysis may be supported by an SMT solver; due to logical complexity or undecidability, however, it may be impossible to complete it within affordable resources (or even to complete it at all).

Before applying the SMT-solver Z3 [9] some preliminary steps are required:

- based on an *n*-tuple $\left(p_{T_1}, \ldots, p_{T_n}\right)$ of thread paths and a scheduling S defining an interleaving $I = \left(\left(p_{T_1}, \ldots, p_{T_n}\right), S\right)$, a corresponding sequence Seq_I of instructions is produced by an interleaving generator developed for this purpose; any branch decision (an *if*-node or a *while*-node) is captured by a corresponding assertion (the branching condition or its negation) indicating the control flow arc to be taken;
- the instruction sequence Seq_I is then symbolically executed by the tool KLEE [5]; in simple cases KLEE is able to determine that the underlying interleaving I is not executable such that it can be excluded from further considerations; otherwise, it produces (E_I, A_I).

For any pair $\left(I_j, I_k\right)$ of interleavings to be checked for potential race conditions the corresponding KLEE outputs $\left(E_{I_j}, A_{I_j}\right)$ and $\left(E_{I_k}, A_{I_k}\right)$ are used to build the predicate $RCC(I_j, I_k)$. Its satisfiability is necessary and sufficient for the occurrence of a race condition revealed by deviating behaviours of I_j and I_k. The SMT-solver Z3 is finally used for the purpose of analysing this satisfiability.

The whole process is summarized in Fig. 1.

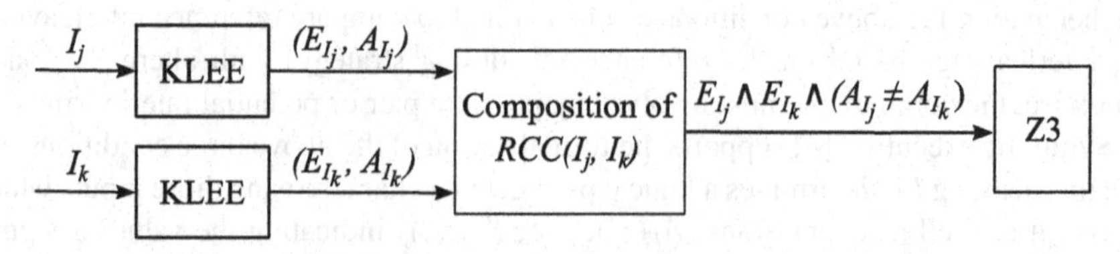

Fig. 1. Structure of approach presented

As already mentioned, the SMT-solver is not guaranteed to terminate in affordable time or to terminate at all. In such cases, a partial evaluation of the overall predicate may provide useful information, e.g.

- if $A_{I_j} \neq A_{I_k}$ results as unsatisfiable, race conditions can be excluded;

- if E_{I_t} $(t \in \{j, k\})$ results as unsatisfiable, I_t is not executable and can therefore be excluded from further comparisons;
- if $E_{I_j} \wedge E_{I_k}$ results as satisfiable, its solutions may serve as input data in the context of a back-to-back testing phase comparing the results of both interleavings;
- if the satisfiability of E_{I_t} $(t \in \{j, k\})$ cannot be evaluated within a tolerable time limit, the analysis may be applied to prefixes of I_t of decreasing length. As soon as a prefix reveals as non-executable, the process can terminate by discharging I_t; on the other hand, as soon as a prefix reveals as executable, heuristic testing (e.g. based on local optimization via evolutionary techniques, s. Subsect. 4.2) may attempt at extending its execution to traverse the complete I_t.

The effort required by considering all possible interleaving pairs can become prohibitive in case of high parallelism; it can, however, be considerably reduced if the acceptable behaviour is provided via a reference scheduling strategy (s. Subsect. 5.3).

5.3 Evaluation

In the present section the approach presented above is evaluated by means of a simple example concerning a fictive store granting a price reduction depending both on the price b of the present purchase and on the amount a of bonus points collected during past purchases. The logic is captured by the code shown in Fig. 2.

It is assumed that originally the application was not planned to include parallel processing of sales involving the same customer ID, such that no analysis of race conditions was required.

In order to avoid undesired side effects due to a successive introduction of multiple family cards, the code may have been adapted such as to enforce atomic regions (i.e. code portions excluding interruptions during their execution) by encapsulating each *if*-condition and its corresponding *then*-part (s. control flow graph shown in Fig. 2).

In the following the approach developed in Subsect. 5.2 is applied for the purpose of verifying the effectiveness of this adaptation. The behaviour to be analysed involves the parallel execution of threads consisting of the same code. In the light of the original system application excluding interleaved executions, the reference scheduling strategy is the sequential one.

Step 1: Restriction of the Search Space
Since all control flow paths access the shared variables a and b, the search cannot be restricted to an input subset.

Step 2: Analysis and Determination of the Search Target Constraint
In this example the control flow of each thread code consists of 6 control flow paths, namely: 1 path consisting of 2 atomic regions, 3 paths consisting of 3 atomic regions, 2 paths consisting of 4 atomic regions.

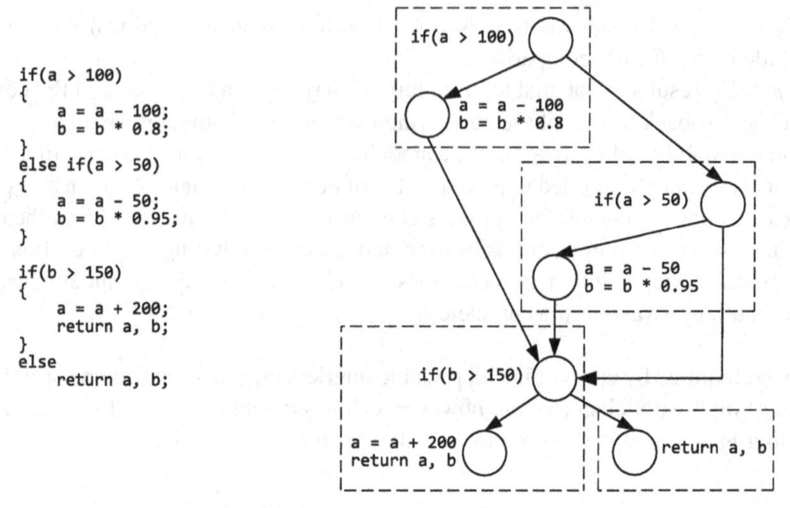

if(a > 100)
{
a = a - 100;
b = b * 0.8;
}
else if(a > 50)
{
a = a - 50;
b = b * 0.95;
}
if(b > 150)
{
a = a + 200;
return a, b;
}
else
return a, b;
if(a > 100)
a = a - 100
b = b * 0.8
if(a > 50)
a = a - 50
b = b * 0.95
if(b > 150)
a = a + 200
return a, b
return a, b

Fig. 2. Code and control flow graph of each thread (dashed lines delimit atomic regions)

As each interleaving $I = \left((p_i, p_j),\ S\right)$ is determined by a particular pair (p_i, p_j) of such paths as well as by a particular scheduling S, the total number of interleavings is given by evaluating the following combinatorial expression:

$$\binom{4}{2} + 3 \cdot \binom{5}{2} + 2 \cdot \binom{6}{2} + 9 \cdot \binom{6}{3} + 6 \cdot \binom{7}{3} + 4 \cdot \binom{8}{4} = 736$$

Among them, 481 were identified by KLEE as non-executable such that they could be excluded from further consideration. The remaining 255 include 12 interleavings based on the reference scheduling strategy. Therefore, the search for a solution of $RCC(I_j, I_k)$ can be limited to $12 \cdot 243 = 2916$ pairs (I_j, I_k) of interleavings, where I_j denotes a reference interleaving and I_k denotes a non-reference interleaving.

Step 3: Search for Scenarios Satisfying the Target Constraint
The search for a system failure due to a race condition is successfully concluded as soon as the constraint $RCC(I_j, I_k)$ is solved for a pair of interleavings (I_j, I_k) by an SMT-solver producing an appropriate scenario. On the other hand, if the constraint $RCC(I_j, I_k)$ results as unsatisfiable for all pairs (I_j, I_k) of interleavings, the absence of race conditions can be implied. The third alternative involving a time-out and demanding for heuristic approaches did not occur in this particular case.

The results concerning the analysis of the constraints and the effort required are summarized in Table 2 and Table 3.

Table 2. Number of interleaving pairs (I_j, I_k) with satisfiable/unsatisfiable $RCC(I_j, I_k)$

Satisfiable $RCC(I_j, I_k)$	Unsatisfiable $RCC(I_j, I_k)$
107	2809

Table 3. Time required for tool execution

KLEE to analyse all interleavings	<23 min
Z3 to detect first race	<5 min
Z3 to analyse all relevant pairs of interleavings	<23 min

As an example, a race condition is triggered by input *(a, b)* = (51, 192) via the interleaved thread execution $(1T_1, 2T_2, 2T_1, 1T_2)$, where nT_i denotes the execution of the next *n* atomic regions of thread T_i ($i \in \{1, 2\}$). The resulting output (401, 182.4) differs from the output (101, 145.92) obtained via the sequential reference scheduling strategy. Although the interleaved execution increases the price, the growth in bonus points (if used to purchase an expensive article) may result in a loss to the store owner.

6 Conclusion and Future Work

This article presented a structured approach supporting the detection of exploitable vulnerabilities of given type. The applicability of the generic pattern proposed was illustrated by considering two different types of software weaknesses among the most relevant ones, namely buffer overflows and race conditions. Depending on the complexity of the application considered, the technique presented may involve undecidable aspects not permitting to rely exclusively on systematic analyses like symbolic execution or SMT-solving. In such cases, the use of heuristics-based testing strategies is considered.

It is planned to devote future work to a refinement of the search for race conditions by optimizing the combination of systematic and heuristic strategies. In order to do so, the insight gained by static analysis should be maximized such as to provide meaningful guidance to the subsequent dynamic phase based on intelligent testing.

Acknowledgment. The authors gratefully acknowledge that the work presented was supported by the German Federal Ministry for Economic Affairs and Energy (BMWi), project no. 1501600C (SMARTEST2).

References

1. Al Sardy, L., Neubaum, A., Saglietti, F., Rudrich, D.: Comparative evaluation of security fuzzing approaches. In: Romanovsky, A., Troubitsyna, E., Gashi, I., Schoitsch, E., Bitsch, F. (eds.) SAFECOMP 2019. LNCS, vol. 11699, pp. 49–61. Springer, Cham (2019). https://doi.org/10.1007/978-3-030-26250-1_4
2. Al Sardy, L., Saglietti, F., Tang, T., Sonnenberg, H.: Constraint-based testing for buffer overflows. In: Gallina, B., Skavhaug, A., Schoitsch, E., Bitsch, F. (eds.) SAFECOMP 2018. LNCS, vol. 11094, pp. 99–111. Springer, Cham (2018). https://doi.org/10.1007/978-3-319-99229-7_10
3. Baldoni, R., Coppa, E., D'Elia, D., Demetrescu, C., Finocchi, I.: A survey of symbolic execution techniques. ACM Comput. Surv. **51**(3), 1–39 (2018)
4. Blanc, N., Kroening, D.: Race analysis for SystemC using model checking. In: IEEE/ACM International Conference on Computer-Aided Design, pp. 356–363 (2008)

5. Cadar, C., Dunbar, D., Engler, D.: KLEE: unassisted and automatic generation of high-coverage tests for complex system programs. In: USENIX Symposium on Operating Systems Design and Implementation, pp. 209–224, USENIX Association (2008)
6. Carnegie Mellon University (SEI): CERT C++ Coding Standard, Concurrency (2016)
7. Common Attack Pattern Enumeration and Classification Community: Overflow Buffers. CAPEC-100 (2020)
8. Cisco: Most Common CWE Vulnerabilities. Annual Cybersecurity Report (2018)
9. de Moura, L., Bjørner, N.: Z3: an efficient SMT solver. In: Ramakrishnan, C.R., Rehof, J. (eds.) TACAS 2008. LNCS, vol. 4963, pp. 337–340. Springer, Heidelberg (2008). https://doi.org/10.1007/978-3-540-78800-3_24
10. Foster, J.C., Osipov, V., Bhalla, N., Heinen, N.: Buffer Overflow Attacks: Detect, Exploit Prevent. Syngress, Rockland (2005)
11. Ganesh, V., Dill, D.L.: A decision procedure for bit-vectors and arrays. In: Damm, W., Hermanns, H. (eds.) CAV 2007. LNCS, vol. 4590, pp. 519–531. Springer, Heidelberg (2007). https://doi.org/10.1007/978-3-540-73368-3_52
12. Goldberg, D.E., Holland, J.H.: Genetic algorithms and machine learning. Mach. Learn. **3**, 95–99 (1988)
13. Holzmann, G.: The power of 10: rules for developing safety-critical code. IEEE Comput. **39**, 95–97 (2006)
14. MITRE Corporation: Common Weakness Enumeration (CWE), Top 25 Most Dangerous Software Weaknesses (2020)
15. MITRE Corporation: Common Vulnerabilities and Exposures (CVE). CVE-2019-3568, CVE-2020-1839, CVE-2021-21006, CVE-2021-21148, CVE-2021-3156
16. Musuvathi, M.: Systematic concurrency testing using CHESS. In: Workshop on Parallel and Distributed Systems: Testing, Analysis, and Debugging. ACM (2008)
17. Nowotka, D., Traub, J.: MEMICS – Memory Interval Constraint Solving of (concurrent) Machine Code. Automotive – Safety & Security, LNI 210, pp. 69–83, Springer (2012)
18. Oulu University (Secure Programming Group): Radamsa (2010)
19. Qadeer, S., Wu, D.: KISS: keep it simple and sequential. SIGPLAN Not. **39**, 14–24 (2004)
20. Schneider Electric Software Security Response Center: Remote Code Execution Vulnerability. Security Bulletin LFSEC00000125 (2018)
21. Tracey, N., Clark, J., Mander, K., McDermid, J.: An automated framework for structural test-data generation. In: Conference on Automated Software Engineering, pp. 285–288. IEEE (1998)
22. Wegener, J., Buhr, K., Pohlheim, H.: Automatic test data generation for structural testing of embedded software systems by evolutionary testing. In: Conference on Genetic and Evolutionary Computation, pp. 1233–1240. Morgan Kaufmann (2002)
23. Yun, I., Lee, S., Xu, M., Jang, Y., Kim, T.: QSYM: a practical concolic execution engine tailored for hybrid fuzzing. In: USENIX Security Symposium, pp. 745–761. USENIX (2018)
24. Zalewski, M.: American Fuzzy Lop (2017)

Formal Modelling of the Impact of Cyber Attacks on Railway Safety

Ehsan Poorhadi(✉), Elena Troubitysna, and György Dán

KTH Royal Institute of Technology, Stockholm, Sweden
{poorhadi,elenatro,gyuri}@kth.se

Abstract. Modern railway signaling extensively relies on wireless communication technologies for efficient operation. The communication infrastructures that they rely on are increasingly based on standardized protocols and are shared with other users. As a result, it has an increased attack surface and is more likely to become the target of cyber attacks that can result in loss of availability and, in the worst case, in safety incidents. While formal modeling of safety properties has a well-established methodology in the railway domain, the consideration of security vulnerabilities and the related threats lacks a framework that would allow a formal treatment. In this paper, we develop a modeling framework for the analysis of the potential of security vulnerabilities to jeopardize safety in communications-based train control for railway signaling, focusing on the recently introduced moving block system. We propose a refinement-based approach enabling a structured and rigorous analysis of the impact of security on system safety.

Keywords: Railway safety · Formal modelling · Event-B

1 Introduction

Modern railway signaling extensively relies on wireless communication technologies for implementing a variety of safety-critical functionalities. The extensive use of standardized communication technologies and the reliance on shared communication infrastructures make the security vulnerabilities inherent to railways. Hence, there is a need to develop techniques that can systematically analyze the impact of cyber attacks on system safety and assess their criticality.

In this paper, we propose a formal approach to analyzing the impact of cyber attacks on ERTMS/ETCS moving block without trackside train detection (TTD) equipment [13], which is a safety architecture for communication-based train control. We formally model the typical data flow between train and trackside and security control mechanisms. Such an approach allows us to model how attackers with different capabilities can circumvent security protection and

I. Habli et al. (Eds.): SAFECOMP 2021 Workshops, LNCS 12853, pp. 117–127, 2021.
https://doi.org/10.1007/978-3-030-83906-2_9

perform an attack aiming at jeopardizing system safety. In our approach, we formally define safety as a set of conditions over the parameters of the function. It allows us to formally assess the impact of various attacks on safety.

Our work relies on modeling and verification in Event-B [14]. Event-B is a formal state-based approach to correct-by-construction system development. It is based on proof-based verification and refinement development technology. The Rodin platform provides mature automated support for modeling and verification in Event-B [10].

In this paper, we focus on studying the moving block functionality. Until now moving block has been studied only from the safety perspective. Hence, we believe the proposed approach to formal modeling of the impact of cyber attacks on the safety of moving block contributes to the study of this emerging reference architecture.

2 Safety and Security Analysis of Moving Block

The ERTMS moving block reference architecture [13] is an enabler of increased railway capacity. It is based on controlling trains' movement through communication between the trains and the radio block centers (RBC). A part of the architecture of the ERTMS moving block is shown in Fig. 1.

Each RBC is responsible for the movement of trains in a specific zone called the area of control. For moving into the RBC's area of control, a train must establish a communication session and send its position to the RBC. When the RBC receives a train's position, it authorizes the train to move in the area up to a specific point called End of Authority (EoA). An EoA could be the rear of a chased train, see Fig. 1. Thus, the RBC needs to accurately estimate trains' position to avoid a hazardous situation. This is done by periodically exchanging position report messages (PRM) sent from trains to the RBC. Such a period for each train defines the control loop for the train. At the end of the control loop of a train, the RBC could send a command that extends the EoA, shortens it, leaves it unchanged, or stops the train. The system-level safety property is to keep the trains' EoA in the rear of the chased trains. One way to capture this requirement is to consider the accuracy of the estimation i.e., ensure that if the RBC's estimate of the rear end of the train is behind its actual location. In this paper, we consider the control loop for the train T_1 in Fig. 1 and investigate if the RBC could always estimate the position of T_1 accurately in the presence of an attacker.

We denote the track by $T = \{1, ..., l\}$, where l is a large enough natural number. Suppose that $P_{real} = (R_{real}, F_{real}) \in T \times T$ is the real location of the train in the beginning of a control loop. Here, R_{real} and F_{real} represent the rear and front end of the train, respectively. At this point, the train sends a PRM to the RBC containing some variables.

The first variable is $front$, which represents the distance that the train travels. We assume that it is greater than or equal F_{real}. The second variable is int, which shows train integrity (if the train is not split). This variable can be

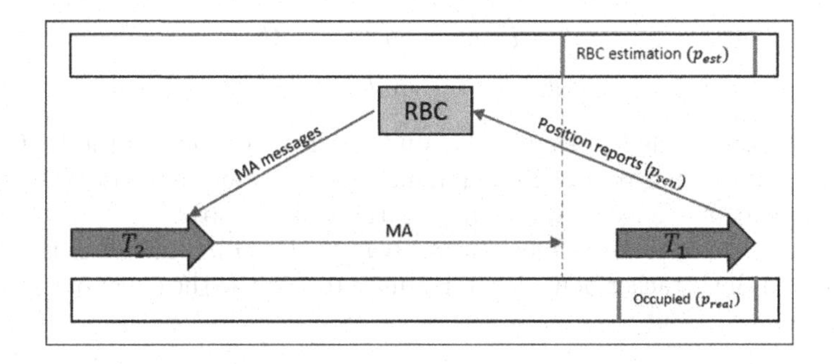

Fig. 1. Illustration of the control loop for ERTMS moving block.

either *confirmed*, *lost*, or *unknown*. In this work, we assume that the train does not split, but it may report its integrity *lost* or *unknown* due to being unable to detect its integrity status. The third variable is len, which represents the distance between the rear end of the train at the time of integrity is confirmed and $front$. If int is *lost* or *unknown*, then len must be zero. By definition, if the train confirms its integrity, $len \geq front - R_{real}$.

When the RBC receives a PRM, it estimates the train's position $P_{est} = (R_{est}, F_{est})$, which depends on the previous estimate and the variables in the PRM as follows. F_{est} will be $front$. If $integrity = confirmed$, $R_{est} = front - len$.; otherwise, $R_{est} = R_{est}^{pre}$. At the end of the control loop, the estimation P_{est} is called accurate if we have,

$$R_{est} \leq R_{real}. \tag{1}$$

Our focus is on cyber attacks on the train-RBC channel targeting data integrity. The attacker tries to change variables in PRMs. Although the communication between the train and RBC is encrypted, we assume that the attacker can break the encryption algorithm as it is discussed in [11]. Thus, the attacker can observe all messages.

Communication between trains and the RBCs is critical for preventing security attacks targeting safety. Hence a pre-shared key is distributed in advance between the entities (RBCs, trains). RBCs and trains use a pre-shared key to authenticate each other and to establish a session key. The session key is used for computing a message authentication code (MAC) for each message. Upon receiving a message, the MAC is verified by the receiver. The message is accepted if the MAC is correct.

After that, the receiver checks whether the values that the message carries are within the expected boundaries i.e., it performs reasonability checks. In this work, we consider the following conditions to define reasonability for each PRM. Let P_{est} be the current estimation of the train's location. A PRM ($front$, int, len) is consider reasonable if,

$$(int = False \Rightarrow len = 0), \tag{2}$$

$$(int = True \Rightarrow len \geq length), \tag{3}$$

$$front \geq R_{est} \ \wedge \ front \in T \ \wedge \ front - len \in T \wedge \ len \in \{0,1,2\}, \tag{4}$$

where *length* is the length of the train known in advance by the RBC. Since the railway system has to be failsafe if a message does not pass any of the checks, the system executes a safe shutdown (i.e., the trains stop).

In Sect. 4, we devise a refinement strategy to develop a model of the system behavior and the attacker behavior in Event-B to analyze the effect of the attacks on Property 1.

3 Event-B

Event-B is a state-based formal method for modeling and verifying a system's properties by theorem proving. It uses set theory as a notation and refinements for modeling a system at different levels of abstraction.

A model in Event-B consists of two components, *Context and Machine.* In the model's context one can introduce constants (numbers, sets, relations, and etc.) and axioms over the constants. Machines model the dynamic behavior of the system as an abstract state machine. A collection of variables represents the model state. A machine models the dynamic behavior of the system. Events represent transitions between system states. Generally, an event has the form

$$any \ a \ where \ G_e \ then \ R_e \ end,$$

where a is the list of local variables, G_e is the list of guards, and R_e is the list of assignments (actions) to some variables. If there exist values a_0 for the local variables a such that G_e is true then the event is enabled, i.e., it can be executed. If the event is executed, the variables obtain the new values according to the assignment R_e. The assignments can be deterministic or non-deterministic. If there are more than one enabled events, an event to be executed is chosen non-deterministically. Events must preserve all invariants of the model. In a machine, the invariant preservation and the correctness of the model are checked by generating and proving theorems called proof obligations. The Rodin platform provides an automated tool for modelling and verification in Event-B [10].

Modelling in Event-B is based on a top-down refinement-based approach to system development. The development starts from an abstract specification of the system that non-deterministically models the essential requirements of the system. Then we reduce non-determinism and introduce detailed design decisions in a sequence of refinement steps. In a refinement step, events and data structures can be refined. Refinement ensures that the model transformations preserve the externally observable behaviour. In case of data refinement, new variables can be added to the model, or some abstract variables replaced with new variables. In the latter case, one has to define a gluing invariant that connects the abstract variable with the concrete one.

```
VARIABLES
    phase, operation, prm, stop, R_real, R_est,
      F_real, F_est
INVARIANTS
    inv7: (phase = checking ∧ prm = TRUE) ⇒
      R_est ≤ R_real
EVENTS
Event Move ⟨⟩ ≙
    any
        m
    where
        grd1: phase = move
    then
        act1: phase := report
        act2: R_real := R_real + m
        act3: F_real := F_real + m
    end
Event Report ⟨⟩ ≙
    when
        grd1: phase = report
    then
        act1: phase := integrity
    end
Event Integrity ⟨⟩ ≙
    where
        grd5: phase = integrity
    then
        act1: phase :∈ {detection, res}
    end
Event Res ⟨ordinary⟩ ≙
    when
        grd1: phase = res
    then
        act1: phase :∈ {detection, est}
    end
Event Est ⟨ordinary⟩ ≙
    where
        grd1: phase = est
        grd4: prm = TRUE ⇒ r_est ≤ R_real
    then
        act1: phase := checking
        act2: R_est := r_est
    end
Event Detection ⟨ordinary⟩ ≙
    when
        grd1: phase = detection
    then
        act1: phase := move
        act2: operation := abnormal
    end
Event Checking ⟨ordinary⟩ ≙
    where
        grd1: phase = checking
        grd3: R_est > R_real ⇔ s = TRUE
    then
        act1: phase := move
        act2: stop := s
    end
END
```

Fig. 2. Modeling the phases of the control loop.

4 Moving Block Development in Event-B

To formally analyze the impact of cyber attacks on safety, we perform several refinement steps to model different aspects of communication between the train and RBC, security control mechanisms, and attacker actions. The specification can be seen in [15]. Our refinement strategy is as follows.

- Modeling the phases of the control loop.
- Modeling PRMs and RBC's estimation.
- Modeling checking data integrity using MAC.
- Modeling the attacker's behavior and tampering attack.

Abstract Specification. In the abstract specification shown in Fig. 2, we model the control loop of the train and obtain a condition under which the RBC estimation is accurate.

The control loop consists of the following phases,

$$\{move,\ report,\ integrity,\ res,\ detection,\ est,\ checking\}.$$

The variable *phase* designates the current phase. The events with the corresponding names model the behavior of the system at each phase. The train movements and sending PRMs are non-deterministically modeled in the phases *move* and *report*. In the phases *integrity* and *res*, the RBC checks the data integrity and reasonability of variables in the PRMs, respectively. If the position report passes all checks, the phase changes to *est*; otherwise, it changes to *detection*. In the phase *detection*, the state of operation changes from *normal* to *abnormal* by defining the variable *operation*. Thus, we deadlock the specification, which corresponds to the system shutdown in our model. In the phase *est*, the RBC computes P_{est}. At this level, the dependency between P_{est} and the

```
MACHINE
VARIABLES
      pl_in, pl_out
INVARIANTS
      inv10: ⟨theorem⟩ (phase = checking ∧ pl_in =
          pl_out) ⇒ R_est ≤ R_real
      inv12: phase = est ⇒ (pl_in(int) ∈ {0,1} ⇒
          pl_in(len) = 0)
      inv13: phase = est ⇒ (pl_in(int) = 2 ⇒
          pl_in(len) ≥ 3)
      inv14: phase = est ⇒ (pl_in(front) ≥
          R_est ∧ pl_in(front) ∈ T ∧ pl_in(front) −
          pl_in(len) ∈ T ∧ pl_in(int) ∈ {0,1,2})
EVENTS
Event Report ⟨ordinary⟩ ≙
refines Report
      where
            grd2: fr ∈ T ∧ le ∈ T ∧ i ∈ {0,1,2}
            grd4: fr ≥ F_real ∧ fr ≤ F_real + 2
            grd5: i = 2 ⇒ (fr − R_real ≤ le)
            grd6: i ∈ {0,1} ⇒ le = 0
      then
            act2: pl_out := {front ↦ fr, int ↦
                i, len ↦ le}
      end
Event Res ⟨ordinary⟩ ≙
refines Res
      where
            grd4: bool(pl_in(int) ∈ {0,1} ⇒
                pl_in(len) = 0) = valid1
            grd5: bool(pl_in(int) = 2 ⇒ pl_in(len) ≥
                3) = valid2
            grd6: bool(pl_in(front) ≥ R_est ∧
                pl_in(front) ∈ T ∧ pl_in(front) −
                pl_in(len) ∈ T ∧ pl_in(int) ∈
                {0,1,2}) = valid3
            grd7: validity = TRUE ⇔ (valid1 =
                TRUE ∧ valid2 = TRUE ∧ valid3 =
                TRUE)
      then
            act1: phase :| (validity = TRUE ⇒
                phase' = est) ∧ (validity = FALSE ⇒
                phase' = detection)
      end
Event Est ⟨ordinary⟩ ≙
refines Est
      where
            grd5: f_est = pl_in(front)
            grd9: pl_in(int) = 2 ⇒ r_est = f_est −
                pl_in(len)
            grd8: pl_in(int) ∈ {0,1} ⇒ r_est = R_est
      then
            act1: phase := checking
            act2: R_est := r_est
            act3: F_est := f_est
      end
END
```

Fig. 3. Modeling PRMs and RBC's estimation.

PRM is represented by a boolean variable prm, which represents status of data integrity of the PRM. If $prm = True$, then the RBC has an accurate estimate. We verify this by adding the invariant,

$$(phase = checking \ \wedge \ prm = True) \Rightarrow R_{est} \leq R_{real},$$

where $phase = checking$ represents the state of the model after RBC's estimation. In the phase *checking*, we check if the Property 1 still holds. If it is violated, then the variable *stop* becomes $True$, and all events become disabled. At this point, the attack is considered successful.

First Refinement. The refinement shown in Fig. 3 models the data flow and allows us to verify system safety in the absence of cyber attacks. Also, the step results in a deterministic specification of reasonability checking. To define sending and receiving a PRM, we define the set

$$Fields = \{front, \ int, \ len\}$$

in the model's context, which represents different fields of a PRM. We define variables p^1_{out}, $p^1_{in} \in Fields \rightarrow \mathbb{N}$, to represent sending and receiving PRMs by the train and the RBC, respectively. Now, we refine the events *Report* to form p^1_{out} and *integrity* to receive a PRM and store it in p^1_{in}. For example, if the train wants to send a PRM which carries $(front_0, \ lost, \ len_0)$, it forms p^1_{out} as $\{(front, \ front_0), \ (int, \ 0), \ (len, len_0)\}$. Note that we model the train integrity status *lost*, *unknown*, and *confirmed* with values 0, 1, and 2, respectively. At this point, we remove the abstract variable prm with the gluing invariant,

$$phase \in \{res, est\} \Rightarrow (prm = True \Leftrightarrow p^1_{in} = p^1_{out}).$$

```
VARIABLES
    p_out, p_in
INVARIANTS
    inv7: phase ∈ {res, est} ⇒ comp_mac(p_in(front) ↦
        p_in(int) ↦ p_in(len)) = p_in(mac)
EVENTS
Event Report ⟨ordinary⟩ ≙
refines Report
    where
        grd10: m = comp_mac(fr ↦ i ↦ le)
    then
        act2: p_out := {front ↦ fr, int ↦
            i, len ↦ le, mac ↦ m}
    end

Event Integrity ⟨ordinary⟩ ≙
refines Integrity
    where
        grd7: comp_mac(rec2(front) ↦
            rec2(int) ↦ rec2(len)) = rec2(mac) ⇒
            acc = TRUE
        grd8: comp_mac(rec2(front) ↦
            rec2(int) ↦ rec2(len)) ≠ rec2(mac) ⇒
            acc = FALSE
    then
        act1: phase :| (acc = TRUE ⇒ phase' =
            res) ∧ (acc = FALSE ⇒ phase' =
            detection)
    end
END
```

Fig. 4. Modeling checking data integrity using MAC.

We also refine the event *Est* to deterministically obtain new P_{est} from p^1_{in} and the current estimation of the train's location. At this point, we prove that,

$$(phase = checking \;\wedge\; p^1_{in} = p^1_{out}) \Rightarrow R_{est} \leq R_{real},$$

which are the formal conditions stating that safety is preserved if the RBC receives an unaltered PRM from the train.

To check reasonability of p^1_{in}, we refine event *Res* by using Conditions 2–4 such that if p^1_{in} satisfies all the conditions then *phase* changes to *est*; otherwise, it will be *detection*. We define the invariant,

$$phase = est \Rightarrow Conditions\, 2\text{–}4,$$

which shows that the RBC accepts an p^1_{in} if it has valid values.

Second Refinement. The excerpt from the second refinement is shown in Fig. 4. We introduce the function,

$$comp_{mac} \in \mathbb{N} \times \mathbb{N} \times \mathbb{N} \rightarrow \mathbb{N}$$

for modelling the computation of the MAC for a PRM. Before sending a PRM which carries $(front_0,\ int_0,\ len_0)$, the train computes,

$$comp_{mac}(front_0,\ int_0,\ len_0),$$

and sends it with PRM. To model this, we add a new set $Fields' = Fields \cup \{mac\}$ and replace the abstract variable p^1_{out} with the concrete variable $p_{out} \in Fields' \rightarrow \mathbb{N}$. Consequently, we replace the abstract variable p^1_{in} with the concrete variable p_{in}. We also define the gluing invariant,

$$\{mac\} \lhd p_{out} = p^1_{out},$$

$$\{mac\} \lhd p_{in} = p^1_{in},$$

where $(\lhd)$ is domain restriction to represent the relation between the abstract and concrete variable.

VARIABLES
tamper
valid_forge
INVARIANTS
inv4: $tamper = FALSE \wedge phase \in \{res, est, checking\} \Rightarrow p\_in = p\_out$
inv3: (theorem) $(tamper = FALSE \wedge phase = checking) \Rightarrow R\_est \leq R\_real$
inv5: $(tamper = TRUE \wedge phase = checking) \Rightarrow R\_real < R\_est$
inv8: $(tamper = TRUE \wedge valid\_forge = TRUE) \Rightarrow phase \neq detection$
inv9: $(valid\_forge = TRUE \wedge tamper = TRUE) \Rightarrow comp\_mac(p\_in(front) \mapsto p\_in(int) \mapsto p\_in(len)) = p\_in(mac)$
inv13: $(tamper = TRUE \wedge phase \in \{res, est, checking\}) \Rightarrow valid\_forge = TRUE$

EVENTS
Event Integrity ⟨ordinary⟩ ≙
refines Integrity
where
grd10: $tamper = FALSE \Rightarrow rec2 = p\_out$
grd9: $tamper = TRUE \Rightarrow rec2 = \{front \mapsto F\_real + 1, int \mapsto 2, len \mapsto F\_real - R\_real, mac \mapsto mac\_forge\}$
then
act3: $valid\_forge :| (tamper = FALSE \Rightarrow valid\_forge' = FALSE) \wedge (tamper = TRUE \Rightarrow valid\_forge' = bool(mac\_forge = comp\_mac(F\_real + 1 \mapsto 2 \mapsto F\_real - R\_real)))$
end
Event Tamper ⟨ordinary⟩ ≙
when
grd1: $phase = integrity$
grd2: $F\_real \neq l$
then
act1: $tamper := TRUE$
end
END

Fig. 5. Modeling the attacker's behavior and tampering attack.

When the RBC receives a PRM m, it checks whether

$$m(mac) = comp_{mac}(m(front),\ m(int),\ m(len)).$$

If it is equal, the PRM is accepted. This results in the refinement of the events *Integrity* and *Report* and in defining the deterministic procedure for entering the phases *res* and *detection*. We prove the invariant,

$$phase \in \{res, est\} \Rightarrow comp_{mac}(p_{in}(front), p_{in}(int), p_{in}(len)) = p_{in}(mac),$$

to show the new PRM is accepted by the RBC only if it has a valid MAC.

Third Refinement. In our previous refinements, we have focused on modeling the behavior of system components, communication, and security control mechanisms. Our next refinement allows us to formally analyze if the specified system is resilient against cyber-attacks. We consider an attacker that monitors and changes PRMs. We assume that the attacker's goal is to violate the safety Property 1.

An excerpt from the specification of the refinement is shown in Fig. 5. To model tampering, we consider a new event *Tamper*. If the attacker decides to tamper with PRMs, the event becomes enabled, and the boolean variable *tamper* becomes *True*. In our model, the attacker changes the message to the following PRM,

$$att = \{(front, F_{real} + 1), (int,\ 2),\ (len,\ F_{real} - R_{real}),\ (mac, m)\},$$

where $m \in \mathbb{N}$. This is an example of a tampering attack that violates the safety property. These sort of injections could be find easily by finding a solution to the following formula.

$$Properties\ 2 - 4 \wedge (integrity = confirmed \Rightarrow front - len > R_{real})$$
$$\wedge\, (integrity \neq confirmed \Rightarrow R_{est}^{pre} > R_{real}.)$$

This immediately follows that *integrity* must be *confiremd* in the injection since we have $R_{est}^{pre} \leq R_{real}$.

Now, we refine the event *Integrity* in which the RBC receives PRMs. In the event, the local parameter *rec*2 used for showing the PRM that the RBC receives becomes restricted by the guards,

$$tamper = False \Rightarrow rec2 = p_{out},$$

$$tamper = True \Rightarrow rec2 = att.$$

We define the following invariants

$$(tamper = False \ \wedge \ phase = checking) \Rightarrow R_{est} \leq R_{real},$$

$$(tamper = True \ \wedge \ phase = checking) \Rightarrow R_{real} < R_{est},$$

which show the effect of the attack on system safety. As can be seen, safety will be violated in the presence of the attacker. Note that, the condition

$$(tamper = True \ \wedge \ phase = checking)$$

implies that m was a valid MAC. Thus, safety is violated only if the attacker can send the message with a valid MAC.

In this section, we have shown the development of moving block specification in Event-B and analyzed the impact of potential cyber attacks on safety. We have followed the refinement-based development strategy and used failed proofs as guidance for understanding which protection measures are required to guarantee safety. Consequently, by modifying the specification to model their effect, we could specify how to protect system safety against cyber attacks.

Reliance on proof-based verification and automated tool support – the Rodin platform – provided a high degree of automation in formal modeling and proof-based verification. It allowed us to consider the behavior of the system in all possible situations without the need to restrict the state space. However, these could be achieved only because we followed the refinement development and created a complex system specification in several incremental model transformations.

5 Related Work

Recently, safety and security interactions have been studied in several works. Authors in [1–3] demonstrate the need for an approach that integrates safety and security aspects.

In some works, techniques used for safety analysis are adopted to perform security analysis. For example, in [4] authors use Boolean logic Driven Markov Processes (BDMP). BDMP is a graphical modeling approach to show failure due to an accident or malicious attack. In [5] authors formulated a security cause-effect chain and adopted failure mode and effect analysis (FMEA) for safety and security analysis. In [6] fault trees and attack trees are combined to integrate safety and security analysis. These approaches help engineers systematically discover security vulnerabilities that can affect safety.

A different line of works tackled the interactions between formal analysis of safety and security requirements [7,8]. Unlike these works, we investigate the effect of the violation of security requirements on safety properties and instead of relying on model checkers, we verify the properties by theorem proving.

In railway domain, in [9] authors formally verified the effect of different types of attacks on the Euroradio protocol used in ERTMS. They used the ProVerif verifier to check security requirements, but they did not consider the effect of attacks on system safety. In [12], the safety analysis of one of the variants of ERTMS/ETCS level 3 was considered without considering security issues. Authors used Event-B to model and specify Hybrid ERTMS level 3, and they managed to find some inconsistency in the specification of the system.

6 Conclusion

In this paper, we have proposed a formal approach to analyzing the impact of cyber attacks on ERTMS moving block safety. We devise a refinement strategy for modeling various aspects of system behavior: communication between the components, data flow within the control cycle as well as mechanisms for security control and the attacker's behavior. We demonstrated how to define safety as an invariant property of the system and rigorously derive the conditions that should be preserved to guarantee that the system is resilient against cyber attacks.

In this paper, we used Event-B as our formal development framework. It allowed us to apply the proposed approach as correctness-preserving model transformations and use proofs as a mechanism for deriving the resilience constraints. The proposed approach has been applied to analyze the resilience against cyber attacks of the moving block reference architecture. It has demonstrated that the current system architecture should be augmented with additional security control mechanisms to ensure safety in the presence of cyber attacks.

In our future work we are planning to develop mitigation schemes and to model those so as to prove safety in the presence of attacks.

References

1. Young, W., Leveson, N.G.: An integrated approach to safety and security based on systems theory. Commun. ACM **57**(2), 31–35 (2014)
2. Paul, S., Rioux, L.: Over 20 years of research into cybersecurity and safety engineering: a short bibliography. Saf. Secur. Eng. **VI**, 335 (2015)
3. Steiner, M., Liggesmeyer, P.: Combination of safety and security analysis - finding security problems that threaten the safety of a system. In: SAFECOMP 2013 - Workshop DECS-2013 (2013)

4. Piètre-Cambacédès, L., Bouissou, M.: Modeling safety and security interdependencies with BDMP (Boolean logic Driven Markov Processes). In: 2010 IEEE International Conference on Systems, Man and Cybernetics (2010)
5. Schmittner, C., Gruber, T., Puschner, P., Schoitsch, E.: Security application of failure mode security application of failure mode and effect analysis (FMEA). In: SAFECOMP 2014, pp. 310–325 (2014)
6. Fovino, I.N., Masera, M., De Cian, A.: Integrating cyber attacks within fault trees. Reliab. Eng. Syst. Saf. **94**(9), 1394–1402 (2009)
7. Brunel, J., Rioux, L., Paul, S., Faucogney, A., Vallee, F.: Formal safety and security assessment of an avionic architecture with alloy. In: ESSS 2014, pp. 8–19 (2014)
8. Troubitsyna, E., Laibinis, L., Pereverzeva, I., Kuismin, T., Ilic, D., Latvala, T.: Towards security-explicit formal modelling of safety-critical systems. In: Skavhaug, A., Guiochet, J., Bitsch, F. (eds.) SAFECOMP 2016. LNCS, vol. 9922, pp. 213–225. Springer, Cham (2016). https://doi.org/10.1007/978-3-319-45477-1_17
9. de Ruiter, J., Thomas, R.J., Chothia, T.: A formal security analysis of ERTMS train to trackside protocols. In: Lecomte, T., Pinger, R., Romanovsky, A. (eds.) RSSRail 2016. LNCS, vol. 9707, pp. 53–68. Springer, Cham (2016). https://doi.org/10.1007/978-3-319-33951-1_4
10. Rodin, Event-B platform. http://www.event-b.org
11. Chothia, T., Ordean, M., Ruiter, J., Thomas, R.: An attack against message authentication in the ERTMS train to trackside communication protocols. In: ACM on Asia Conference on Computer and Communications Security (2017)
12. Hansen, D., et al.: Using a formal B model at runtime in a demonstration of the ETCS hybrid level 3 concept with real trains. In: Butler, M., Raschke, A., Hoang, T.S., Reichl, K. (eds.) ABZ 2018. LNCS, vol. 10817, pp. 292–306. Springer, Cham (2018). https://doi.org/10.1007/978-3-319-91271-4_20
13. Moving Block System Specification (2019). http://projects.shift2rail.org/download.aspx?id=a81c93c2-36a5-46cf-8bd8-4924ae612dd7
14. Abrial, J.R.: Modeling in Event-B. Cambridge University Press, Cambridge (2010)
15. https://github.com/Poorhadi/Moving-Block

LoRaWAN with HSM as a Security Improvement for Agriculture Applications - Evaluation

Reinhard Kloibhofer[1(✉)], Erwin Kristen[1], and Afshin Ameri E.[2]

[1] AIT Austrian Institute of Technology GmbH, Giefinggasse 4, 1210 Vienna, Austria
{reinhard.kloibhofer,erwin.kristen}@ait.ac.at
[2] MDH Mälardalen University, Högskoleplan 1, 722 20 Västerås, Sweden
afshin.ameri@mdh.se

Abstract. The future of agriculture is digital and the move towards it has already started. Comparable to modern industrial automation control systems (IACS), today's smart agriculture makes use of smart sensors, sensor networks, intelligent field devices, cloud-based data storage, and intelligent decision-making systems. These agriculture automation control systems (AACS) require equivalent, but adapted, security protection measures. Last year, the agriculture-related security theme was addressed in a presentation [1] on DECSoS '20 workshop of the SAFECOMP 2020 conference. In the workshop a simple soil sensor prototype with wireless communication system was presented. The sensor was used to demonstrate the improvements to operational security of field devices through employing cyber security protection techniques. As a continuation of the last year contribution, this paper presents the evaluation of the technologies presented in real-life deployment of AACS. It also describes operational scenarios and experiences with the implementation of security measures for AACS, e.g.: the implementation of a four-layer cyber security architecture, the signalling concept of alarms and notifications in the event of a cyber-attack, the assessment of security measures, the costs of security, and an outlook of upcoming future security requirements for wireless IoT devices which are specified by the European Commission through the European Radio Equipment Directive (RED).

Keywords: Agriculture Automation Control Systems (AACS) · Hardware Security Module (HSM) · Internet of Things (IoT) · Cyber-Physical Systems (CPS) · Safety & Security · Agriculture · LoRaWAN · European Radio Equipment Directive (RED)

1 Introduction

Digitalization is making a rapid progress in agriculture. Ground vehicles are becoming more and more intelligent with integrated control electronics that support the farmer in doing their daily work in a more precise and simple manner. The vehicles themselves are constantly generating operational and environmental data that can be used for future maintenance and mission decisions. Some of this data is stored on removable data storage

I. Habli et al. (Eds.): SAFECOMP 2021 Workshops, LNCS 12853, pp. 128–140, 2021.
https://doi.org/10.1007/978-3-030-83906-2_10

media, which are manually transferred to a data repository after completion of the task. However, often such data are transferred directly through a wireless connection to an online cloud-based data center.

Agricultural companies, such as fertilizer and pesticide suppliers and the machinery manufacturers also benefit from the continuous data collection. In such cases, all production data is recorded during task execution, which enables cooperation with the machinery supplier and/or other agriculture companies for task optimization. This in turn can lead to an increase in product output and quality.

While machinery suppliers exchange data, for example, via DataConnect (used by John Deere, Claas, CNH, New Holland and Steyr), the farmers collaborate on platforms, like 356FarmNet [2] and Agrirouter [3].

Using different deep learning approaches and statistical evaluations, the information gathered can be used to discover connections between various data sets. Such approaches can result in recommendations for process improvements in the agriculture farm. They will help to reduce the use of energy resources, especially the fossil fuels and water and to reduce the emission of environmentally harmful greenhouse gases such as carbon dioxide.

Apart from ground vehicles and machinery, other hardware involved in agriculture (i.e. ground sensors, drones, aircrafts, etc.) can also produce data sets. All together they span a dense data collecting sensor network over the agricultural operation area.

Despite all these new and fascinating digitalization technologies, data security and security of ensuring the trustworthiness of the data must never be forgotten in design and system architecture considerations.

The authors of this paper presented a data security approach in the paper "LoRaWAN with HSM as a Security Improvement for Agriculture Applications" [1] at the SAFECOMP - DECSoS workshop 2020. In this paper, a Long Range Wide Area Network (LoRaWAN) end node, built around a soil sensor, was presented, which was connected to the cloud data repository by a LoRaWAN [4] network. While the LoRaWAN data communication standard already offers a high level of data security measures, the sensor prototype used has been expanded with additional security improvements to provide protection against tampering, theft, misuse and malicious use. In last year's paper, the focus was on the technology behind the integration of a so-called Hardware Secure Module (HSM). The sensor with this HSM is part of the Security Evaluation Demonstrator (SED). This year, as a continuation, the evaluation of the implemented security features and the lessons learned during the installation will be presented.

This paper is divided into the following parts: a brief overview of the functions of the HSM, which are described in detail in the document mentioned above [1]. A layer structure of the security improvement features is also presented. The next chapter shows the alarm and notification properties. Security improvement features realized on the SED, will be presented in their own sub-chapter with an explanation of the evaluation steps, lesson(s) learned, and experiences gained during the implementation.

The paper is closed by conclusions, where the costs of extended security and the higher power requirements of future field devices are discussed. In the outlook, the expected upcoming European Radio Equipment Directive (RED) [5] will be presented.

The RED describes what must be provided by product manufacturers and product integrators in the future for wireless network IoT devices.

2 Security Features Provided by an HSM

A Hardware Security Module (HSM) is a module with a security controller and a cryptographic processor that can be added to a system to generate, manage, and securely store cryptographic keys. See [1] for more details. For this demonstrator, a HSM from Zymkey [6] is used. It offers the following security and general features:

- Multi Device ID and Authentication.
- Data Integrity, Encryption & Signing.
- Key Security, Generation & Storage.
- Physical Tamper Detection.
- Real Time Clock (RTC).
- Ultra-Low Power Operation.
- Secure Element Hardware Root of Trust.

In combination with further securing methods and system condition supervision mechanisms, the HSM allows a multi-layered active cyber security protection architecture (CSPA). As shown in Fig. 1, the CPSA of the SED implementation consists of four layers.

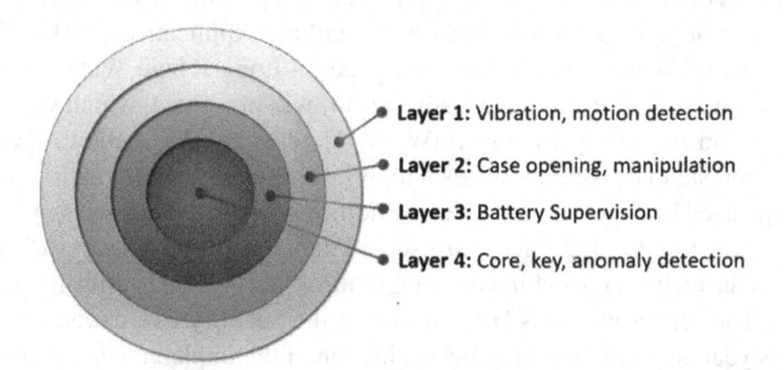

Fig. 1. Four-layer cyber security architecture

Layer 1 defines the outmost protection mechanism. It detects unusual vibrations caused by forcibly removing and tilting the sensor from its location. This event triggers the build-in Global Positioning System (GPS) tracker, which begins to continuously send the current geographical position to the system supervision. Also, the sensor sends an alarm message to the Alarm Processing & Reporter (APR), a functionality in the cloud-based middleware of the system control center. From there the alarm will be visualised in the main Mission Management Tool (MMT), a graphic screen equipped control center. It informs the operator about the event and notifies them that something irregular is happening to the sensor. Additionally, a Short Message Service (SMS) message is generated and warns a selected person in the field about the situation. The person in the field can get more detailed information on his mobile MMT, which could be a tablet or the information screen in a field vehicle. With this information, the first steps can be taken to prevent theft by quickly intervening. Layer 1 is the first barrier, which can help with theft attempts on field devices. If Layer 1 security measure fails, Layer 2 provides a further break-in barrier. It detects the physical intrusion by an unauthorised person who tries to open the device case. The HSM provides alarm wires which are embedded in the housing material. These wires will be broken if the housing is opened improperly. In such cases, the HSM can, if activated, destroy the device firmware to prevent reverse-engineering of the device firmware software and its reuse.

These first two layers are mechanical security protection measures. The next layer, Layer 3, provides the battery supervision function. This is an immensely important function to ensure correct functioning of the device components with a good energy supply. The battery monitor reports the current battery level and initiates charging or replacement of the battery before the battery power falls below a defined power level.

Finally, Layer 4 provides the core security functions, such as software encryption, and key storage, which cannot be read out or manipulated. The HSM allows, when activated, the destroying of the main security key to prevent code manipulation and reverse engineering of the code. This last layer is the software-related security function and represents the last possibility to prevent device abuse. The onion-like security architecture, by means of a coordinated layer structure, represents the so-called Super Security Solution Protection (S^3P).

3 Event Signalisation: Alarms and Notifications

A field device performs in diverse operational modes. These are, for example, the normal, the update and the emergency modes of operation. In the normal mode of operation, the field device both receives commands from a control instance or transmits data to a gateway located nearby, which transfers the data to a data center. In the update mode of operation, the field device enters an operation condition in which, the on-board firmware can be overwritten by a new software version. In the emergency operating mode, the field device enters a self-protecting operating state if an abnormal operating condition or an anomaly in the data communication has been detected. These events trigger the generation of alarm messages that are send to the control center to inform the operator of the condition. Another type of message, which includes the status of the field device, is the notification message. In contrast to the alarm message, this message type can be generated in all three of the operating modes mentioned above. A good example is the low battery message, which informs the operator to charge or replace the battery in the field device.

Alarms are used in the SED to notify the operator of various abnormal operating conditions caused by illegal acts and handling of the field device. The supported security protection measures are described in detail in the next chapter.

When an alarm is triggered, a SMS message is generated to inform a predetermined person about the emergency condition. The SMS is provided by the on-field edge node for a fast and immediate reaction. The alarm message is handled and acknowledged at the operator center. A hand-held device, such as a tablet or mobile phone provides the person on site with more details and the selected countermeasures to fix the alarm situation.

Mission Management Tool (MMT) is the software responsible for receiving and visualizing the alarms and notifications to the operator. MMT also is used to provide visualizations of different sensor readings on the farm. Apart from that, MMT can be used to plan, execute and supervise different missions on the farm involving tractors, drones and other systems. MMT's planning, execution and supervision features have been demonstrated earlier in underwater scenarios [7].

In this work, MMT is run at the command and control center and also on handheld devices. This way, MMT's planning functionalities can be used at the command and control center, while the operators on the field can use it for visualization of required sensor values and receiving alarms and notifications. Sensor alarms and notifications are presented on MMT's panels in two different ways: (1) through the "Alarms" panel which lists all the alarms and notifications received and (2) on the map. In the latter cases, if a sensor reports a notification or raises an alarm, this sensor is marked with red colour on the map and hovering the mouse over the marker shows extra information regarding the received alarm/notification (Fig. 2).

Fig. 2. MMT displaying a tamper attempt alarm on the map and the alarm panel.

4 SED Evaluation of Security Measures

In last year's DECSoS workshop paper [1], the implemented security mechanism of SED architecture was presented and described. In short, SED consists of the battery-powered sensor, which is located in the field. This sensor is part of the SED and includes the HSM. Components like the data gateway, edge computer, interface to the cloud environment and MMT are part of the SED, as well. In this paper, the successful integration of all SED components, especially the HSM and an advanced battery supervision in the sensor are described. Since the SAFECOMP conference last year, the following evolution of the correct functioning of the security mechanisms were done and shall be documented in this chapter. The following security mechanisms are implemented to improve the protection of IoT devices against cyber security attacks:

- Detection of unauthorised moving of the sensor.
- Ensuring the physical integrity of the sensor.
- Inhibit unauthorised reuse of manipulated sensors.
- Prevent manipulation of the sensor communication data.

These mechanisms are implemented and evaluated because they give a broad coverage of security features from theft to manipulation. Further security features can be added at any time. Each of these cyber security functions is explained using an application example and the personal experiences, learned during the implementation work.

4.1 Detection of Unauthorised Moving of the Sensor

Security Application: External manipulation. The sensor is placed in the field, self-powered with a built-in battery, and transfers the measurement sensor data over a wireless LoRaWAN communication link to the LoRaWAN gateway. The gateway collects the data from several LoRaWAN IoT devices and transmits it to a cloud-connected edge computer. A built-in shock sensor detects the theft of the IoT device by a series of strong shock events and sets the IoT node in the emergency mode. In this mode, a GPS receiver is started to determine the current geographical position, which is continuously sent to the gateway. The edge node incorporates a Long Term Evolution (LTE) interface, which sends a SMS alarm to a designated person. Sensor movement is reported in the MMT. The MMT can also run on a mobile device like a tablet or a light version of MMT on a mobile phone for persons working on the field to receives GPS position updates from the sensor through the cloud infrastructure.

Event Signalisation: ALARM (theft), pre-information with an emergency SMS, GPS position.

Evaluation Steps:

- Place the sensor in the field. The sensor automatically determines the reference position and starts normal operation. This is the initial state.
- The sensor is removed from its place.
- The sensor enters the emergency state.
- The sensor generates an alarm message.
- The edge computer sends a SMS message.
- If the sensor leaves the non-alarm area, the theft is confirmed.
- The GPS receiver starts to continuously transmit the current position.
- The SMS is received by the designated person.
- The movement of the sensor is displayed on the MMT.

Implementation Experience: Challenges to setup a GPS module for an application.

To get geographical position data, a GPS module is needed. There are many GPS modules on the market. Standard modules either have a Universal Asynchronous Receiver Transmitter (UART) interface or a provide UART through a Universal Serial Bus (USB) interface. The standard configuration of these modules is to send a data packet each second via a NMEA (National Marine Electronics Association) standard. For using this data, the UART interface can be polled or read out via interrupts. Reading and processing the UART data is time consuming. A better solution in Linux-based systems is to use a so-called GPS demon (GPSD), all data access work is done in the background. Once the GPSD is configured, the GPS position data can be retrieved via a port access and a time out.

However, use of GPS modules can lead to some problems. On the LoRaWAN/GPS module used in our field demonstrator, there is a GPS implementation based on the Mediatek MT3339 [8]. The data can be read from the controller via UART in standard mode. If the GPS antenna is stationary small variations in measured GPS position are expected. This is due to variations in radio wave propagation conditions of the different tracks between GPS satellite and GPS antenna. But data read from the GPS module appears frozen, even if GPS time is changed as expected. The reason for this behaviour is that the module is set up in a so called "tracking mode". This means that GPS position data is only changed if the measured speed of the module is higher than a certain threshold. For car navigation this mode is helpful because the displayed position seems to be constant for low speed. In the case of sensor tracking, where the position varies with low speed, this function is a problem. This "tracking mode" of the module can be switched off with a command from the microcontroller to the GPS module. For the demonstrator, a USB module with a LEA-6S GPS-chip from the company "u-blox" [9] is used. This module has an active antenna which is very sensitive.

The accuracy of GPS position data varies due to many parameters. These are the most important errors for GPS reception:

- Atmospheric effects: changes in ionosphere and troposphere.
- Timing errors of GPS satellites.
- Multipath effects of the received signal.
- Number of GPS satellites received.
- Dilution of precision due to satellite position.

From a user aspect, buildings and trees surrounding the GPS receiver are important. The receive condition is lower if the receiver is placed in a street canyon and worse if trees are shadowing some GPS satellites. Cloudy weather, fog, rain and snow attenuate the signal and influence the accuracy of the measured position. Moving objects near the receiver such as vehicles driving on the street change the multipath situation and therefore the accuracy. To estimate the practical computed accuracy some measurements were done in different situations. In Fig. 3 measurements are shown in an unfavourable situation (high buildings and trees near the receiver, cloudy weather) and in a good receiving situation (free view, cloudless weather).

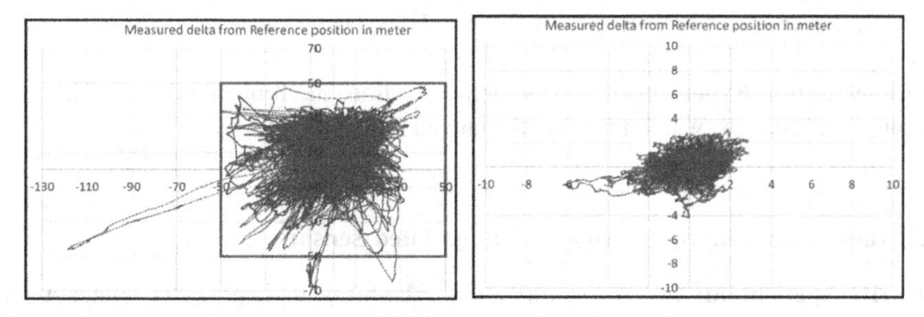

Fig. 3. GPS accuracy (a) unfavourable and (b) good receiving situation

It can be concluded that the position error most of the time is lower than ±50 m from the reference point also in the case of unfavourable receiving situation. If the position drifts out of this area it is only for a short time and can be eliminated by a rule that the measurements are out of the area for a specific time. Only in this case a movement of the reference position is detected.

To determine the reference position, hundreds of measurements are taken, and the average is used.

For higher accuracy differential GPS can be used. The component cost would be higher, and the power consumption would increase. For a theft detection normal GPS is sufficient.

4.2 Ensuring the Physical Integrity of the Sensor

Security Application: External manipulation. The electronics of the SED are protected by a safe housing. Violent and unauthorized opening of the housing interrupts tamper detection wires and activates alarm switches. This is detected by the HSM and an unauthorised opening, an intrusion, is assumed. This event will be notified to the control center by an alarm message. The housing alarm can be activated by a dedicated magnetically activated switch or via software by an activation command. The activation is acknowledged by a blinking light-emitting diode (LED). Deactivation is done with a magnet.

Event Signalisation: ALARM (Tamper detection), pre-information with an emergency SMS.

Evaluation Steps

- The sensor housing is opened without prior authentication.
- The tamper detection wires get broken, the HSM detects this.
- The sensor enters the emergency state.
- The sensor generates an alarm message.
- The edge computer sends a SMS message.
- The alarm message is received and displayed on MMT.

Implementation Experience: For the demonstration, the broken wires are emulated by plugs and sockets, which interrupt the tamper detection wires.

4.3 Inhibit Unauthorised Reuse of Manipulated Sensors

Security Application: Internal manipulation. The firmware is protected against unauthorized reading and updating through encryption. The encryption keys are stored in the HSM and cannot be changed or read out. An unauthorised access attempt is detected by wrong credentials. After a predefined number of tries the access interface is blocked for a given time period and after another unsuccessful attempt, access is completely locked.

Event Signalisation: ALARM (unauthorised access), pre-information with an emergency SMS.

Evaluation Steps:

- The attacker connects a terminal to the access interface.
  - The sensor asks for authentication.
  - For each wrong credential the sensor generates a notification message.
  - After a predefined number of access attempts with wrong credentials, the sensor blocks the access interface.
- The sensor enters the emergency state.
- The sensor generates an alarm message.

- The edge computer sends a SMS message.
- The alarm message is received and displayed on MMT.
- Reading out the firmware from external memory for reverse engineering and manipulation does not make sense because the firmware is encrypted.

Implementation Experience: The HSM must be handled carefully. There are two types of activation of the security mechanism of the HSM. Development mode and permanent mode. If the permanent mode is activated, it cannot be cleared again. The module can no longer be removed without losing the software.

4.4 Prevent Manipulation of the Sensor Communication Data

Security Application: Internal manipulation. The field device protects the stored communication keys in such a way that these keys cannot be read out by an unauthorised person and used in a different device for (manipulated) data communication in the agriculture sensor network.

Event Signalisation: ALARM, pre-information with an emergency SMS.

Evaluation Steps

- Connect the sensor to the network normally.
- Try to read communication keys from ongoing communication with a sniffer → not possible.
- Try to read out the communication keys memorised in the HSM → not possible without successful authentication.
- If the sensor stops working and does not send any data to the cloud, the system center triggers an alarm that the sensor has been lost.

Implementation Experience: In common implementations of LoRaWAN end nodes, the communication keys are stored in a memory which can be read out with much effort. With this prototype sensor, the keys are stored in the HSM and the keys cannot be read out even with great effort.

5 Cost of Security

Improving security of sensors used in outdoor environments comes at its own cost. Part of these costs are on the development side in the form of time required for developing such solutions and the cost of the necessary hardware components. There are ongoing costs such as System Control Centre and maintenance. Another part of these costs lies on the sensor itself in the form of increased power consumption which is needed to run the extra security components.

For a prototype such as the SED prototype presented here, the development costs are not the essential factor. For industrial production the cost of development must be divided by the number of devices expected to be sold.

In the SED two special hardware components are used: the HSM and a GPS receiver with an external active antenna. Additionally, the case of the device must be constructed with a wire loop that is broken if the case is opened. The HSM costs about 40€ for single items, the GPS module about 40€ and the active antenna about 15€. The wire loop in the enclosure has no significant cost. The connection of these components to the sensor is not complex and in general plug and play. All components together can be purchased for less than 100€.

The running costs for these safety functions are also low, since the System Control Center is normally be used without these safety improvement functions and there are very low additional costs. Mobile devices such as cell phones or tablets are typically used for general maintenance of field devices and do not need to be purchased for these security functions.

As many sensors in the field are battery powered, the power consumption of the additional security components is of high importance. Especially LoRaWAN sensors are designed to work for up to several years from a built-in battery, related to the sensor component itself. The HSM used in our demonstrator has a coin cell as a power supply and can work for years. The wire loop in the enclosure also requires only very low current and is not significant for the overall power consumption. However, the power consumption of a GPS receiver with active antenna is much higher. It is in the range of 50 mA in receive mode with active antenna. To decrease this power consumption, the GPS receiver is in general in power down modus or shut off and only activated by a tilt or acceleration sensor. Standard acceleration sensors such as the ones in mobile phones consume very low current. With the use of a passive antenna, the power consumption can be lowered as well. A GPS receiver with a passive antenna has a lower performance but it is suitable for many outdoor environments.

In general, the additional cost for the security features comes mainly from the additional GPS module and its greater power supply requirements.

6 Conclusions

For the demonstration of cyber security protection improvements on field devices an implementation for a soil sensor was selected. Such a sensor represents one of the smallest field devices used in the agriculture domain. Regardless of the size, these field devices need the same cyber security measures as large field devices, such as tractors and field operation machines. In the future, field devices will become more and more powerful, equipped with a powerful 32-bit microcontroller, with large firmware and large data memories as well as with a wireless broadband communication interface such as 4G or 5G. These devices will become part of the multitude of IoT devices in our future digital world. But the increasing distributed computing power of these devices, which are spread across the region, will also awaken the interest of cyber criminals. For example, the Mirai [10] IoT Botnet attack has shown what criminals are able to establish, when the cyber security protection of IoT devices is very weak. In 2016, Mirai disturbed several high-profile services via massive DDoS (Distributed Denial of Service) attacks with a data bandwidth of 1 Tbps (Tera bits per second), by compromising IoT devices such as routers and edge nodes.

Today it is not so simple to integrate both an HSM, a 4G/5G data modem and a GPS receiver of small size into one sensor and at low unit costs. But these functions will become available together in one small system module for the mass market in near future.

The evaluation of the SED showed that the security measures are working efficiently. The cost for components is not very high and the power consumption can be trimmed to a good level.

7 Outlook

Securing networked IoT devices against cyber-attacks and unauthorized use is becoming increasingly important. Modern electronics, the basis of digitalization, must not be operated unprotected. The technology offers a highly interesting operating platform for cyber-attacks if strong security measures are not taken.

In the next few months, the European Commission will clarify Article 3.3 of the European Radio Equipment Directive (RED) [5]. Manufacturers and distributors of IoT devices that exchange data via wireless interfaces must fulfil security requirements that are very difficult to meet. These are the "Essential Requirements" - specifications that are binding but have not yet been implemented. For example, in Article 3.3 the following security requirements are relevant for future developments of wireless devices:

Radio equipment within certain categories or classes shall be so constructed that it complies with the following essential requirements:

e) radio equipment incorporates safeguards to ensure that the personal data and privacy of the user and of the subscriber are protected;

f) radio equipment supports certain features ensuring protection from fraud;

g) …

i) radio equipment supports certain features in order to ensure that software can only be loaded into the radio equipment where the compliance of the combination of the radio equipment and software has been demonstrated.

Excerpt from the EU-Guideline [5]

The distributors are obliged to integrate the necessary cyber security protection measures and to prove efficient function. There are security improvement efforts necessary in any domain of the digitalization, such as residential, automotive, industrial and agriculture technology.

Acknowledgments. This project has received funding from the ECSEL Joint Undertaking (JU) under grant agreement No. 783221 (AFarCloud). The JU receives support from the European Union's Horizon 2020 research and innovation programme and Austria, Belgium, Czech Republic, Finland, Germany, Greece, Italy, Latvia, Norway, Poland, Portugal, Spain, Sweden.

Parts of this work were funded by the Austrian Research Promotion Agency (FFG) and BMK (Austrian Federal Ministry for Climate Action, Environment, Energy, Mobility, Innovation and Technology).

Federal Ministry
Republic of Austria
Climate Action, Environment,
Energy, Mobility,
Innovation and Technology

References

1. Kloibhofer, R., Kristen, E., Davoli, L.: LoRaWAN with HSM as a security improvement for agriculture applications. In: Casimiro, A., Ortmeier, F., Schoitsch, E., Bitsch, F., Ferreira, P. (eds.) SAFECOMP 2020. LNCS, vol. 12235, pp. 176–188. Springer, Cham (2020). https://doi.org/10.1007/978-3-030-55583-2_13
2. Website of 365FarmNet platform. https://www.365farmnet.com/en
3. Website of Agrirouter platform. https://my-agrirouter.com/en/
4. LoRa Alliance. LoRaWAN 1.1 Specification. http://lora-alliance.org/lorawan-for-developers. Accessed 02 May 2020
5. Directive 2014/53/EU of the European Parliament and of the Council of 16 April 2014. https://eur-lex.europa.eu/legal-content/en/TXT/?uri=CELEX:32014L0053
6. https://community.zymbit.com/c/zymkey/22
7. Ameri, E.A., et al.: Planning and supervising autonomous underwater vehicles through the mission management tool. In: Global Oceans 2020: Singapore–US Gulf Coast. IEEE (2020)
8. https://www.mediatek.com/products/locationintelligence/mt3339, https://community.zymbit.com/c/zymkey/22
9. https://www.u-blox.com/en/product/lea-6-series?lang=de
10. Inside the infamous Mirai IoT Botnet: A Retrospective Analysis, 14 December 2017. https://blog.cloudflare.com/inside-mirai-the-infamous-iot-botnet-a-retrospective-analysis/

2nd International Workshop on Dependable Development-Operation Continuum Methods for Dependable Cyber-Physical System (DepDevOps 2021)

2nd International Workshop on Dependable Development-Operation Continuum Methods for Dependable Cyber-Physical Systems (DepDevOps 2021)

Miren Illarramendi[1], Haris Isakovic[2], Aitor Arrieta[1] and Irune Agirre[3]

[1] Software and Systems Enginering, Mondragon Unibertsitatea, Mondragon-Arrasate, Spain
{millarramendi,aarrieta}@mondragon.edu
[2] Computer Engineering, Cyber-Physical Systems, Technische Univeristat Wien, Vienna, Austria
haris@vmars.tuwien.ac.at
[3] Dependable Embedded Systems , Ikerlan Research Centre, Mondragon-Arrasate, Spain
iagirre@ikerlan.es

1 Introduction

In recent years it has become evident that the use of software to perform critical functions is on the rise. As a result, dependable embedded systems are getting more intelligent and automated. For instance, the automotive industry is a clear witness of this trend, where more and more Advanced Driver-Assistance Services (ADAS) are already embedded in cars. This results in a dramatic increase of software complexity, which also requires hardware platforms with higher computing power. All these trends hinder the safety certification, as it is increasingly difficult to guarantee at design time that system errors can be prevented or controlled in such a way that there will be no unreasonable risk associated with the electrical/electronic system component at operation time. These challenges are leading to the need for new development practices that reduce the overall system development time and costs without compromising safety and certification.

The rise of new connection technologies (e.g., 5G) brings new opportunities in terms of the download of frequent software updates of new (improved) releases and the retrieval of operation-time information for fixing bugs and enhancing the design. Advances in new development practices like DevOps have shown effectiveness in software development while reducing overall development costs. The DevOps paradigm aims at having seamless methods for the Design-Operation Continuum of software systems. This paradigm has shown promising results in different domains, including web and mobile engineering. Its practices can bring several advantages to dependable CPSs, including bug fixing based on operational data, inclusion of new functionalities, etc.

However, in the context of dependable CPSs, several challenges arise, requiring DevOps paradigms to have adaptions from several perspectives: the environment in which the CPS operates needs to be considered when updating the software, dependability of software needs to be ensured to a certain level, software faults might lead to severe damages, etc. Furthermore, the safety-critical industry has well established safety-lifecycles dictated by safety standards and adopting the DevOps paradigm has several open research challenges.

The International Workshop on Dependable Development-Operation Continuum Methods for Dependable Cyber-Physical Systems (DepDevOps) is dedicated to exploring new ideas on dependability challenges brought by over-the-air-software updates to the critical domain, with special focus on safety, security, availability, and platform complexity of emerging dependable autonomous systems. This is a fundamental step for the adoption of DevOps approaches in dependable embedded systems. Over-the-air updates can bring several benefits to dependable cyber-physical-systems, like solving security vulnerabilities, adding new functionalities, or bug fixing, and they are a key enabler for improving the design based on operation time data. In addition to this, the workshop aims to identify novel tools and architectures that enable the developers to implement a streamlined and automatic workflow that makes methods and tools to be seamlessly used during design phases as well as in operation.

The second edition of DepDevOps was held as part of the 40th International Conference on Computer Safety, Reliability, and Security (SAFECOMP 2021).

2 H2020 Projects: Dependable DevOps

The DepDevOps project has been organized by researchers from two H2020 projects that are in line with the workshop:

- Adeptness: Design-Operation Continuum Methods for Testing and Deployment under Unforeseen Conditions for Cyber-Physical Systems of Systems (https://adeptness.eu/)
- UP2DATE: New software paradigm for SAfe and SEcure (SASE) Over-the-Air software updates for Mixed-Criticality Cyber-Physical Systems (MCCPS) (https://h2020up2date.eu/)

which means that the topics of the workshop are in line with the research objectives of these projects and as both projects are in their second year, the papers presented during the workshop will be considered as inputs and inspiration for the next stages.

3 Acknowledgments

As chairpersons of the workshop, we want to thank all authors and contributors who submitted their work, Friedemann Bitsch, the SAFECOMP publication chair, and the members of the International Program Committee who enabled a fair evaluation through reviews and considerable improvements in many cases. We want to express our thanks to the SAFECOMP organizers, who provided us the opportunity to organize the workshop at SAFECOMP 2021. In particular, we want to thank the EC and national public funding authorities who made the work in the research projects possible. We hope that all participants will benefit from the workshop, enjoy the conference and accompanying programs, and will join us again in the future!

International Program Committee

Erwin Schoitsch	AIT Austrian Institute of Technology, Austria
Friedemann Bitsch	Thales Deutschland GmbH, Germany
Jon Perez	Ikerlan, Spain
Leonidas Kosmidis	Barcelona Supercomputing Center, Spain
Shaukat Ali	Simula Research Laboratory, Norway
Paolo Arcaini	National Institute of Informatics, Japan
Mikel Azkarate-Askasua	Ikerlan, Spain
Blanca Kremer	Ikerlan, Spain
Francisco J Cazorla	Barcelona Supercomputing Center, Spain
Aitor Agirre	Ikerlan, Spain
Goiuria Sagardui	Mondragon Goi Eskola Politeknikoa, Spain
Wasif Afzal	Mälardalen University, Sweden
Kim Gruettner	OFFIS, Germany
Simos Gerasimou	University of York, UK

Towards Continuous Safety Assessment in Context of DevOps

Marc Zeller(✉)

Siemens AG, Otto-Hahn-Ring 6, 81739 Munich, Germany
marc.zeller@siemens.com

Abstract. Promoted by the internet companies, continuous delivery is more and more appealing to industries which develop systems with safety-critical functions. Since safety-critical systems must meet regulatory requirements and require specific safety assessment processes in addition to the normal development steps, enabling continuous delivery of software in safety-critical systems requires the automation of the safety assessment process in the delivery pipeline. In this paper, we outline a continuous delivery pipeline for realizing continuous safety assessment in software-intensive safety-critical systems based on model-based safety assessment methods.

Keywords: Safety assessment · Agile · DevOps · Continuous delivery

1 Introduction

DevOps mixes the development and operations phases of a software product by promoting high frequency software releases which enable continuous innovation based on feedback from operations [12]. DevOps uses continuous integration and test automation to build a pipeline from development to test and then to production (so-called *continuous delivery*). While companies are already implementing agile practices and continuous delivery in "non-critical" software development, safety-critical software is nowadays still developed using classical waterfall or V-model-based development processes. Safety-critical systems must meet regulatory requirements and shall comply to safety standards (such as IEC 61508). This requires (re-)certification processes for safety compliance after each change of the system. Thereby, introducing continuous delivery of software in this context requires to solve specific issues [27]. However, also in the area of safety-critical systems, the need for accelerating the delivery of software is essential to reduce the time-to-market for new features and to respond faster to changing customer/market demands or technical concerns like deploying security patches. Hence, there is an increasing need to build a *"continuous safety assessment machine"* which enables continuous assessment and delivery also for software in safety-critical systems.

This paper provides an overview of the challenges to enable continuous delivery of software in the context safety-critical systems. These challenges are a summary

I. Habli et al. (Eds.): SAFECOMP 2021 Workshops, LNCS 12853, pp. 145–157, 2021.
https://doi.org/10.1007/978-3-030-83906-2_11

of [27]. In contrast to [27], this paper also outlines how model-based approaches can cope with these challenges. Moreover, it illustrates a pipeline which allows continuous safety assessment based on these approaches as a first vital step towards continuous delivery in software-intensive safety-critical systems.

The remainder of this paper is organized as follows: In Sect. 2 related work is summarized. Afterwards, the challenges that architects and developers face in the context of continuous delivery for safety-relevant systems are outlined. In Sect. 4, model-based approaches to cope with these challenges are sketched. Based on these concepts, a pipeline to enable continuous safety assessment in Sect. 5 is illustrated. Afterwards, additional steps towards DevOps for safety-critical systems (Sect. 6) are discussed. The paper is concluded in Sect. 7

2 Related Work

R-Scrum [2] and SafeScrum [5] are existing approaches to develop safety-critical systems using agile methods, but do not show how to build a continuous delivery pipeline. Also [22] only presents challenges how to enable agile development of safety-critical systems in large organizations. All of these papers name traceability and continues safety validation or compliance, which is an integrated part of all sprints, as a necessary step to realize agile development in regulated environments. First ideas to realize a continuous delivery pipeline for safety-critical software are outlined in [23,24]. This delivery pipeline includes iterative safety analysis approaches as well as automated safety test generation execution integrated in the agile software engineering process. However, the ideas w.r.t. methods and tools to realize the delivery pipeline are very abstract and immature. The authors in [24] propose a combination of component-based design, the use of contracts, modular assurance cases, and agile practices to realize continuous delivery in the development of safety-critical systems. In contrast to existing work, this paper addresses all phases of the safety engineering life-cycle and sketch a concrete delivery pipeline for continuous safety assessment based on existing model-based safety analysis approaches and model-connected assurance cases.

3 Challenges in Enabling Continuous Delivery for Safety-Critical Systems

DevOps is an approach that shortens the gap between software development and software operation in the production environment. Continuous delivery of software is based on flexible and scalable product definitions that focus on feature-wise clustered components which are loosely coupled and can be upgraded independently (e.g. using microservice architectures and patterns). It uses continuous integration and test automation to build pipeline from development to test and then to production (so-called *continuous delivery pipeline*).

While continuous monitoring and measurement (often called supervision in the safety domain) is a standard design feature of safety-critical systems, continuous delivery is challenging in safety-related applications with strict regulatory requirements and safety guidelines. Since safety is a system-level property, the continuous delivery process must be lifted from software to system level and people from different engineering disciplines must be included in the continuous delivery pipeline. Moreover, additional development steps are required in safety-critical systems, such as the HARA, the safety analysis of the architecture, the safety argumentation, and the certification (see Fig. 1). These steps are required by any safety standards (e.g. IEC 61508) and must be performed within the delivery pipeline.

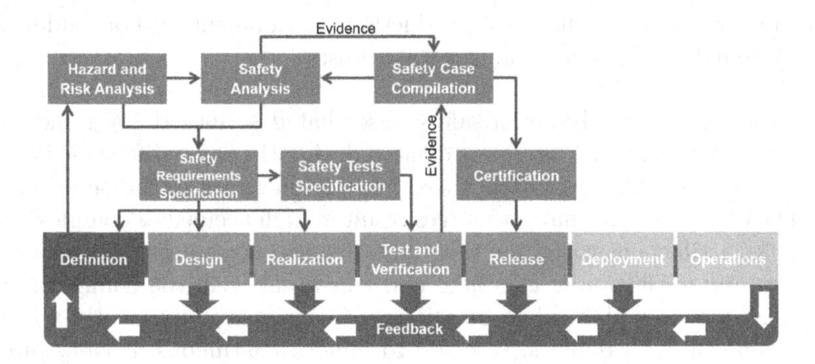

Fig. 1. Continuous delivery pipeline with additions for safety-critical systems development

Thereby, the following challenges related to the safety engineering activities need to be addressed in order to realize a delivery pipeline:

Hazard Analysis and Risk Assessment: Today, the HARA is a manual process which requires the assessment of potential system hazards typically documented in spreadsheets. To enable continuous delivery for safety-critical systems, we need to speed-up the HARA process. This means, that we need automation for both the identification of hazards and the assessment of the risk associated to hazards. Moreover, it must be possible to identify new hazards and to adapt the risk associated with known hazards of the system.

Safety Analysis: The safety analysis techniques used today in industrial practice, such as *Fault Tree Analysis (FTA)* and *Failure Mode and Effect Analysis (FMEA)*, are performed manually by experts on system level. To enable continuous delivery, we need to speed-up the safety analysis process by increasing the level of automation. The changes of the software and their influence on the system safety must be reflected in the analysis automatically. Failure modes specific to highly integrated software-intensive systems such as failures due to feature interactions, emerging features, or not-wanted interactions must be integrated in the safety analysis. Moreover, the safety analyses must be linked to the system (or software/hardware) design so that changes to the system (or software/hardware) architecture can be synchronized with the safety analysis models.

Safety Tests: Since the predictable behavior of a system in the presence of faults is crucial to its safe operation, testing of safety-critical systems addresses questions such as: Are fault reactions correctly and effectively implemented? Is the timing of these reactions sufficient? Consequently, safety-critical systems require the testing of safety mechanisms that consider the fault models of the system components. Typically, tasks related to the testing of safety mechanisms are manually done by experts. In the context of a continuous delivery pipeline, the synthesis of tests for the specified failure mitigation mechanisms must be automated. Moreover, we need the possibility to inject faults into the productive system under test in virtual/real production environment without side-effects and to automatically execute the generated tests.

Safety Argumentation: Today, a safety case that describes the argumentation and references all the work products created during the safety life-cycle is manually captured in documents. Often, these work products are spread across various tools. The disparate information sources result in high accidental complexity and keeping the artifacts consistent is time-intensive and error-prone. Checking that the safety argumentation is complete and consistent with the configuration of the system, which is planned to be released, is expensive and mostly a manual process done through reviews. In order to build a continuous delivery pipeline for safety-critical systems, the creation and maintenance of the safety argumentation must be automated to reduce the time and costs involved in evolving the safety arguments. Therefore, detailed traceability between safety arguments and evidences, created during the development, must be provided.

Orchestration of Different Disciplines: Since safety is a system-level property, the assessment of safety-critical software in terms of safety must be conducted on system level. Therefore, not only software engineers must be involved in the continuous delivery process, but also experts from other engineering disciplines as well as safety experts. Moreover, external assessors must be incorporated into the agile development process. In this way, at the end of each sprint/iteration, not only correct software but also the necessary assets and documentation for the independent safety assessment and certification can be delivered.

4 Methods to Enable Continuous Delivery for Safety-Critical Systems

Automation of the safety assessment process of a safety-critical software-intensive system is a viable first and mandatory step to realize DevOps in a safety-relevant context. To automate the safety assessment process, we need to automate the activities of the safety engineering life-cycle.

Model-based safety engineering and assurance approaches can help to cope with the challenges described in Sect. 3. Similar to model-based engineering, model-based safety assurance [8] provides the foundation to automate tasks related to the safety engineering life-cycle. Artifacts of the safety engineering life-cycle are represented by models. With today's document-driven safety engineering practices, establishing automation is hardly possible, because information is documented in natural language text and images. Documents do not provide fine-granular traceability of information created during the system and the safety engineering life-cycle. In contrast, models provide the possibility to capture information in a machine-readable way and enable the interlinking of different kinds of elements to establish traceability. By capturing the results of different activities performed during the safety engineering life-cycle, in the form of models, their interdependencies can be captured, and consistency and traceability can be established. Moreover, the safety engineering models can be integrated with system design artifacts provided by model-based system engineering. As a consequence, safety cases are represented as model artifacts that reference other fine-granular model elements created during the system development and safety engineering life-cycle [17]. Based on the model-based approach, it is possible to provide methods to automate activities within the safety engineering life-cycle, e.g. when updates of the safety-relevant system are performed.

Potential approaches to solve the challenges described in Sect. 3 by automating the respective tasks using model-based techniques are outlined in the following:

Dynamic Hazard and Risk Analysis: Detailed knowledge about the safety-relevant system, its environment, and the interfaces between the system and its environment is required to address the aforementioned challenges. For instance, in a robot-based industrial automation system, the robot's trajectories must be known, the complete environment of the robot within a plant, and the tasks that the robot can perform in order to judge which hazards may occur in such a system and how to assign the associated risks to each of the identified hazards.

A possible solution is to simulate the system including its different subsystems (software, electronics, mechanics, etc.) to identify the effects of component failures on the behavior of the system. However, in order to determine if a failure results in a hazard, the context (or the environment) in which the system operates (e.g. the driver of a car, roads, traffic, etc.) must also be simulated. Hence, the behavior of a system at its interfaces can be observed and the consequences of a system's action on the environment can be identified and assessed in terms of occurrence and severity. Therefore, a sophisticated simulation of the socio-technical system which consists of co-simulation of various kinds of simulations (1D, 3D, physics, human behavior, etc.) is necessary. By simulating the technical system and its environment and by including the effects of failures on humans, hazards can be identified, and the risks of the hazards can be assessed. An example from the manufacturing domain to dynamically perform risk assessment based on a 3D simulation of the system is presented in [15].

Automated Safety Analysis: With *Component Fault Trees (CFTs)*, there is a model- and component-based methodology for fault tree analysis [6,9,10]. In CFTs, every system component is represented by a CFT element. Each element has its own in-ports and out-ports that are used to express propagation of failure modes through the tree. Similar to classic fault trees, the internal failure behavior that influences the output failure modes is modeled by Boolean gates.

A library, which contains CFT elements for all system components, supports reusability by allowing stakeholders to create different CFTs by changing the assembly of the CFT elements according to the system architecture. Based on the methods described in [13,14], it is possible automate the composition of CFTs. Hence, by automatically generating mappings between the input and output failure modes, system-wide safety analysis models can be automatically created. Moreover, the safety analysis can be automatically adapted when making modifications to a system's architecture. Also analysis models for software can be automatically generated [9,11,26].

Automate Side-Effect Free Fault Injection Tests: To ensure that the safety mitigation mechanisms are correctly implemented and executed on the target hardware, by performing side-effect free fault injection tests in a system.

Therefore, it is possible to combine the model-based safety analysis and the CFTs with the concept of *data probes* [4]. CFTs represent the failure behavior of a system and is the basis for the generation of system-level test cases for failure mitigation mechanisms. Data probes are special purpose library components that monitor, trace, check, and optionally, manipulate one or more variables in one or more functional components of one or more process components. Data probes can be programmed. A probe program selects probe points and tells a data probe what to do with the selected probe points in accordance with an analysis goal. Thus, data probes enable non-intrusive monitoring and side-effect free manipulation of data during the operation of software-intensive systems.

Based on the failure behavior descriptions, using CFTs for model-based safety analyses allows to automatically generate test data at system-level [25]. The resulting test cases show that the actual implementation of a system is compliant to its defined failure behavior and can be used to test the specified failure mitigation mechanisms. Furthermore, with test case generation based on CFTs (with the data probes concept), it is possible to perform non-intrusive, side-effect free fault-injection tests and make reliable statements about the behavior of safety-relevant systems in the presence of software faults and component failures [4]. These tests are derived automatically from the system's failure behavior for the detection and handling of value and time errors. Thereby, the derived test cases provide the input, which is used to manipulate data using the probes at the input of a component and to configure monitors via the probes to observe the output of a component that is being tested.

Model-Based Safety Argumentation: A fundamental problem in today's safety engineering processes is that safety argument models are not formally integrated with the evidence models supporting the claims of the argumentation. This lack of integration hampers the effective automation of the safety assessment [17]. Concrete examples of evidence models include hazard and safety analysis models as well as the dependability process execution documentation. These artifacts refer to the same system and they are interrelated with each other. In order to solve these challenges, the artifacts created during the safety engineering life-cycle as well as their relationships should be a part of the system's model-based reflection. In this context, the concept of the *Digital Dependability Identity (DDI)* [21] can be used to capture the various artifacts of the safety life-cycle in a model-based way and to establish a relationship between the argumentation and the supporting evidence models. By establishing traceability across the artifacts, DDIs represent an integrated set of safety data models ("What is the evidence data?") that are generated by engineers and are reasoned upon in safety arguments ("How is the evidence data supporting the claim?"). A DDI provides traceability between a safety argumentation captured in form of the *Structured Assurance Case Metamodel (SACM)* [16] and safety-related evidence models, namely, hazard and risk analysis, functional architecture, safety analysis, and safety design concept. SACM provides the assurance case backbone for creating the required traceability. The DDI meta-model formalizing the traceability and evidence semantics is the so-called *Open Dependability Exchange (ODE)* meta-model[1].

5 Continuous Safety Assessment

In this section, the model-based approaches presented in Sect. 4 are integrated to enable continuous safety assessment in a delivery pipeline as depicted in Fig. 2.

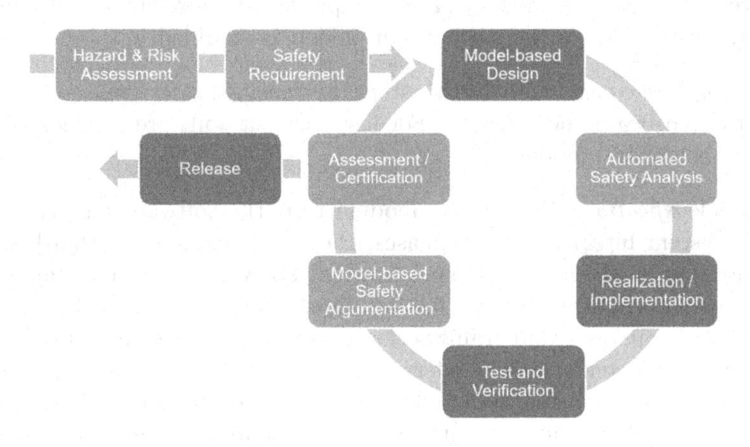

Fig. 2. Continuous safety assessment pipeline for safety-critical systems

[1] https://github.com/Digital-Dependability-Identities and https://github.com/DEIS-Project-EU.

Prerequisites: We assume that the Hazard & Risk Assessment for the system has been conducted (manually). Thus, potential hazards have been identified and the associated risk was assessed. Moreover, safety goals have been derived from these hazards and the safety goals have been refined into a set of safety requirements. The safety requirements are the input for the continuous safety assessment pipeline.

Model-Based Design: The system architecture is defined using a *Model-Based System Engineering (MBSE)* approach (such as SysML). This includes a structural description of the system architecture including the hardware and the software architecture as well as their interactions. The safety requirements are allocated to these system elements. Moreover, the model-based design comprises of a behavioral description of the software elements, for instance in the form of UML Statecharts, Continuous Functions Chars (CFCs) or Simulink models.

Automated Safety Analysis: Based on the model-based system description, safety analysis models in the form of CFT elements can be derived automatically for the software components when these components are defined as UML Statecharts [11], CFCs [26] or Simulink models [9]. For the hardware components, the CFT elements can be composed automatically, if a library with CFT element for all necessary parts (e.g. resistors, transistors, contactors) is predefined. Hence, a circuit diagram of a safety-critical system can be transformed automatically in a CFT model by assembling predefined CFT elements according to a given model of the circuit diagram. The system-level CFT can be built according to [7] based on the CFT models derived for the software, the CFT model of the different hardware parts, and the allocation of software to hardware components. Hence, the safety analysis for a system design can be conducted in an automated way. Moreover, the safety analysis can be adapted and re-executed automatically when the system design (i.e. the system model) is modified.

Realization/Implementation: In this step, the specified safety-relevant system is realized in hardware and software. For instance, the software is generated from the given behavioral models.

Test and Verification: Within the module test, the software integration test, and the system integration test phases, the developed system (hardware and software) is verified against the specification. For verification techniques (such as source code verification, unit testing, integration testing, etc.) various test automation techniques and frameworks are existing and can be used according to the recommendations in safety standards. In addition, testing of system safety mechanisms must automated. Since test data can be derived from the safety analysis models in form of CFTs, the required test data to ensure the correctness of the implemented safety mechanisms at system-level can be generated automatically [25]. The tests can be executed as Software-in-the-Loop (SiL) tests using the generated code running in a simulation environment [20] or as Hardware-in-the-loop (HiL) tests where the test cases can be executed non-intrusively and side-effect free using the data probe concept [4].

Model-Based Safety Argumentation: Finally, the safety-relevant system is validated, and a so-called *Safety Case* is compiled to argue that the system is safe. Based on the DDI concept, it is possible to automate the creation of the safety case. Since safety requirements are provided as an input to the continuous safety assessment pipeline, the basic structure of the safety argumentation (i.e. the claims derived from the safety requirements of the systems and regulatory rules) can be defined manually and only needs to be adopted if requirements are changing. Moreover, the evidences to prove that the claims are fulfilled are crated automatically in the pipeline (in the safety analysis and test & verification steps). According to [18] the relationships between the claims in the safety argumentation and evidences provided by the artifacts represented in an ODE model can created automatically. The resulting safety case can then be checked w.r.t. consistency and completeness [1,3]. The DDI concept also eases the effort in case of modifications, since the impacts of changes in the safety argumentation or the evidences can be identified automatically due to the traceability established by the DDI [19].

Assessment/Certification: The entire artifacts, processes, and tools of the system development life-cycle are subject to independent reviews and independent assessments. The documents necessary for the assessment or certification process (at least the technical part) can be generated from the model-based safety case created by the continuous safety assessment pipeline. Hence, assessors can be easily integrated in the development process and can receive the latest information in a well-established form at the end of the sprint or at dedicated points in time during the development without additional overhead for the developers to create the necessary documentation.

6 Towards DevOps for Safety-Critical Systems

Although continuous safety assessment is the first steps towards DevOps in the context of safety-relevant systems, there are additional challenges which must be solved:

Culture Change in Assessment and Certification Organizations: To fully leverage the benefits of DevOps in the context of safety-relevant systems, the certification process must be integrated into the continuous delivery process. Therefore, the culture of a so-called *"big-bang"* all-at-once certification before the release [22] needs to be changed and assessors should review the status of the project in terms of functional safety frequently. Hence, in context of DevOps, there is a need to integrate assessor as stakeholders into the continuous delivery process. In [2,5], it is recommended that the assessor is incorporated into the agile development process, such that they are able to review the status of the project in terms of functional safety frequently. Therefore, we propose to explicitly add the role of the "Independent Safety Assessor" in a framework such as the Scaled Agility

Framework (SAFe)[2]. Hence, providing the independent assessor the possibility to continuously review documents or models and enabling transparency of the sprint and release planning. Thereby, it is possible to (1) make the assessor aware of ongoing activities and (2) ensure that the assessor can plan appropriate time slots for the review of the safety case and the approval or formal certification of the release.

Automated Change Impact Analysis (CIA): Since safety standards require a thorough Change Impact Analysis (CIA) for every change, automating CIA is essential to realize continuous delivery in safety-relevant systems. To automatically perform a CIA, all the steps of the safety engineering life-cycle (e.g. HARA, safety analyses, etc.) which need to be updated must be identified. Afterwards, the required updates must be applied to the system.

To automate CIA, we must enable strict traceability between the various artifacts of the system, software, and safety engineering life-cycle to determine the elements with the artifacts that are affected by change (e.g. in the source code or in a requirements document). This requires a holistic meta-model to capture all the artifacts and their relationships created during a system's development life-cycle. Moreover, tool support must be provided to architects and developers for storing the information without any overhead [17]. The concept of the DDI provides a first version of a holistic meta-model for providing strict traceability and enabling an automated CIA [19].

Qualification of the Tool Chain: In order to be able to develop a safety-relevant system and to deliver a sound safety case in conjunction with each new release of the product, also the environment used to create the project requires specific attention. Hence, tool qualification is mandatory if the results of a tool are not reviewed by humans (e.g. a compiler). However, the qualification of tools is domain, company, business unit and even project specific. Domain specific check lists or tool qualification kits already exist, but a generalization of these approaches is currently not possible. Therefore, we propose to add a dedicated role of the "Certification Engineer/Manager" in a framework such as the Scaled Agility Framework (SAFe). The Certification Engineer must be responsible for establishing and maintaining the regulatory compliance of the development environment (including the required documentation) within the agile development project.

7 Conclusions and Future Work

Safety-critical systems must meet regulatory requirements and shall comply to safety standards. Moreover, the continuous delivery process of a new software version must be lifted from software to system level and people from different engineering disciplines must be involved. Thus, applying agile development

[2] https://www.scaledagileframework.com/.

and the concepts of continuous delivery in context of safety-critical, software-intensive systems requires to solve specific challenges. In this paper, the challenges which need to be solved in order to realize continuous delivery of software in safety-critical systems are summarized. Moreover, potential approaches how to overcome the mentioned challenges based on existing model-based safety analysis approaches (such as Component Fault Trees) and model-based assurance cases in form of Digital Dependability Identities (DDIs) are sketched. Thus, it is possible to create a pipeline for continuous safety assessment for safety-critical software-intensive systems.

Furthermore, additional steps towards DevOps for safety-critical systems beyond the automation of the safety assessment process are discussed. This includes the mandatory qualification of the tools since delivery pipeline itself becomes a subject to regulatory compliance, the integration of assessors as stakeholders into the continuous delivery process, and the automation of change impact analysis for changes of the software after the system is released (feedback for Ops). These topics will be addressed in future research as well as the integration of the continuous safety assessment pipeline in the Scaled Agility Framework (SAFe). Moreover, we will investigate the extension of Machine Learning Operations (MLOps) w.r.t. safety-critical environments.

References

1. Cârlan, C., PetriŞor, D., Gallina, B., Schoenhaar, H.: Checkable safety cases: enabling automated consistency checks between safety work products. In: 2020 IEEE International Symposium on Software Reliability Engineering Workshops (ISSREW), pp. 295–302 (2020). https://doi.org/10.1109/ISSREW51248.2020.00088
2. Fitzgerald, B., Stol, K., O'Sullivan, R., O'Brien, D.: Scaling agile methods to regulated environments: an industry case study. In: 35th International Conference on Software Engineering (2013)
3. Foster, S., Nemouchi, Y., Gleirscher, M., Wei, R., Kelly, T.: Integration of formal proof into unified assurance cases with Isabelle/SACM (2020)
4. Fröhlich, J., Frtunikj, J., Rothbauer, S., Stückjürgen, C.: Testing safety properties of cyber-physical systems with non-intrusive fault injection – an industrial case study. In: Skavhaug, A., Guiochet, J., Schoitsch, E., Bitsch, F. (eds.) SAFECOMP 2016. LNCS, vol. 9923, pp. 105–117. Springer, Cham (2016). https://doi.org/10.1007/978-3-319-45480-1_9
5. Hanssen, G.K., Stålhane, T., Myklebust, T.: SafeScrum®-Agile Development of Safety-Critical Software. Springer, Heidelberg (2018). https://doi.org/10.1007/978-3-319-99334-8
6. Höfig, K., et al.: Model-based reliability and safety: reducing the complexity of safety analyses using component fault trees. In: 2018 Annual Reliability and Maintainability Symposium (RAMS) (2018)
7. Höfig, K., Zeller, M., Heilmann, R.: Alfred: a methodology to enable component fault trees for layered architectures. In: 41st Euromicro Conference on Software Engineering and Advanced Applications (SEAA), pp. 167–176 (2015)
8. Joshi, A., Miller, S.P., Whalen, M., Heimdahl, M.P.E.: A proposal for model-based safety analysis. In: 24th AIAA/IEEE Digital Avionics Systems Conference (2005)

9. Kaiser, B., et al.: Advances in component fault trees. In: Proceedings of 28th European Safety and Reliability Conference (ESREL), pp. 815–823 (2018)
10. Kaiser, B., Liggesmeyer, P., Mäckel, O.: A new component concept for fault trees. In: Proceedings of the 8th Australian Workshop on Safety Critical Systems and Software, pp. 37–46 (2003)
11. Kaukewitsch, C., Papist, H., Zeller, M., Rothfelder, M.: Automatic generation of RAMS analyses from model-based functional descriptions using UML state machines. In: 2020 Annual Reliability and Maintainability Symposium (RAMS) (2020)
12. Loukides, M.: What is DevOps? O'Reilly Media Inc., Sebastopol (2012)
13. Möhrle, F., Bizik, K., Zeller, M., Höfig, K., Rothfelder, M., Liggesmeyer, P.: A formal approach for automating compositional safety analysis using flow type annotations in component fault trees. In: Proceedings of the 27th European Safety and Reliability Conference (ESREL) (2017)
14. Möhrle, F., Zeller, M., Höfig, K., Rothfelder, M., Liggesmeyer, P.: Automating compositional safety analysis using a failure type taxonomy for component fault trees. In: Proceedings of ESREL 2016, pp. 1380–1387 (2016)
15. Moncada, D.S.V.: Dynamic safety certification for collaborative embedded systems at runtime. In: Model-Based Engineering of Collaborative Embedded Systems, pp. 171–196. Springer, Cham (2021). https://doi.org/10.1007/978-3-030-62136-0_8
16. Object Managemnet Group (OMG): Structured Assurance Case Metamodel 2.0 (SACM) (2018)
17. Ratiu, D., Zeller, M., Killian, L.: Safety.Lab: model-based domain specific tooling for safety argumentation. In: Koornneef, F., van Gulijk, C. (eds.) SAFECOMP 2015. LNCS, vol. 9338, pp. 72–82. Springer, Cham (2015). https://doi.org/10.1007/978-3-319-24249-1_7
18. Reich, J., Zeller, M., Schneider, D.: Automated evidence analysis of safety arguments using digital dependability identities. In: Romanovsky, A., Troubitsyna, E., Bitsch, F. (eds.) SAFECOMP 2019. LNCS, vol. 11698, pp. 254–268. Springer, Cham (2019). https://doi.org/10.1007/978-3-030-26601-1_18
19. Reich, J., Frey, J., Cioroaica, E., Zeller, M., Rothfelder, M.: Argument-driven safety engineering of a generic infusion pump with digital dependability identities. In: Zeller, M., Höfig, K. (eds.) IMBSA 2020. LNCS, vol. 12297, pp. 19–33. Springer, Cham (2020). https://doi.org/10.1007/978-3-030-58920-2_2
20. Reiter, S., Zeller, M., Höfig, K., Viehl, A., Bringmann, O., Rosenstiel, W.: Verification of component fault trees using error effect simulations. In: Bozzano, M., Papadopoulos, Y. (eds.) IMBSA 2017. LNCS, vol. 10437, pp. 212–226. Springer, Cham (2017). https://doi.org/10.1007/978-3-319-64119-5_14
21. Schneider, D., Trapp, M., Papadopoulos, Y., Armengaud, E., Zeller, M., Höfig, K.: Wap: digital dependability identities. In: IEEE International Symposium on Software Reliability Engineering (ISSRE), pp. 324–329 (2015)
22. Steghöfer, J.-P., Knauss, E., Horkoff, J., Wohlrab, R.: Challenges of scaled agile for safety-critical systems. In: Franch, X., Männistö, T., Martínez-Fernández, S. (eds.) PROFES 2019. LNCS, vol. 11915, pp. 350–366. Springer, Cham (2019). https://doi.org/10.1007/978-3-030-35333-9_26
23. Vost, S., Wagner, S.: Keeping continuous deliveries safe. In: IEEE/ACM 39th International Conference on Software Engineering Companion (2017)
24. Warg, F., Blom, H., Borg, J., Johansson, R.: Continuous deployment for dependable systems with continuous assurance cases. In: IEEE International Symposium on Software Reliability Engineering Workshops (2019)

25. Zeller, M., Höfig, K.: Confetti - component fault tree-based testing. In: Proceedings of the 25th European Safety and Reliability Conference (ESREL), pp. 4011–4017 (2015)
26. Zeller, M., Höfig, K., Schwinn, J.P.: Arches - automatic generation of component fault trees from continuous function charts. In: Proceedings of the 2017 IEEE 15th International Conference on Industrial Informatics (INDIN), pp. 577–582 (2017)
27. Zeller, M., Ratiu, D., Rothfelder, M., Buschmann, F.: An Industrial Roadmap for Continuous Delivery of Software for Safety-critical Systems. In: 39th International Conference on Computer Safety, Reliability and Security, Position Paper (2020)

The Digital Twin as a Common Knowledge Base in DevOps to Support Continuous System Evolution

Joost Mertens[1,2](✉) and Joachim Denil[1,2]

[1] Faculty of Applied Engineering, University of Antwerp, Antwerpen, Belgium
{joost.mertens,joachim.denil}@uantwerpen.be
[2] Flanders Make @ University of Antwerp, Antwerpen, Belgium

Abstract. There is an industry wide push for faster and more feature rich systems, also in the development of Cyber-Physical Systems (CPS). Therefore, the need for applying agile development practices in the model-based design of CPS is becoming more widespread. This is no easy feat, as CPS are inherently complex, and their model-based development is less suited for agile development. Model-based development does suit the concept of digital twin, that is, design models representing a system instance in operation. In this paper we present an approach where the digital twins of system instances serve as a common-knowledge base for the entire agile development cycle of the system when performing system updates. Doing so enables interesting possibilities, such as the identification and detection of system variants, which is beneficial for the verification and validation of the system update. It also brings along challenges, as the executable physics based digital twin is generally computationally expensive. In this paper we introduce this approach by means of a small example of a swiveling pick and place robotic arm. We also elaborate on related work, and open future challenges.

Keywords: DevOps · Digital twin · Cyber-physical system

1 Introduction

In multiple industries, there is a drive to develop faster and more feature-rich systems. That is also the case for Cyber-Physical Systems (CPS) that integrate computation, communication and physical processes in a single system. Examples of these systems can be found in trains, vehicles, industrial machines, etc. The software of CPS is usually very complex and tries to control the system in uncertain environments. However, at design time it is hard to predict all the possible circumstances where the system needs to operate in. Furthermore, user requirements of the system may change during the operation of the system.

Joost Mertens is funded by the Research Foundation - Flanders (FWO) under strategic basic research grant 1SD3421N.

I. Habli et al. (Eds.): SAFECOMP 2021 Workshops, LNCS 12853, pp. 158–170, 2021.
https://doi.org/10.1007/978-3-030-83906-2_12

This results in the demand for more agile approaches to CPS development and operations [3,5] that bridge these phases of the system lifecycle.

DevOps (Development and Operations) is a set of practices that looks to link design and operations of a system with each other in a continuum. Through continuous integration, delivery and deployment, the common idea is to achieve faster lead times for system updates [12]. Within the context of CPS, which are mostly designed in a Model-Based manner, the application of DevOps is feasible but challenging [8].

Given the model-based engineering approach of CPS, performing the most accurate design iteration necessitates a synchronization of the existing design models with real-life. In contrast with software engineering, physical systems age and undergo revisions independently from changes to the design models. To achieve the synchronization, operational data must be incorporated back into the design models. Because CPS are usually engineered using a set of models (architecture, behavior, etc.), the idea of continuously synchronizing a model with the real-world is not far-fetched, but feasible. In fact, this idea closely resembles a digital twin, which is often known as an executable physics-based model that represents a physical system. It can be used in several scenarios, from monitoring in which case it is more accurately called a digital shadow [13], to product life cycle management [10] and prediction, where it becomes clear that a link can be made to the initial question of how to incorporate operational data back to design models.

In [16], the uses of a digital twin throughout a DevOps cycle are made explicit. The digital twin serves as the enabler for carrying operational data back over to the development part of the DevOps cycle, but is also employed in other phases, e.g. in the test phase for testing against the most correct representation of the current system, or in the build phase, to virtually commission the system. We share this vision, and like to consider a digital twin as the common knowledge base that can be used throughout the entire DevOps cycle.

One additional point of interest is that of system variants. At design time, designers can account for a multitude of variants in a deployed system, yet at run time system variants develop naturally, for example hardware revisions of parts or degradation of electrochemical components such as batteries. In principle, such variations are detectable events for a digital twin. To clarify, at design time, the interpretation of variants is the same as in software product lines: a system configuration with specific features. However, at run time, we additionally deem a system that is no longer accurately represented by it's design time model a variant, in which case the model requires updating.

In [15], we shared a vision and the challenges related to using a digital twin as a means to bridge the design and operational phases of a system lifecycle. In this paper, we start by introducing a small example of a robotic arm making a pick and place swivel movement. Next, we elaborate on our approach of combining a digital twin with agile methods to update systems. We apply a subset of the approach on a system update of the small example as a demonstration. We then elaborate on some challenges, and discuss our proposed approach to tackle them. Afterwards we discuss related work, and lastly a conclusion summarizes the paper.

2 Example System

In this example, an electric motor performing a repetitive movement is studied, more specifically a swiveling robot arm performing a pick and place move. The arm is deployed in various locations around the world, were temperatures and humidity levels differ. In the example, we update the controller to perform a similar, but more rapid movement. Figure 1 illustratively shows this system.

Note that the physical system does not actually exist. Instead, we work on a simulated environment, with a different parameter set for a warm, cold and "lab" environment. This yields 3 models, one development model, and two "real-life" models. Those "real-life" models are conveniently reused as digital twins in this case. In a real setup this would not be the case, and the "real-life" models would be the physical system, instead of a model.

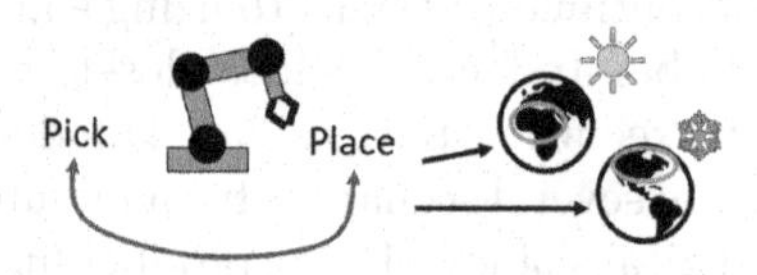

Fig. 1. Iconized example of the example system.

The system is modeled in Simulink®, shown in Fig. 2. The pick and place controller consists of a chain of PI(D) controllers that control the position of the arm and, the angular velocity and current of the motor. The controller generates the control signal to drive the inverter connected to the motor. The rotational movement is modeled using the SimScape library as a brushless DC motor that is driven by an inverter. The effect of the environment being cold or warm is incorporated as a viscous friction component in the rotational part of the BLDC model. In a cold environment, this friction component is higher than the in a warm environment, for example due to tighter tolerances or viscosity of lubricants. In the model this is characterized by the Coulomb friction. The only requirements for the system are for the pick and place movement to happen at a speed 2 Hz, and with a steady state error of less than 0.75°. This is briefly summarized in Table 1.

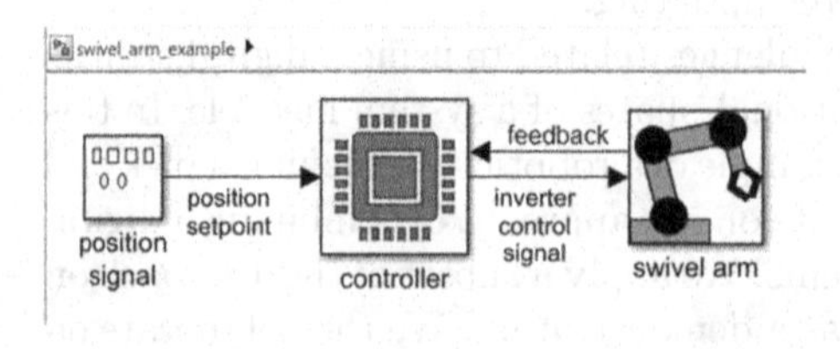

Fig. 2. Simulink®model of the system.

Table 1. Model information.

Description	Value
Pick & place move	−90° to 90°
Frequency	2 Hz
Error	<0.75°

During the development the friction component remains a question for the modeller, but we can assume that lab conditions suffice to tune the controller, that is, regular room temperature. After tuning and "deploying" the controller on the real system, that is, testing it on the models where the correct environmental conditions are applied, we find that the friction component matters little, and the swivel motion happens as expected. The movement of the swivel arm can be seen in Fig. 3. Model calibration reveals that to be entirely correct, the Coulomb friction must be 0.5 Nm in the cold climate and 0.2 Nm in the warm climate.

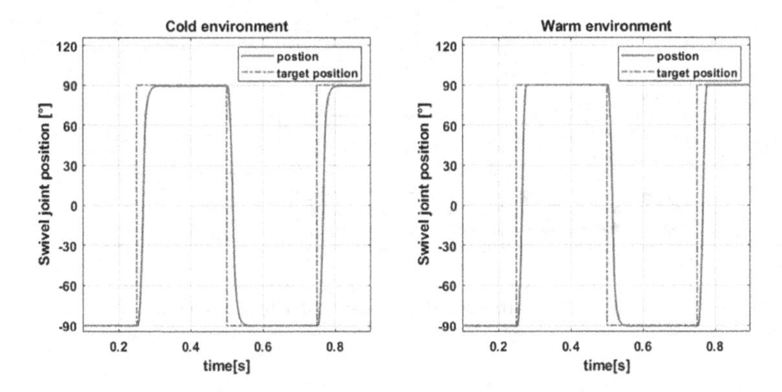

Fig. 3. Traces for the initial controller in both environments.

As can be seen in Fig. 3, the swivel arm reaches the setpoint reasonably quickly, and, inspecting the difference at steady yields worst case absolute errors of 0.46° and 0.53° for the cold and warm environments respectively. In other words, the controller tuned under lab conditions suffices for both the actual in-the-field conditions.

3 Approach

Our approach combines digital twins with agile methods to update systems, with a specific interest in the verification and validation of the updates. The foundation of the approach is multi-paradigm modeling [19]. Multi-paradigm modeling advocates the explicit modeling of all parts of a system, at the most appropriate level of abstraction, and in the most appropriate formalism. In other words, in our approach we thus look to model the various parts of the system: real-world representative system models (digital twins), deployment models, architectural models of the whole, networking models for communication. This allows to perform different processes on a deployed system, as the architectural overview in Fig. 4 depicts. Next we explain the process of release management and variant detection in more detail.

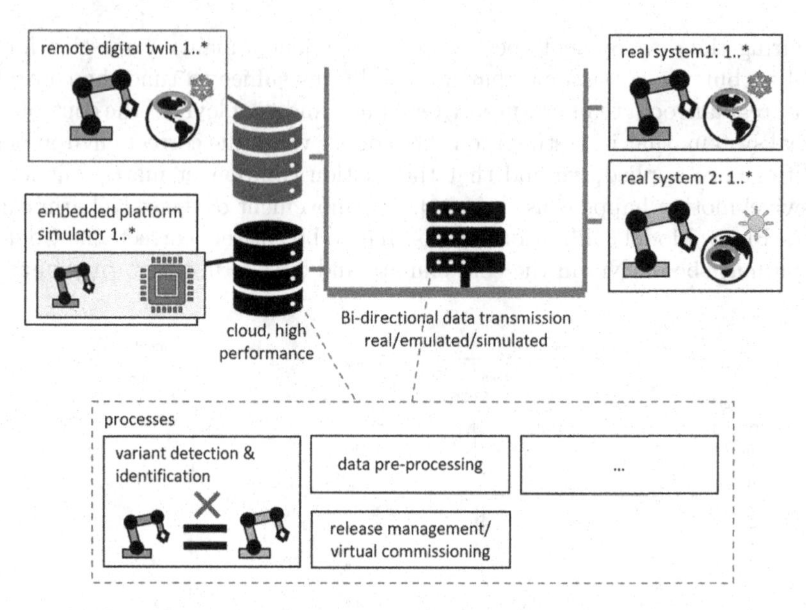

Fig. 4. Architectural overview of the approach.

3.1 Variant Detection

Our approach for the detection of new, undocumented variants relies on the digital twin. Figure 5 shows the process of discovering new variants of the system configuration and environments. The first activity is the detection of deviations between the real system and the digital twin within a certain tolerance. The means to detect a deviation are dependent on the digital twin technology. For example, when using a Kalman filter-based digital twin, the observation over the Kalman filter gain is a possible means to detect deviations between the model and the real world. The deviation can be purely parameter-based. In such a case, the structure of the model is still valid. This results in a new calibration of the model using the data from the real-world. If the re-calibration is successful, a new variant is detected in the real world.

If the calibration fails, the model structure is not sufficient to describe the state of the real system. This can happen when a component of the real-system has changed (e.g. a different DC motor is used because of a repair). In this case, another model is needed to describe the state of the system. When a library of models is available (with different alternative models), the variant detection searches within the library for possible configurations. We rely on the validity frame [1,4,17,18] to describe the validity range of the model. The Validity Frame covers the general concept of validity of a simulation model. In [4], four uses of the Validity Frame concept are analysed: (i) defining the validity of a model, (ii) model/component discovery, (iii) calibration of a model, (iv) defining the experimental process. In the context of this paper, (ii) and (iii) are of interest. When no valid model can be found, the developers need to review the data and find possible system configurations that explain the deviation from the data. This can happen when a non-standard component was used for the repair of the system or because of wear and tear of the system.

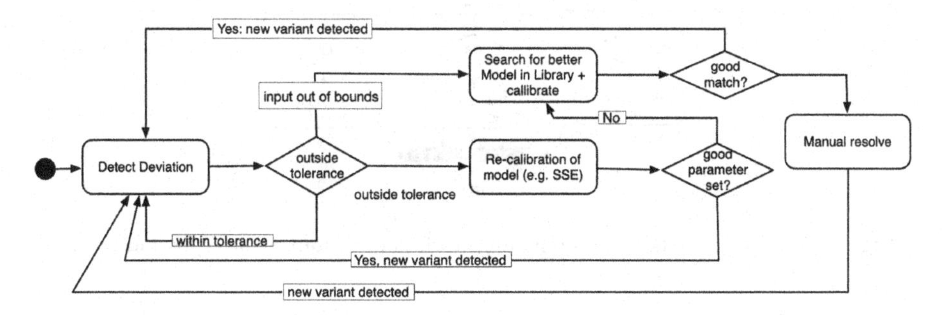

Fig. 5. Workflow for detection system variants using digital twins.

Similarly, another deviation happens when the input of the real system is outside the valid input bounds of the model. The same procedure with the validity frame is used to detect and resolve such a deviation. When no model is available, the environment in which the system now operates is not taken into account during design of the system. It is therefore necessary to resolve the deviation manually.

In each of the cases, a new variant is added to the library of variants that is used by the release management and virtual commissioning component.

3.2 Release Management and Virtual Commissioning

Figure 6 shows the simplified workflow for the virtual commissioning of the system update on the different variants. From the variant detection mechanism, the virtual commissioning component knows which different variants exist of the system. For each of the different variants, an automatic verification and validation experiment is set up. If the experiment passes for the variant, the software update can be deployed onto the systems that are represented by that specific variant. Note that for most systems, in the interest of safety, a human intervention is still required to sign off on the release. This can be provided by showing the proof of the V&V activity to the human to make a decision on deployment of the update.

Because of the nature of the system-under-test, the experimentation environment is build from a co-simulation setup using FMI co-simulation [2]. This is necessary as a typical CPS system is created with different models in different languages. However, some of the requirements can only be checked at the system level, for which the different sub-systems must be integrated to evaluate the system-level behavior, thus requiring co-simulation.

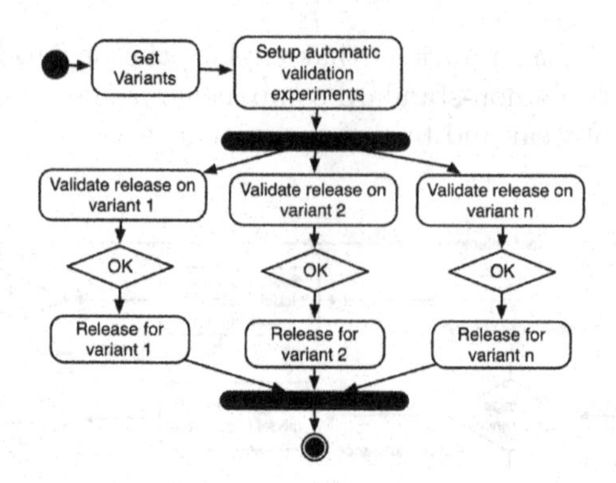

Fig. 6. Simplified virtual commissioning workflow without error reporting paths.

We also expect that system requirements are available that can be automatically checked. Simulation typically outputs traces of the signals in the simulated system. The requirements specified over these traces should be evaluated. For temporal properties, we rely on monitors generated from temporal logics such as Signal Temporal Logic [6]. Other properties, such as energy performance, typically require an extra step to be computed from the traces (e.g. the integration of the signal in time).

4 Application of the Approach

As a proof-of-concept for the approach, we show a new release for the control mechanism on the arm. The user of the system wants more performance and demands for a more rapid movement 10 Hz instead 2 Hz. This is shown in the system information in Table 2. We study the system update in two cases, one with and one without a DevOps loop with a Digital Twin as knowledge base.

Table 2. Updated system information.

Description	Value
Pick & place move	−90° to 90°
Frequency	2 Hz
Error	<0.75°

Updating without DevOps. In this first update, we assume no DevOps with a digital twin as common knowledge base is used. The concept is illustrated in Fig. 7a, which shows that the system in operation is monitored, but no feedback is given for future developments. The new controller is therefore still created

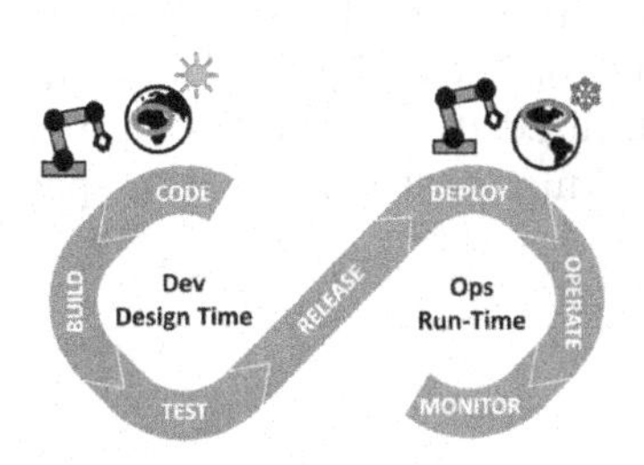

(a) Development and deployment without feedback.

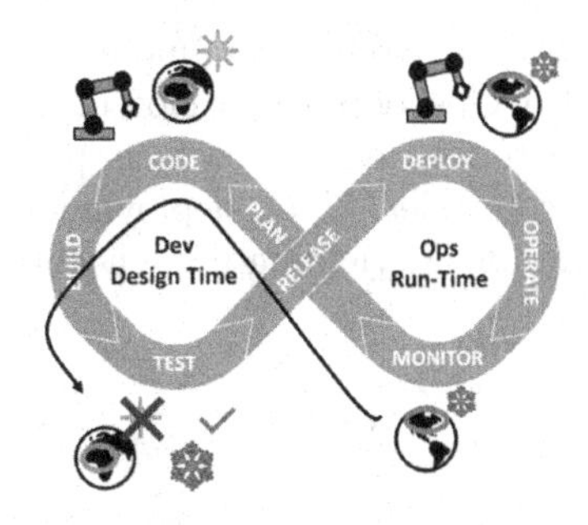

(b) DevOps assuming digital twin as common knowledge base.

Fig. 7. Update schemes for the system update.

for, and tested in lab conditions, even though it may be deployed in different conditions.

Inspecting the results in Fig. 8, at first glance both systems seem to pass the requirements, but on closer inspection, in the cold environment the arm fails on the maximum allowable absolute error. It hits a worst case number of $-1.57°$, which is far past the $0.75°$ allowed value. The arm in the warm environment performs alright, with an error of only $0.61°$.

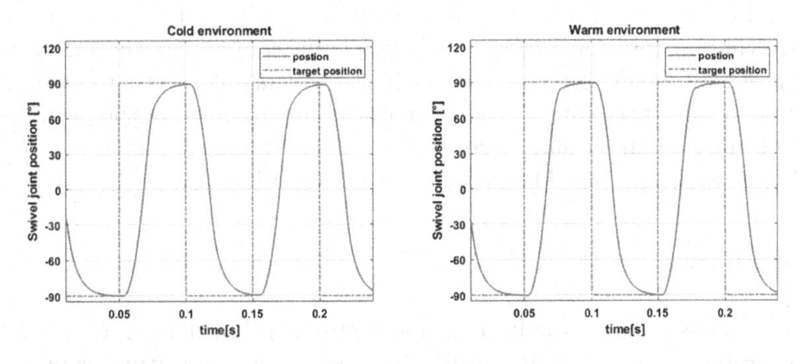

Fig. 8. Traces for the updated controller in both environments.

Updating with DevOps and Digital Twin. In this second case, we assume a digital twin (or digital shadow), of the arm exists. Since the digital twin is continuously calibrated with the real system, the specific environment in which the robotic arm is located is identified. As per the flow in Fig. 5, a deviation detection notifies us of a difference between the model and the real world. In this case, given that structurally the model is valid, a re-calibration of the model suffices. After re-calibration, a proper new value for the friction in the cold environment is found and the software update verified and validated. Figure 7b shows this system where monitored data at runtime is used at design time. Since the DevOps loop is now closed, instead of relying on the design models only, the characterized environment is processed and utilized during the testing of the new controller. Upon deployment, each robot arm is tested in its own specific environment, yielding more accurate results.

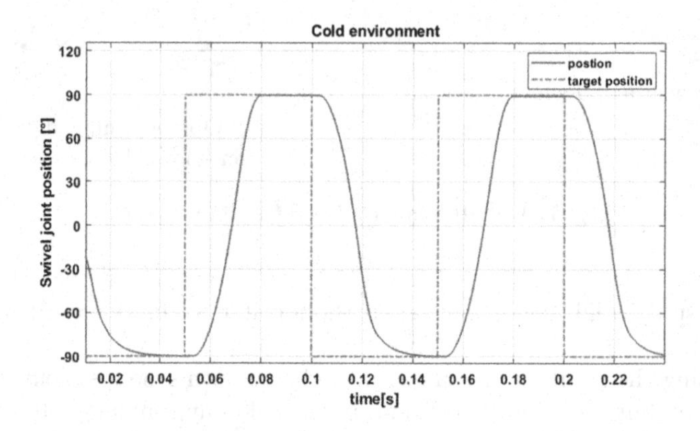

Fig. 9. Traces for cold environment, with corrected controller.

For the warm environment nothing changes, except that, if needed the controller can be adjusted even better to the actual condition. For the cold environment, testing against the calibrated digital twin would show the failed results from Fig. 8, and the designer is able to adjust the controller to that environment. Indeed, with some tweaking of the differential gain parameter of the position controller, the maximum absolute error in the cold environment can also be brought to 0.62°, passing the test. Those results are plotted in Fig. 9.

5 Discussion

Although the example is small, it demonstrates well how having feedback from existing systems can aid in the verification process of new feature or update for already deployed systems. We rely on DevOps in Fig. 7b since it the most complete set of practices for continuous feedback. One might argue that the approach does not specifically need the use of digital twins to detect the different environments, and that the problem could have been prevented by more thoroughly defining tests in Sect. 2, but that is a moot argument, since one cannot account for all possible variations. Alternatively, instead of varying the environmental conditions, we could also have studied a case in which physical components such as the motor are replaced by equivalent but not 100% identical parts. From the approach, the example shows two possibilities that stem from using a digital twin as knowledge aggregator for data:

- Detecting undocumented variants in deployed systems or their environments. Once detected, those variants can be used to aid in the updating of existing or the development of new systems.
- Detecting run-time problems. A physics-based, executable digital twin, allows to detect run-time problems that cannot be identified on the deployed system itself, e.g. by being too computationally heavy.

Besides these possibilities, the approach brings along some new challenges:

C1 When using a digital twin as knowledge base, where should its executable parts be ran and where should its data be stored? Various options are possible, such as machine-local, on a centralized server in the cloud, or using some decentralized scheme of edge computing per operational site.

C2 Is it feasible to detect and uncover every potential variant in a computational sense, and test for every discovered variant? That's to say, is this scalable to larger number of systems, such as in a fleet of vehicles. Can heuristics be used to make this approach more scalable, e.g. by aggregating data from multiple systems into a single representative digital form. Distribution towards the edge and local system might help.

Some parts of the approach do remain unexplored by the example, more specifically the embedded platform simulator and along with it a more realistic process of release management/virtual commissioning. Our main solution for handling both challenges is by modeling these explicitly. Performance models are under construction to explicitly reason over the deployment issues. Sensitivity analysis on the different input and parameters will be used to automatically reason over lumping together individual variants. Furthermore, as certain subsystem tests are shared between variant classes, an analysis of the test set might increase the scalability of the approach. Finally, the challenges are not independent, as evaluation can be distributed based on the type of feature. Features that are much more customized benefit from distribution towards the edge and system, while features that are very common might benefit from centralization. This remains future work.

6 Related Work

Papers such as [3,5,20] make note of the ever faster recurring development cycle for CPS. In the CPS domain, where model-based engineering is widely applied due to its abstraction of complexity, there is thus a need for practices such as DevOps. In [3], the challenge of transferring data back from the running system to the models is specifically noted. In [14], it is clearly identified that model-based design practices and digital twin practices integrate well with eachother, and in [16], a description of how the digital twin can be used throughout the DevOps stages is elaborated on. That is also why we like to call the digital twin the common-knowledge base in the DevOps loop, and such a knowledge base can consist of executable models but also non-executable models. Specifically in the verification and validation stage, it is noted in [16] how the digital twin can be used to ensure the system is represented as best as possible, though no note is made on the identification variants, which is where we believe we can further contribute. On the topic of testing, we must note that this remains one of the larger challenges for the adoption of DevOps in the development for CPS, as noted in [21]. Indeed, the development of CPS relies on thorough testing, often followed by safety and/or security certification. In this respect, the potential contributions of the presented approach will be limited to identifying problems earlier on, since we remain in a modeled world, and certification happens on a codebase. Models can however help with producing certifiable code more easily. It's also on the model aspect that we differentiate from other variant management approaches such as [11], which operate on a code level, and uses emulation. This is interesting from the certification point of view, but less interesting from the early testing point of view. In [7] however, it is stated that for continuous integration chains, fast transaction level simulation is proven to work for the simulation of virtual hardware, and is more scalable than testing with hardware. Another aspect is that of the run-time validity monitoring. Where the runtime monitoring of Validity Frames is currently limited to input/output value/datatype and relation monitoring, for correct variant detection, it might be needed to incorporate other monitors as well. In this regard in [9], the concept of multilevel monitors is presented for detecting attacks or faults. A distinction is made between Data Monitors, Functional monitors and Network monitors. The data monitoring and functional monitoring are similar to the I/O monitoring in Validity Frames, but the overlapping monitors and network monitoring presents interesting insights that can be used to extend the monitoring capabilities of Validity Frames.

7 Conclusion

In this paper we elaborated on our vision of utilizing the digital twin of a system in the agile deployment of new system updates. The approach builds on the idea of using the digital twin as a common knowledge base throughout the DevOps cycle. We demonstrated a subset of the approach on a small example study, from which we elaborated some open challenges that remain to be researched. Lastly, we also elaborated on how related work matches our vision, and how our approach can contribute to or differs from that existing work.

References

1. Acker, B.V., Meulenaere, P.D., Denil, J., Durodie, Y., Bellinghen, A.V., Vanstechelman, K.: Valid (re-)use of models-of-the-physics in cyber-physical systems using validity frames. In: 2019 Spring Simulation Conference (SpringSim), pp. 1–12 (2019). https://doi.org/10.23919/SpringSim.2019.8732858
2. Blochwitz, T., et al.: The functional mockup interface for tool independent exchange of simulation models. In: Proceedings of the 8th International Modelica Conference, pp. 105–114. Linköping University Press (2011)

3. Combemale, B., Wimmer, M.: Towards a model-based DevOps for cyber-physical systems. In: Bruel, J.-M., Mazzara, M., Meyer, B. (eds.) DEVOPS 2019. LNCS, vol. 12055, pp. 84–94. Springer, Cham (2020). https://doi.org/10.1007/978-3-030-39306-9_6
4. Denil, J., Klikovits, S., Mosterman, P., Vallecillo, A., Vangheluwe, H.: The experiment model and validity frame in m&s. In: SpringSim (2017)
5. Denil, J., Salay, R., Paredis, C., Vangheluwe, H.: Towards agile model-based systems engineering. In: MODELS (Satellite Events), vol. 2019, pp. 424–429. CEUR Workshop Proceedings (2017). http://ceur-ws.org/Vol-2019/
6. Donzé, A., Ferrère, T., Maler, O.: Efficient robust monitoring for STL. In: Sharygina, N., Veith, H. (eds.) CAV 2013. LNCS, vol. 8044, pp. 264–279. Springer, Heidelberg (2013). https://doi.org/10.1007/978-3-642-39799-8_19
7. Engblom, J.: Continuous integration for embedded systems using simulation. In: Embedded World 2015 Congress (2015)
8. Garcia, J., Cabot, J.: Stepwise adoption of continuous delivery in model-driven engineering. In: Bruel, J.-M., Mazzara, M., Meyer, B. (eds.) DEVOPS 2018. LNCS, vol. 11350, pp. 19–32. Springer, Cham (2019). https://doi.org/10.1007/978-3-030-06019-0_2
9. Gautham, S., Jayakumar, A.V., Elks, C.: Multilevel runtime security and safety monitoring for cyber physical systems using model-based engineering. In: Casimiro, A., Ortmeier, F., Schoitsch, E., Bitsch, F., Ferreira, P. (eds.) SAFECOMP 2020. LNCS, vol. 12235, pp. 193–204. Springer, Cham (2020). https://doi.org/10.1007/978-3-030-55583-2_14
10. Grieves, M.: Origins of the digital twin concept. In: Working Paper, pp. 1–7. Florida Institute of Technology, August 2016. https://doi.org/10.13140/RG.2.2.26367.61609
11. Guissouma, H., Lauber, A., Mkadem, A., Sax, E.: Virtual test environment for efficient verification of software updates for variant-rich automotive systems. In: 2019 IEEE International Systems Conference (SysCon), pp. 1–8 (2019). https://doi.org/10.1109/SYSCON.2019.8836898
12. Kim, G., Humble, J., Debois, P., Willis, J.: The DevOps Handbook: How to Create World-Class Agility, Reliability, and Security in Technology Organizations. IT Revolution (2016)
13. Kritzinger, W., Karner, M., Traar, G., Henjes, J., Sihn, W.: Digital twin in manufacturing: a categorical literature review and classification. IFAC-PapersOnLine **51**(11), 1016–1022 (2018). https://doi.org/10.1016/j.ifacol.2018.08.474, 16th IFAC Symposium on Information Control Problems in Manufacturing INCOM 2018
14. Madni, A.M., Madni, C.C., Lucero, S.D.: Leveraging digital twin technology in model-based systems engineering. Systems **7**(1), 7 (2019)
15. Mertens, J., Denil, J.: Digital twins for continuous deployment in model-based systems engineering of cyber-physical systems. CEUR Workshop Proceedings, vol. 2822, pp. 32–39 (2020). http://ceur-ws.org/Vol-2822/
16. Ugarte Querejeta, M., Etxeberria, L., Sagardui, G.: Towards a DevOps approach in cyber physical production systems using digital twins. In: Casimiro, A., Ortmeier, F., Schoitsch, E., Bitsch, F., Ferreira, P. (eds.) SAFECOMP 2020. LNCS, vol. 12235, pp. 205–216. Springer, Cham (2020). https://doi.org/10.1007/978-3-030-55583-2_15

17. Van Acker, B., Oakes, B.J., Moradi, M., Demeulenaere, P., Denil, J.: Validity frame concept as effort-cutting technique within the verification and validation of complex cyber-physical systems. In: Proceedings of the 23rd ACM/IEEE International Conference on Model Driven Engineering Languages and Systems: Companion Proceedings. MODELS 2020, New York, NY, USA. Association for Computing Machinery (2020). https://doi.org/10.1145/3417990.3419226
18. Van Mierlo, S., Oakes, B.J., Van Acker, B., Eslampanah, R., Denil, J., Vangheluwe, H.: Exploring validity frames in practice. In: Babur, Ö., Denil, J., Vogel-Heuser, B. (eds.) ICSMM 2020. CCIS, vol. 1262, pp. 131–148. Springer, Cham (2020). https://doi.org/10.1007/978-3-030-58167-1_10
19. Vangheluwe, H., De Lara, J., Mosterman, P.J.: An introduction to multi-paradigm modelling and simulation. In: Proceedings of the AIS 2002 Conference (AI, Simulation and Planning in High Autonomy Systems), Lisboa, Portugal, pp. 9–20 (2002)
20. Warg, F., Blom, H., Borg, J., Johansson, R.: Continuous deployment for dependable systems with continuous assurance cases. In: 2019 IEEE International Symposium on Software Reliability Engineering, WoSoCer workshop. IEEE Computer Society (2019)
21. Zeller, M., Ratiu, D., Rothfelder, M., Buschmann, F.: An industrial roadmap for continuous delivery of software for safety-critical systems. In: 39th International Conference on Computer Safety, Reliability and Security (SAFECOMP), Position Paper (2020)

1st International Workshop on Multi-concern Assurance Practices in Software Design (MAPSOD 2021)

1st International Workshop on Multi-concern Assurance Practices in Software Design (MAPSOD 2021)

Jason Jaskolka[1], Brahim Hamid[2], and Sahar Kokaly[3]

[1] Systems and Computer Engineering, Carleton University, Ottawa, ON, Canada
jason.jaskolka@carleton.ca
[2] IRIT, University of Toulouse, Toulouse, France
brahim.hamid@irit.fr
[3] General Motors, Markham, ON, Canada
sahar.kokaly@gm.com

1 Introduction

Complex software systems have become increasingly entwined in a wide variety of systems such as critical infrastructure, industrial control systems, medical devices, automobiles, airplanes, and spacecraft. Assuring the security and safety, as well as other dependability aspects such as availability, robustness, and reliability, of these software-intensive systems remains among the top priorities for governments and providers of critical systems and services. Manufacturers, owners, and operators of the components and devices that make up these software systems strive to ensure that they have adequately addressed emerging concerns related to safety hazards, security threats, and performance challenges, among others. For this reason, there is a need to address these various concerns within the architecture and design of these systems in the context of the subjective and often contradicting, competing, and conflicting needs and beliefs of stakeholders, and to do so with a level of confidence that is commensurate with the tolerable loss consequences associated with each of these objectives.

2 This Year's Workshop

This is the first edition of the MAPSOD workshop co-located at SAFECOMP 2021. MAPSOD 2021 is centred on rethinking the concept of assurance and certification, taking into account the nature and the full range of design concerns of software-intensive systems. The topics provide coverage of architecture and design trends supporting multi-concern assurance of software-intensive systems of high relevance to software practitioners. Special emphasis has been devoted to promoting discussion and interaction between researchers and practitioners focused on addressing concerns related to safety, security, reliability, availability, and robustness within the architecture and design of software-intensive systems.

The workshop comprised three presentations:

1. An Accountability Approach to Resolve Multi-stakeholder Conflicts, by Yukiko Yanagisawa and Yasuhiko Yokote.
   This paper presents an approach to resolve conflicts among multiple stakeholders based on the notion of an accountability map. The approach involves identifying events requiring accountability and the risks that come from conflicting stakeholder viewpoints and claims to incorporate the recommendations from IEC 62853. It also involves describing the resolution procedures to support various risk responses.
2. Architecture-Supported Audit Processor: Interactive, Query-Driven Assurance, by Sam Procter and Jerome Hugues.
   This paper presents an interactive tool called the Architecture-Supported Audit Processor (ASAP). ASAP focusses on system safety and extracts safety-specific viewpoints from system architecture models expressed using AADL. The viewpoints are dynamically generated as diagrams and tables. Using these viewpoints, the paper discusses integrations with Systematic Analysis of Faults and Errors (SAFE) and System Theoretic Process Analysis (STPA).
3. Towards Assurance-Driven Architectural Decomposition of Software Systems, by Ramy Shahin.
   The paper discusses the limits of using abstraction techniques to handle the complexity of computer systems from design and analysis perspectives. The proposed approach combines three architectural views (computation, coordination, stateful) and a meta-programming envelope to decompose software systems. The resulting 4-dimension meta-architecture allows the capture of assurance expectations, early on in the design stage. The proposed decomposition method contributes to the reconciliation between the two different visions, cultures, and the produced artifacts related to SDLC and assurance activities.

The program was completed with a keynote talk given by Barbara Gallina (Mälardalen University) on “Multi-concern Assurance: Current Practices, Challenges, and Ways Forward”. Moreover, the workshop closed with a discussion session about the future of the workshop.

3 Acknowledgements

As chairpersons of the workshop, we want to thank all authors and contributors who submitted their work, Friedemann Bitsch, the SAFECOMP publication chair, Simos Gerasimou and Erwin Schoitsch, the SAFECOMP workshop co-chairs, and the members of the International Program Committee who provided a fair and thorough evaluation through reviews of the submitted papers. We extend our thanks to the SAFECOMP organizers, and their chairperson John McDermid, who provided us the opportunity to organize the workshop at SAFECOMP 2021 as an online event.

It is our sincerest hope that all participants will enjoy and benefit from the workshop and will join it again in the future!

International Program Committee

An Accountability Approach to Resolve Multi-stakeholder Conflicts

Yukiko Yanagisawa and Yasuhiko Yokote(✉)

Advanced Data Science Project, RIKEN Information R&D and Strategy Headquarters, RIKEN, Tokyo, Japan
{yukiko.yanagisawa,yasuhiko.yokote}@riken.jp

Abstract. A problem that has been increasingly emerging recently in software development is that the interests of stakeholders with different positions and roles can lead to system failures due to exacerbation of conflicts among them. This research considers the degradation of software reliability due to conflicts of interest among multiple stakeholders as one of the challenges of multi-concern assurance, and describes a method to resolve stakeholder conflicts and improve software dependability. We propose an accountability map that incorporates the approach to achieving accountability outlined in international standard IEC 62853. We identify events that require accountability, identify risks that arise from exacerbating stakeholder conflicts, and describe resolution procedures that support and resolve risk responses. The results of this paper are valuable in that increased trust among stakeholders will enable software development to be responsive to accountability issues.

Keywords: Accountability · Accountability map · Multi-stakeholder conflict · Multi-concern assurance · Dependability · IEC 62853

1 Introduction

In recent years, the number of stakeholders involved in software development has been increasing due to the growing scale and complexity of systems and the shift to system of systems and essentially to open systems. Since stakeholders have different backgrounds, positions, and roles, conflicts of interest are likely to occur, reflecting gaps in their expectation regarding the requirements, lack of mutual understanding, different priority to objectives, cultural differences and capabilities to achieve project goals. In addition, many cases arise in which the boundaries of stakeholder responsibilities are unclear. Where multiple parties are trying to protect their own positions while fulfilling their respective duties and responsibilities, it is obvious that conflicts of interest will arise. Obara [1] refers to this as multi-stakeholder conflicts and observes that the negative communication cycle can further worsen the conflicts, turning them into risks of damage such as increased costs due to work delays.

This research recognizes multi-stakeholder conflicts as one of the challenges of multi-concern assurance and proposes a method to resolve such conflicts from the perspective

I. Habli et al. (Eds.): SAFECOMP 2021 Workshops, LNCS 12853, pp. 175–186, 2021.
https://doi.org/10.1007/978-3-030-83906-2_13

of overall software dependability. This paper is organized as follows: Sect. 2 describes the background of this work in relation to accountability, Sect. 3 explains a method to derive an accountability map (abbreviated as A-map) and shows how to schematize events that may cause multi-stakeholder conflicts using specific examples, and Sect. 4 outlines how to verify the A-map created. Section 5 discusses on resolution of conflicts of interest among stakeholders in terms of complexity and satisfaction on accountability. We finally summarize this paper in Sect. 6.

2 Background

The authors' interests are implementation of lifecycle processes to perform data-intensive medical science with AI technologies accordance with the open systems dependability standard, IEC 62853 (abbreviated as 62853). Based on the concept of a process view [2], which summarizes the processes related to each common stakeholder interest, this standard defines four process views—consensus building, accountability achievement, failure response, and change accommodation [3]. Matters related to stakeholder response are mentioned throughout all four process views in 62853. Two of them are crucial to implement it in practically and effectively satisfying its requirements. One of them is on the consensus building process view, which a concrete implementation method has been proposed in [4]. The other is on the accountability achievement process view, which requires anticipation of events that require accountability and implementation of corresponding actions. These actions include providing information to stakeholders and providing remedies. Here, the crucial issue is how to manage conflicts of stakeholders' interest and for stakeholders to be assured its implementation.

Methods for analyzing and resolving multi-stakeholder conflicts have been studied in the fields of project management and governance. In recent years, multi-stakeholder processes (MSPs) have been introduced as a concept to deal with multi-stakeholder conflicts. MSPs are communication processes in which three or more stakeholder groups participate on an equal footing to build consensus and make decisions [5]. As "a new model of governance to support sustainable development," [6] they are being implemented by the United Nations, governments, and non-profit organizations. MSPs are a means of moving forward in addressing difficult issues such as the Sustainable Development Goals (SDGs) with all participants connected by trust, rather than relying on one party to lead the way.

However, neither MSPs nor 62853 include discussion on specific ways of resolving conflicts that this paper focuses on. The perception gap that causes conflicts among stakeholders arises from the ambiguity of each other's responsibilities. To narrow the gap, it is necessary to define the boundaries first. Disclose each other's demands, take them in to evaluate based on one's own interests, and then clarify the boundaries of responsibility. When any two parties can see each other's boundaries, it is possible to find a problem that might be overlooked because of wide gap between them and left out from the scope of responsibilities on both sides. If such a problem is found, further consultations can be made to adjust the scope of responsibility and close the gap.

Methods for clarifying the scope of responsibilities among stakeholders are being studied. According to Bovens [7], accountability has four components: "to whom," "by whom," "about what," and "why". By identifying these four factors among stakeholders, the scope of responsibility can be documented and articulated in a consultative manner. Yamamoto [8] has diagrammed these four components and presented them as an information model. These four components recapitulate the multiple stakeholders into two-party relationships and identify the accountability that occurs between them. Using the two-party information model that includes the above four components of accountability, we schematize the events that can cause multi-stakeholder conflicts. By incorporating the concept of 62853, the events that require accountability can be identified as risks, and stakeholders can consider and agree on the options to handle them, such as avoidance, taking, mitigation, elimination, sharing, and retention. By clarifying its scope and specific responsibilities to fulfil among stakeholders in the process of achieving accountability, we can eliminate potential causes of the occurrence of multi-stakeholder conflicts that cause such conflicts, as well as improve the handling of risks.

3 A Proposal of Accountability Map

In order to schematize the events that can cause multi-stakeholder conflicts, we will attempt to construct an information model that can grasp the entire picture of events that can cause multi-stakeholder conflicts by investigating specific cases and examining the components of accountability and the relationships between the components. In this study, the information model is called the accountability map. The specific example to be used is the COCOA failure described in next paragraph.

An example of failure resulting from a multi-stakeholder conflict is the critical glitch in the COVID-19 Contact-Confirming Application (abbreviated as COCOA) distributed by Japan's Ministry of Health, Labor and Welfare (MHLW), which was discovered in February 2021. This glitch, in which smartphones running the Android OS were not notified of contact with an infected person, was left unresolved for more than four months. The subsequent investigation reported findings that "the importance of testing was not fully recognized among ministry officials involved" and that there was "an ambiguous division of roles between the ministry and the developer" [9]. This is a case in which the risk of multi-stakeholder conflicts eventuated in a failure. The fact that software intended to help protect the health of the public did not fulfill its function over a long period has significantly lowered the dependability of the software.

We first explain the flow of deriving the A-map using the structural modeling method, and then we explain the procedure of executing accountability using the map. Finally, we summarize the verification method for the A-map and the procedure to achieve accountability.

3.1 Development of the Accountability Map

When assessing what kind of information is necessary for the act of an accountable party to give a specific explanation, we consider the four components ("to whom," "by whom" "about what," "why") and the relationships among them, as identified in Bovens' and Yamamoto's work, to be the basic structure of this information model, because they limit the events that require accountability.

The four components capture the events that require accountability in the relationship between two parties, the *explainer* and the *explainee*. To determine the scope of the problem in the specific case of the COCOA failure, the events that require accountability occurring between the two parties are extracted using the following steps: 1) identification of stakeholders; 2) identification of relationships between stakeholders; 3) identification of interactions occurring between the two stakeholders; and 4) identification of the direction of the interactions. In addition, events that require a specific action by the explainer in the exchange between the two parties are identified as events that require accountability.

According to a report by the MHLW [10], the direct cause of the problem with COCOA's functionality as deployed was a bug that failed to take into account the difference in calculated values depending on the operating system. However, the report emphasized that the underlying problems were: (1) releasing the product without conducting the final acceptance test on the actual device; and (2) the lack of shared understanding about the maintenance of the test environment and the purpose of the software itself. For this reason, the scope of the problem was defined as the testing process related to the release of COCOA. Furthermore, according to the Nikkei newspaper, the business in charge of implementation and operation was responsible for functional testing of COCOA, while the business in charge of project management and quality control was responsible for regression testing, and the MHLW was supposed to conduct acceptance testing [11]. This fact shows that there were three stakeholders. The events that require accountability extracted from this information are shown in Table 1.

Table 1. Scope of specific cases and events requiring accountability

Events for which accountability is required	Stakeholders	Interactions
Functional tests	The businesses in charge of implementation and operation	Reporting of test results to higher-level processes
	The businesses in charge of project management and quality control	Checking the test results
Regression tests	The businesses in charge of project management and quality control	Reporting of test results to higher-level processes
	MHLW	Checking the test results
Acceptance tests	MHLW	Release report to society
	End user	Update COCOA app

Next, for each of the above three events requiring accountability, we will apply the respective information to the four components described above. The functional tests in Table 1 are used as a sample to explain how they are assigned to the components. In this case, the business in charge of implementation and operation of COCOA is responsible for conducting functional tests for the relevant release range and reporting the test results to the business in charge of project management and quality control, which is in charge of regression testing, as an upstream process. Such events are classified as organizational or hierarchical accountability [8], and require accountability achievement.

The role of explainer is assigned to the business in charge of implementation and operation, and the role of explainee is assigned to the business in charge of project management and quality control. The explainee, which is a higher-level process, requests what it wants the explainer to do in order to implement the regression test that the explainee is in charge of. We set the following as typical requirements that are considered in general software development: we want the functions in the scope to be released to work correctly, we want the bugs in the scope to be released to be fixed, we want the fixed program to be free of new bugs, and we want the functional tests to be conducted. These requirements are assigned to the "why" accountability component. The explainer receives requests from the explainee and considers accordingly how to respond to those requests in order to comply with the internal business rules, guarantees that the developed content will work, guarantees that functional tests will be conducted to prove that the content works, and protect its own position. On the basis of that consideration, the explainer presents the explanation to the explainee. This is the "about what" of accountability, namely the assumption that the scope of the relevant release was correctly grasped, that test specifications were prepared, that a test environment was in place to enable testing of actual equipment, that functional tests were conducted, and that the results were correctly grasped and reported, as shown in Fig. 1.

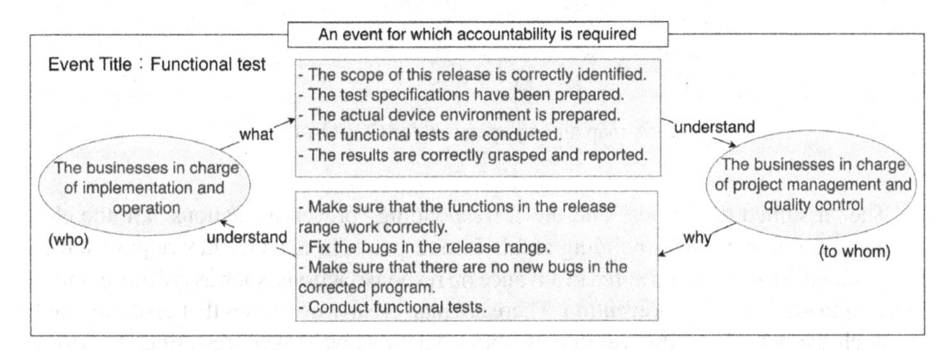

Fig. 1. The four components involved in functional testing of COCOA app

The same considerations as above are repeated, and the four components are applied to the other two accountability-requiring events: regression testing and acceptance testing. As a result, we can confirm that the four components identify the events that require accountability.

Next, we add the following elements to satisfy the requirements of 62853: "details of accountability implementation," "assumed deviations," and "preventive actions" to indicate the response to the possible deviations. In the example in Fig. 1, the business unit in charge of development and operation that conducts functional testing indicates the scope of the release, presents the test specifications within that scope, and explains the development of the test environment and test results to the business in charge of project management and quality control, which are higher-level processes. However, this alone cannot be said to have fulfilled accountability requirements. This is because the explainee has not conducted a regression test to prove that the downstream process test is acceptable overall after the test specifications are checked and test results presented by the explainer. There must be evidence that the explainee has reviewed the test specifications and test results, understands their contents, and has accepted them. When both the evidence of execution by the explainer and the evidence of confirmation by the explainee are present, accountability has been properly achieved, and a relationship of trust based on mutual agreement has been maintained.

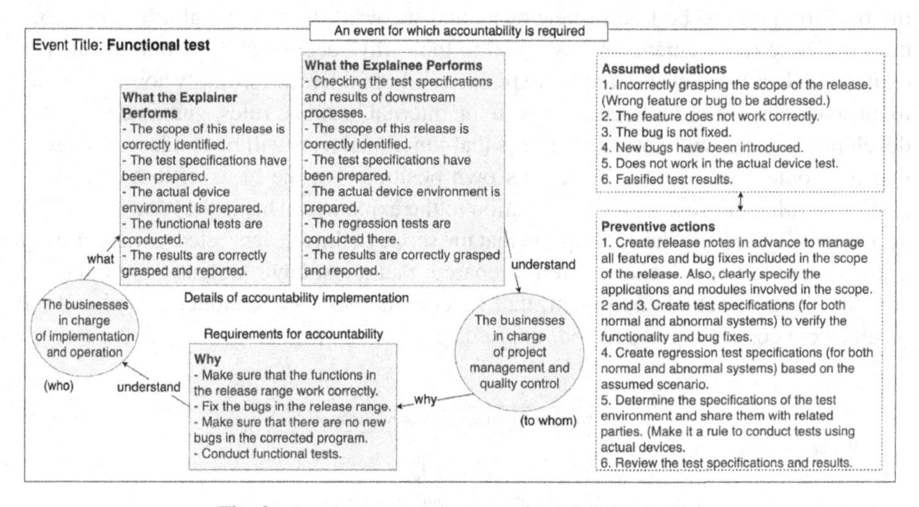

Fig. 2. A-map among operators for COCOA testing

The "assumed deviations" and the corresponding "preventive actions" are the elements related to consensus building and failure response in 62853. They require stakeholders to identify risks and agree in advance on response actions, such as risk mitigation, risk elimination, and risk prevention. There are many real-world cases that could be cited in which the potential risks were known but no actions was taken, resulting in serious harm. Failure to properly fulfill the actions taken to achieve accountability is a deviation from the state of agreement between the two parties. In 62853, a condition in which stakeholder consensus is maintained is a condition in which dependability is maintained. Therefore, it is essential to: (1) identify the risk of deviation from the agreed condition in advance; (2) be able to detect the deviation; and (3) determine specific actions to take immediately in response to detection of the deviation in order to minimize its impact. Figure 2 shows an A-map with the aforementioned three additional elements. By applying specific information to the seven elements of the A-map for each of the three events that require accountability, we can verify that the A-map can identify the accountability that exists between two parties.

3.2 Identify Steps to Implement Accountability Using Accountability Maps

The A-map can be used to schematize the events that can cause multi-stakeholder conflicts throughout examination of the work done in the process of developing the A-map. Stakeholders agree on the content of the A-map to be carried out as described. They then move on to the stage of implementing the specific items written in the "what" component of accountability. A mechanism to prevent the occurrence of possible deviations is implemented, along with a mechanism to monitor whether deviations are occurring. If a deviation is identified as having occurred, accountability is achieved through performing the failure response and the change accommodation process views defined in 62853, including implementation of recurrence prevention. 62853 states that repeating these process views over and over, starting with risk identification, will reduce the number of recurrent failures and improve the dependability of the system. Therefore, after action is taken to prevent recurrence, the procedure should be repeated iteratively, once again reviewing the A-map and, if necessary, the actions and stakeholders that identify them. The results of the derived procedure are summarized below.

1. Identify two stakeholder groups.
2. Identify the actions of the two parties.
3. Identify events for which accountability is needed.
4. Create an A-map for each event.
5. Link multiple, related A-maps.
6. Agree on an A-map between the two parties.
7. Execute what is to be done.
8. Monitor for possible deviations.
9. Respond to failures and changes in case of deviations.
10. Achieve accountability.
11. Based on the results of the failure response and change accommodation process views, return to step 1 and review the A-map.

If we look at explainers and explainees, there are cases where the person who is the explainer in one event requiring accountability becomes the explainee in another related event. Taking Fig. 2 above as an example, the business in charge of project management and quality control conducts a regression test based on the results of the functional test. Explaining, in turn, these results to the MHLW is a higher-level process. It can be seen that the business in charge of project management and quality control, which was the explainee in the event of functional testing in Fig. 2, becomes the explainer in the next event, regression testing, and therefore another accountability process occurs with the MHLW as the explainee. In this way, two or more A-maps are linked together to represent one large event, which we call an *accountability chain.*

4 Verification of Accountability Map

The aforementioned report on the COCOA failure will be used to verify whether the A-map and its methods are effective or not. The verification will be conducted by creating A-maps for the remaining two events requiring accountability (regression testing and acceptance testing), as per the aforementioned accountability execution steps 1–5, and checking the validity of A-maps and their procedures against the report on the COCOA failure published by the MHLW.

The A-map between the MHLW and the business related to the regression test shows a case in which the stakeholder that was the explainee in Fig. 2 becomes the explainer. The explainer is designated to be the business in charge of project management and quality control that conducts the regression testing. The explainee is designated to be the MHLW, which will perform the acceptance test. The explainer's task is to conduct regression testing. The explainee's task is to confirm the results of the regression test by means of the acceptance test that the explainee is in charge of, the results of which therefore encompass the results of the regression test. The content of the execution and possible deviations will be the same as in the tests of the lower-level process between the two businesses, except for a change in the type of test.

The third A-map shows the case in which the MHLW, which is the explainee in the second A-map, becomes the explainer. In this case, the explainee is the end user who uses the application which has been downloaded to their device. The content to be performed by the explainer is the acceptance test. The end user, as the explainee, updates the release of COCOA on their device based on the results of the acceptance test by the MHLW, which designates the actual use of the application by the end user as the content of the "what" component. The results are shown in Fig. 3. The three completed A-maps will be linked together as an accountability chain and checked against the COCOA failure report to verify whether the 62853 assumed deviations could have prevented the failure in question.

According to the MHLW's COCOA failure report [10], the parties involved expressed the following perceptions. "One of the managers stated that he had received a report from the person in charge that although no tests had been done, it was okay because it was only a change in logic, and that at the time he was aware that the public health center had requested a response to the problem of frequent push notifications and that the modification needed to be done quickly." On the other hand, according to the COCOA update information on the MHLW's website, "The application has been modified so that the determination of contact when displaying a push notification that there has been contact with a positive person is consistent with the determination of contact when displaying that there has been contact with a positive person in the contact confirmation application, and the accuracy of contact detection has been improved. In addition, we have made improvements to the internal processing in order to optimize the accuracy of contact detection" [12]. In other words, the contents of the release included improvements to the push notification of contact with a positive person and improvements to the display of contact with a positive person in the application. However, the fact that the MHLW never tested the release because it thought the only change was in logic and that it was in a hurry to release the application means that the MHLW did not understand the contents of the release correctly. If it had understood the contents of the release correctly, it would have tested that push notifications were sent on the actual device and displayed in the application. If it had tested the actual device, it could have found this bug. Another reported reason for not conducting acceptance tests was the fact that the testing environment was not in place at the time. In this regard, it is assumed that the recognition of the necessity of testing on the actual device would have been enhanced by deciding, agreeing, and sharing the specifications of the testing environment in advance when the three parties set up the testing arrangements.

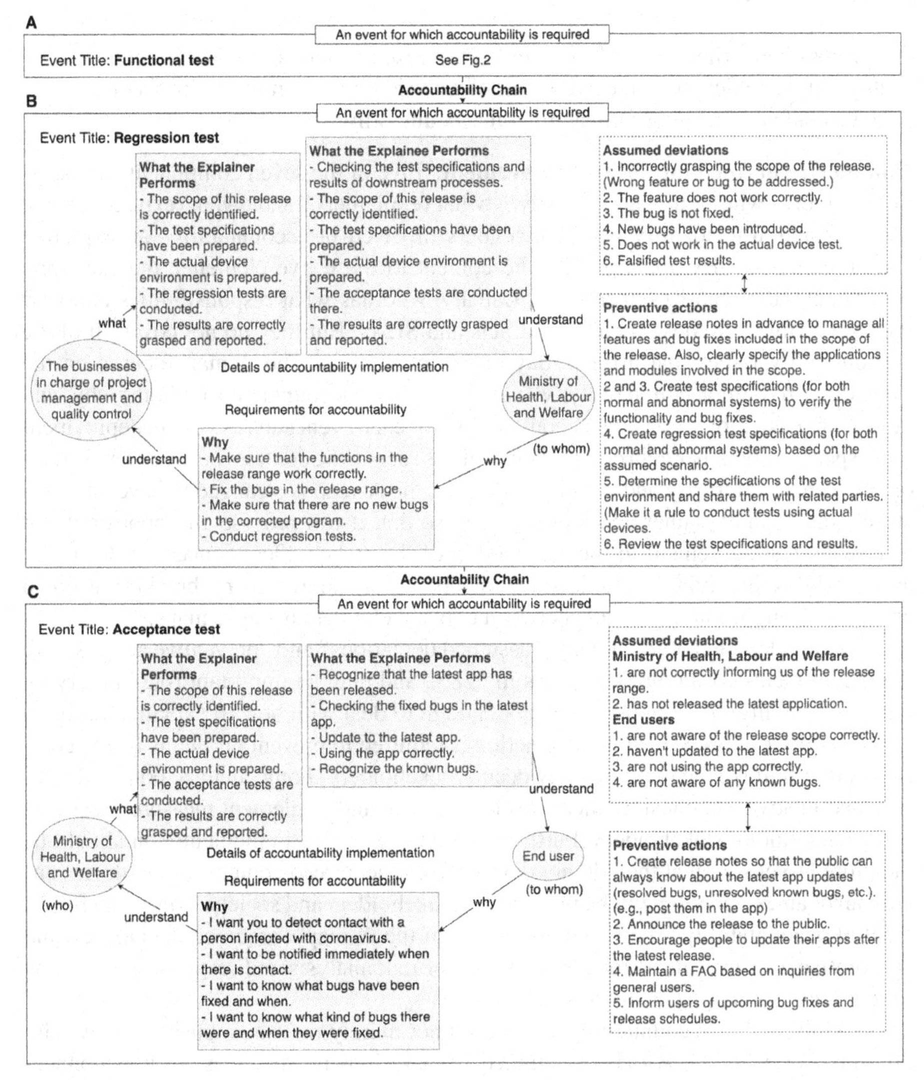

Fig. 3. A: A-map among operators for COCOA testing; B: A-map between the business unit and the MHLW for testing COCOA; C: A-map between the MHLW and the general public for testing COCOA.

A comparison of the A-map with the report about the COCOA failure confirms that the map can clearly organize multi-stakeholder conflicts. It is also possible to identify the deviations that are expected to cause multi-stakeholder conflicts. If such an A-map had been created in advance, failures could have been prevented or detected at an early stage.

5 Discussion

We discuss here whether the A-map in Sect. 3 resolves the conflicts of interest among multiple stakeholders from the two points of view: handling complexity in accountability achievement and reaching satisfaction on accountability.

Complexity in Accountability Achievement. Five of the seven components in the A-map, namely "who," "to whom," "why," "what (explainer)," and "what (explainee)" are the information to identify two stakeholders involved in accountability, an explainer and an explainee. By applying specific content to these five elements, the execution of accountability can be concretized. Usually, who has what responsibilities and how they are fulfilled are described in contracts and SLAs. But they do not cover all of the accountability issues that arise in day-to-day operations. The A-map focuses on the events that require accountability, visualizing who is accountable to whom, why, and how. A-map is a tool to consider accountability for each event such as system deployment more specifically rather than comprehensively as seen in the contract between enterprises or between an enterprise and a user. In addition, the A-map is arranged to have only two stakeholders, an explainer and an explainee, so that it can provide an opportunity for discussion and agreement to the satisfaction of the stakeholders concerned. Complex accountability involving multiple stakeholders can be addressed by breaking it down and simplifying it into the events between two stakeholders using A-maps.

The two elements of the A-map, "assumed deviations" and "preventive actions," are the requirements for failure responses in 62853 and play an important role in carrying out accountability. Accountability is not enough to deal with how to explain a problem to society after it has occurred. Prior action is required to prevent problems from occurring or to minimize the impact if they occur. Stakeholders should analyze potential risks assumed in advance, consider measures to be taken, and implement measures based on agreement among stakeholders. Furthermore, the results of implementation should be monitored to ensure that they do not deviate from the agreed content. If any deviations are found, an explanation will be provided to stakeholders and society. This is the fulfillment of accountability. The two components of the A-map, "assumed deviations" and "preventive actions," allow for consideration of risk analysis and responses, and allow for consensus among stakeholders.

In addition, the accountability chain does not merely indicate a business flow relationship. For example, A-maps are linked from various perspectives, such as connections involving the same stakeholders, connections involving the same requirements, and connections involving the same risks. Complex accountability involving multiple stakeholders can be broken down into multiple detailed A-maps, but the accountability chain from different perspectives makes it more practical to use as a single event for each perspective.

Satisfaction on Accountability. Multi-stakeholder conflicts are exacerbated by a negative cycle of failing communication, misunderstanding, stress built and growing mistrust

[1]. It is obvious that stakeholders will not be able to understand each other's demands if they are not clear about the scope of their interests and responsibilities. What is the state of non-conflict among stakeholders? It is a state in which stakeholders are satisfied. Stakeholder satisfaction is achieved by understanding the demands of the other stakeholders, agreeing with the other stakeholders in advance on how to meet the other stakeholders' demands while protecting our own interests, and correctly executing what has been agreed upon. It is a state of accountability achievement.

The A-map provides an opportunity for both stakeholders to consider, validate, and agree on actions for accountability based on the other stakeholders' requirements. The opportunity solves the problem of not having a clear definition of what to agree upon in order to carry out accountability. 62853 requires that deviations from the agreement be monitored. Problems can be detected early by having a mechanism to discover the state that is being accountable as agreed upon. Early detection and early response can shorten the time to resolution, thereby reducing stakeholder stress and distrust, and preventing the situation from developing into a multi-stakeholder conflict. The A-map is a general-purpose tool that can schematize events that can cause multi-stakeholder conflicts, ensure mutual stakeholder satisfaction, and improve dependability.

6 Conclusions

The accountability map organizes the events where multi-stakeholder conflicts can occur. It also identifies the risks that could result from exacerbating conflicts and supports risk response, thereby improving stakeholder satisfaction and system dependability. The accountability map is based on the concept of open systems dependability, as required by 62853, and can be used in the life cycle of various systems.

We recognize that there are two items that have not been discussed so far. The first is the verification of the effectiveness of the implementation procedure for accountability enforcement. At this stage, we have been able to demonstrate the effectiveness of the A-map by applying past examples, but we have not been able to demonstrate the effectiveness of the implementation procedure as a whole. We are planning to experiment with this issue by implementing the procedure for actual projects related to data intensive medical science applications.

The second is the development of a support tool to manage A-maps. 62853 requires that the identified information be recorded in a traceable manner. Based on the procedure for achieving accountability in this research result, we are developing a support tool to input and output A-maps using graphical and conversational UIs. This support tool consists of a database recording traceable information such as A-maps themselves, derived assurance cases, and their supplementary documents. Thus, stakeholders can achieve their accountability they need while referring to A-map and recording evidence of agreement implementation.

Acknowledgments. This paper is partially supported by the Innovation Platform for Society 5.0 of Japan's Ministry of Education, Culture, Sports, Science and Technology.

References

1. Obara, S.: Reality research to multi stakeholder conflicts issue in project business: reframing agenda by psychology, standards, organization based triad approach. J. Int. Assoc. P2M **6**(2), 79–97 (2012). https://doi.org/10.20702/iappmjour.6.2_79. (in Japanese)
2. ISO/IEC/IEEE 15288: 2015 Systems End Software Engineering - System Life Cycle Processes (2015)
3. IEC 62853: 2018 Open Systems Dependability (2018)
4. Yanagisawa, Y., Yokote, Y.: A new approach to better consensus building and agreement implementation for trustworthy AI systems, In: Fourth International Workshop on Artificial Intelligence Safety Engineering, York (to appear) (2021)
5. Hemmati, M., Dodds, F., Enayati, J., McHarry, J.: Multi-Stakeholder Processes for Governance and Sustainability, 1st edn., p. 19. Earthscan, Oxfordshire (2002)
6. Cabinet Office of Japan: https://www5.cao.go.jp/npc/sustainability/concept/index.html. Accessed 30 Apr 2021. (in Japanese)
7. Bovens, M.: Analysing and assessing accountability: a conceptual framework. Eur. Law J. **13**(4), 447–468 (2007)
8. Yamamoto, K.: Thinking About Accountability: How Did it Become Accountability?, 1st edn., p. 52. NTT Publishing, Tokyo (2013). (in Japanese)
9. Japan Times: Limited Testing Leaves COVID-19 App Glitches Overlooked. The Japan Times Ltd. (2021). https://www.japantimes.co.jp/news/2021/04/16/national/virus-app-glitches/. Accessed 16 Apr 2021
10. Ministry of Health, Labor and Welfare: Investigation into the Circumstances Surrounding the Occurrence of Problems with the Contact Confirmation Application "COCOA" and Consideration of Ways to Prevent Recurrence (2021). https://www.mhlw.go.jp/content/000769774.pdf. (in Japanese)
11. Gen, T.: COCOA Serious Bugs Left Unattended, MHLW's Sloppy Testing Practices. Nikkei Inc. (2021). https://www.nikkei.com/article/DGXZQOFK107S60Q1A210C2000000. Accessed 30 Apr 2021. (in Japanese)
12. Ministry of Health, Labour and Welfare: (COCOA) COVID-19 Contact-Confirming Application (2021). https://www.mhlw.go.jp/stf/seisakunitsuite/bunya/cocoa_00138.html. Accessed 30 Apr 2021. (in Japanese)

Towards Assurance-Driven Architectural Decomposition of Software Systems

Ramy Shahin(✉)

University of Toronto, Toronto, Canada
rshahin@cs.toronto.edu

Abstract. Computer systems are so complex, so they are usually designed and analyzed in terms of layers of abstraction. Complexity is still a challenge facing logical reasoning tools that are used to find software design flaws and implementation bugs. Abstraction is also a common technique for scaling those tools to more complex systems. However, the abstractions used in the design phase of systems are in many cases different from those used for assurance. In this paper we argue that different software quality assurance techniques operate on different aspects of software systems. To facilitate assurance, and for a smooth integration of assurance tools into the Software Development Lifecycle (SDLC), we present a 4-dimensional meta-architecture that separates computational, coordination, and stateful software artifacts early on in the design stage. We enumerate some of the design and assurance challenges that can be addressed by this meta-architecture, and demonstrate it on the high-level design of a simple file system.

Keywords: Software design · Assurance · Decomposition

1 Introduction

Computer systems are so complex, so they are usually designed and analyzed in terms of layers of abstraction. An operating system typically runs on top of the bare hardware. A run-time system possibly comes next, and then the sets of abstractions introduced to programmers by a programming languages. Each layer can restrict the power of the layer underneath, but it can not be more powerful. For example machine instruction sets allow for arbitrary jumps to instruction addresses, while many high-level programming languages do not allow that.

Abstractions allow software designers to manage the inherent complexity of systems. Complexity is still a challenge facing logical reasoning tools that are used to find software design flaws and implementation bugs. Scalability (or lack thereof) of many such tools (e.g., model checkers) make them only suitable

I. Habli et al. (Eds.): SAFECOMP 2021 Workshops, LNCS 12853, pp. 187–196, 2021.
https://doi.org/10.1007/978-3-030-83906-2_15

for relatively simple systems or toy examples. Abstraction, again, is a common technique for scaling those tools to more complex systems. However, abstractions used in system design are in many cases different from those used for assurance.

In this paper, we argue that the software abstractions used for assurance can also guide software designs, making them simpler, easier to understand, and readily suitable for automated reasoning. In particular, we present a *4-dimensional software meta-architecture* that separates the computational, coordination, stateful, and meta-programming aspects of a software system early on in the design process. We argue that this systemic separation allows different assurance tools and techniques to be applied to each of the meta-architecture dimensions.

The rest of this paper starts with outlining some of the challenges of quality assurance of monolithic systems (Sect. 2). The 4-dimensional meta-architecture is then presented in Sect. 3. This meta-architecture is then demonstrated on a simple file system design (Sect. 4). Related work is then briefly discussed in Sect. 5, and we finally conclude and outline some future directions in Setc. 6.

2 Challenges of Assurance of Monolithic Systems

In this paper, we refer to systems that do not separate computational, coordination, and state abstractions as *monolithic systems*. This section outlines some of the challenges primarily caused by this lack of separation.

2.1 Automated Logical Reasoning

Integrating automated reasoning tools into the Software Development Life Cycle (SDLC) can significantly improve the quality of software systems by finding design/implementation flaws early in the process. Due to the inherent complexity of software systems, many automated reasoning techniques and tools can efficiently operate only on *abstractions* rather than the concrete system artifacts. For example, model checkers operate on transition systems capturing abstract representations of system behavior. Theorem provers on the other hand are typically used to prove the correctness of abstract representations of algorithms.

Abstraction is a prevalent design technique allowing software designers to manage system complexity. However, design abstractions are in many cases different from the abstractions used in automated reasoning. For example, objects, classes, and interfaces are typical abstractions used in object-oriented design. On the other hand, when applying a software model checker to a program, program state is abstracted using bounded unrolling of loops. As a result, automated reasoning abstractions are *synthetic* in many cases. One serious consequence is abstractions becoming out of sync with concrete system artifacts, leaking flaws into the system even when they are proven not to exist on abstractions. Decomposing system artifacts into different categories suitable for different reasoning and/or assurance techniques early in the design process provides a synergy between design and reasoning abstractions.

2.2 Partial Correctness

The axiomatic verification literature carefully, and rightfully, distinguishes Total Correctness from Partial Correctness. Total Correctness is accomplished when given a guarantee of the correctness of a precondition, an implementation is proven to satisfy a given postcondition. Partial Correctness on the other hand describes the same ternary relationship between a precondition, a postcondition and an implementation only if an implementation terminates. Proving whether an implementation will always terminate is undecidable, so Partial Correctness is a weaker guarantee than Total Correctness.

Separating terminating computations from non-terminating coordination artifacts allows for designing modeling languages of computational algorithms that always terminate. In his seminal book The Art of Computer Programming, Knuth characterizes the notion of an algorithm in terms of five attributes, the first of which is finiteness. In his definition of algorithms, Knuth explicitly states that "an algorithm must always terminate after a finite set of steps" [22]. Making termination explicit when modeling computational algorithms thus does not limit algorithm designers.

2.3 Performance Assurance

Performance analysis has always been an important aspect of software engineering. Theoretical asymptotic analysis of algorithm time and space complexities [22] has been an established field of Computer Science, but its practical counterpart has not been fully materialized yet. We typically use profilers to measure the execution time or memory consumption of a specific implementation under a specific workload, but we cannot do that statically. Profilers are analogous to dynamic type checkers, potentially reporting problems at run-time rather than compile-time, and potentially missing problems that are not covered by the workload used for analysis.

Real-time systems in particular would benefit a lot from statically analyzing the performance of programs to make sure they meet their real-time constraints. Embedded systems with limited memory and processing resources would also benefit a lot from checking the performance characteristics of a system before being deployed. Power consumption has also been an important performance metric of programs running on handheld systems. Reusable library designers and users can leverage resource consumption contracts on modules to declaratively and soundly distinguish high-performance components from other components meeting the same functional requirements but consuming more processing time, memory and/or power.

But automatic performance analysis is theoretically impossible in general, simply because almost all performance analysis problems reduce to the halting problem, which is undecidable. Several heuristic algorithms have been proposed to address the problems of termination (e.g. [12]) and resource consumption. However, those are heuristics that can not be used in a logically sound and complete analysis.

Modeling algorithms using only terminating constructs would enable new performance analysis scenarios:

- Time Analysis: With algorithm models terminating by design, the compiler becomes responsible of making sure recursion is bounded and loops all have an upper bound on the number of iterations. Since the compiler needs either to infer those bounds or to require the model to make them explicit, then worst case asymptotic complexity can be directly calculated. Asymptotic time complexity can be also added to the contract of an algorithm. The implementation will be checked against that contract, and clients using that algorithm will be "taxed" that upper bound on time complexity when their individual complexities are calculated. Those contracts can then be used by performance analysis tools to find performance bottlenecks statically.
- Space Analysis: Similar to time complexity, space complexity can be also calculated directly. This will be an asymptotic analysis as well because algorithms usually abstract away platform-specific details for portability. Still, given those asymptotic bounds, lower-level compilation phases can come up with more accurate space analyses as they generate platform-specific code.
- Power Analysis: Power consumption is at least as important as time and space in handheld systems. Time complexity, individual instructions, using specific peripheral devices and network bandwidth are among the factors affecting power consumption. Since we can at least asymptotically quantify each of those factors, we can again statically calculate an asymptotic bound on power consumption given performance contracts.
- Bandwidth Analysis: Similar to Power, network bandwidth analysis can be performed based on time analyses, and the different levels of overhead added at different layers of a communication stack.

3 A 4-Dimensional Meta-architecture

Given the differences in the nature of different software artifacts, we propose the meta-architecture in Fig. 1. This is a meta-architecture in the sense that it is system-independent, and can be instantiated for different systems. This is also sometimes referred to as an *architectural pattern* [3]. The three orthogonal models we identify here are *computation*, *coordination*, and *state*. The coordination model interfaces with the others, accessing only constructs publicly exported. For example, a message handler in the coordination model might need to perform a computation. This can be achieved by calling a function exported from the computation model. Similarly, a transition from one object state to another might involve calling a computational function. In such case, the function is called by a process in the coordination model, and the result is used to atomically update the stateful object.

To allow for direct integration with program reasoning tools, the exported constructs of all three models form a meta-programming envelope. Those constructs need to export programmable interfaces for models to integrate with each other, and also to be used by tools (e.g., verification, test-case generation,

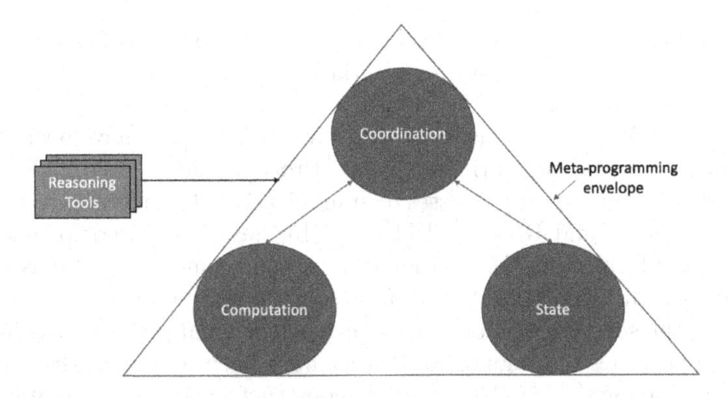

Fig. 1. A 4-dimensional meta-architecture (architectural pattern) decomposing architectural artifacts into four models: coordination, computation, state, and meta-programming artifacts.

performance analysis). The three models presented earlier, in addition to the meta-programming envelope, form a *4-dimensional meta-architecture* that can guide the design of software systems.

Table 1. Examples of aspects of the different models within the 4-dimensional meta-architecture.

	Computation	Coordination	State
Underlying Logic	Hoare logic, Constructive logics	Linear/Temporal logics	Description logics
Calculus	λ-calculus	Process calculi (e.g., π-calculus)	Relational calculus
Semantics	Denotational	Operational	Axiomatic
Type systems	System F and dependent types	Session types	Type-state
Verification	Contract-based	Model checking	Symbolic model checking

The main advantage of splitting the architecture into multiple models is liberating software designers to use the formalisms that best suite the responsibilities of each of the models, instead of having to stick to only one set of formalisms throughout the design of the whole system. Table 1 presents a taxonomy of formalisms, with examples of particular formalisms that might be more suitable for particular models than others.

Software reasoning tools are usually based on an underlying logic. For computation, Hoare logic [19] has been widely used to reason about sequential programs, and truth judgments are usually defined in constructive logics. Coordination on the other hand is more about causality and timing constraints. Temporal logics [29] and Linear logics [18] are capable of expressing those judgments.

Description logics [16] have been commonly used to formalize data representation, and thus might be best suited for stateful object models.

Similarly, different kinds of formal calculi are applicable to the different modeling languages. Variants of λ-calculus (e.g., λC, λP, λD [25]) have been designed with computation as their primary focus. Process calculi [4,20,23,24] are all about modeling concurrent processes, communication channels and interactions. Relational calculus and Relational algebra [13] have been used for decades as the underlying formalisms for relational data management in databases. Many relational concepts are applicable to in-memory stateful objects.

Defining the semantics of language constructs is what gives models meaning. Different approaches to language semantics have been used over the years [26]. Denotational semantics model language constructs using mathematical functions, so they are a natural fit for computational models. Operational semantics model the operational effects resulting from the evaluation of constructs. Coordination is effectful, and we naturally tend to think about coordinating systems operationally. Axiomatic semantics focuses on defining logical invariants that are to be maintained across evaluations. This is exactly what stateful objects and their integrity invariants and constraints are to be defined on top.

Typed languages usually integrate their type systems together with their formal calculi. System F of polymorphic types [28] and dependent types [25] associates types (with varying expressive powers) with expressions. Concurrent systems need a different sort of type systems (e.g., Session types [9]). Stateful objects are themselves treated as types in many language paradigms. Elaborate Typestate-based systems have been designed though to track the dynamic interfaces (types) of objects subject to logical object state [2,14,31].

There are several approaches to program verification, and the assurance of program properties in general. Again, different approaches would fit better than others to different models. Correctness of computations can be verified based contracts (pre-conditions and post-conditions). Model checking approaches are best-suited to verifying temporal properties of systems defined in process calculi [6,7,33]. With invariants being first-class constructs of stateful objects, symbolic model checking [8,11] can be used to verify that object state transitions do not violate object invariants.

4 Example: A Simple File System

To demonstrate the concepts presented in this paper, we use a simple file system as an example (Fig. 2). A file system can be thought of as a process communicating with other processes. In addition to the user-mode processes that need to read and write data from/to files, a file system also interacts with a storage system that manages a block-based storage device (e.g., disk or tape), and possibly with other operating system processes. Different processes communicate with each other through message passing. Processes and their communication channels are modeled as blue model elements in Fig. 2. Their dependencies on stateful objects and computations are modeled as dashed arrows. A process can

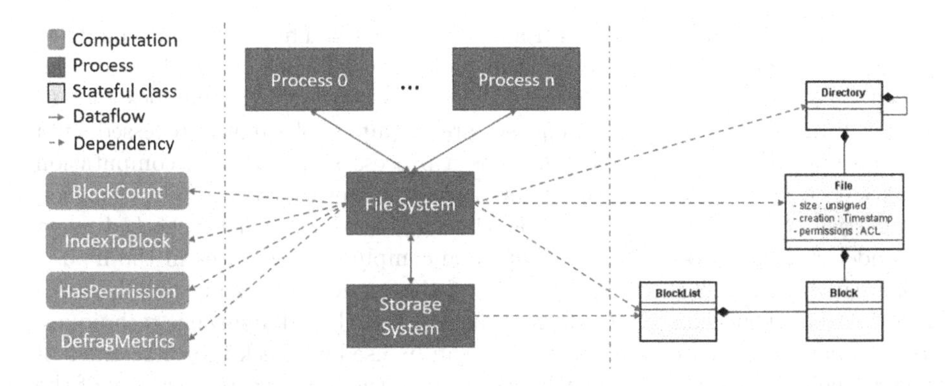

Fig. 2. A simple file system example. Diagram is split into three sub-models: stateful classes/objects (yellow), processes (blue), and computations (green). (Color figure online)

call a computational function, and can also query or update the state of an object. Computations and stateful objects do not directly interact though.

The stateful objects managed by the file system are modeled as a UML class diagram in Fig. 2 (yellow model elements). Stateful objects include files, directories, and a list of storage blocks that can be allocated to different files. A file exclusively contains the set of blocks where its contents are stored. A directory contains a set of files and possibly sub-directories. Stateful objects have to preserve a set of invariants. For example, a block is either free, or belongs to strictly one file. Also the size of a file is a function of the number of blocks it contains. Structurally, the directory structure has to be acyclic, with each file/directory having at most one parent (i.e., directories form a tree).

In addition to communication between processes, and state transitions of stateful objects, a file system needs to compute the values that are to be passed across processes, or used to determine which state an object should transition to. Computations are modeled as green elements in Fig. 2. For example, when a file is created or appended, the number of blocks needed has to be computed using the *BlockCount* function. Files do not necessarily occupy contiguous blocks on a storage device, so given a per-file block table, accessing a particular byte within a file, the index of that byte needs to be translated into a pair of values: a block identifier, and an index within that block. Those are computed using the *IndexToBlock* function. The Access Control List (ACL) of a file encodes access permissions to that file, and determining whether a user, a group or a process has access to that file typically involves a computation (*HasPermission*). Whenever possible, storing file contents in contiguous blocks improves access time due to locality patterns, especially for sequential file access. Defragmentation is a process where file contents are moved to unused blocks that are physically closer to other blocks used by the file. Deciding whether to defragment a volume is usually subject to several metrics that also need to be computed (*DefragMetrics*).

File systems, pretty much like other operating system components, are usually cited as examples of non-terminating software systems. A file system has to continuously respond to requests from user processes. This is typically modeled as an *event loop*, where a system waits indefinitely for an external event, and when that event arrives it is processed by the system, which then goes back to the waiting state. Read/write requests are examples of events processed by a file system. Modeling this event loop as a process rather than a computation makes it easier to assure the safety and correctness of the system. Coordination properties of processes (e.g., deadlock/livelock freedom) can be checked using a model checker without having to include computational states in the model. This can highly reduce the state-space of the model, improving scalability of existing model checkers. At the same time, contract-based assurance techniques, or axiomatic tools based on Hoare-logic, can be used to check the correctness of sequential computations without having to take the inherent concurrency of the system into consideration.

5 Related Work

Systematic decomposition of software systems into smaller units has been the driving force behind several software engineering paradigms. Seminal work by Parnas [27] suggests hiding each design decision in a separate module. Decomposing systems statically into functions [1], or objects [5], or deployment-time services [15], are among the most commonly used paradigms. Hybrid decomposition techniques have been also suggested [10]. In addition, cross-cutting concerns inspired multi-dimensional decomposition techniques [32], such as Aspect-Oriented Programming (AOP) [21], and Feature-Oriented Programming (FOP) [30].

The aforementioned techniques and paradigms base their decomposition decisions upon either problem domain abstractions, or encapsulation of design decisions. This paper on the other hand suggests an orthogonal dimension of decomposition, taking assurance techniques and their abstractions into consideration. This can be thought of as a generalization of multi-dimensional separation of concerns [32], adding an explicit assurability dimension.

Separation of computation and coordination aspects of software systems has been argued for by Gelernter back in the early 1990s [17]. In this paper we follow that argument, and it is one of the inspirations behind the 4-dimensional meta-architecture. It is unfortunate though that almost 30 years later, monolithic system architectures are still the norm.

6 Conclusion and Future Work

In this paper we argued that different software quality assurance techniques operate on different aspects of software systems. To facilitate assurance, and for a smooth integration of assurance tools into the Software Development Lifecycle (SDLC), we presented a 4-dimensional meta-architecture that separates computational, coordination, and stateful software artifacts early on in the design stage. We enumerated some of the challenges that can be addressed by this meta-architecture, and demonstrated it on a simple file system design.

For future work, we plan to study the adequacy of existing software modeling tools, and potentially provide tool support for the 4-dimensional meta-architecture. Integrating modeling with logical reasoning tools (e.g., model checkers, theorem provers, SMT solvers), and effectively combining the results computed by reasoning tools are two future research directions as well. Tooling support might involve the design of notations/languages suitable for the different aspects of the meta-architecture. Integration of results from different reasoning tools would involve proving that this integration preserves soundness.

Acknowledgments. The author thanks the anonymous reviewers for their feedback and insightful suggestions.

References

1. Abelson, H., Sussman, G.J.: Structure and Interpretation of Computer Programs, 2nd edn. MIT Press, Cambridge (1996)
2. Aldrich, J., Sunshine, J., Saini, D., Sparks, Z.: Typestate-oriented programming. In: Proceedings of the 24th ACM SIGPLAN Conference Companion on Object Oriented Programming Systems Languages and Applications. OOPSLA 2009, New York, NY, USA, pp. 1015–1022. Association for Computing Machinery (2009)
3. Bass, L., Clements, P., Kazman, R.: Software Architecture in Practice. Addison-Wesley Professional, 3rd edn. (2012)
4. Bergstra, J.A., Klop, J.W.: ACPτ a universal axiom system for process specification. In: Wirsing, M., Bergstra, J.A. (eds.) Algebraic Methods 1987. LNCS, vol. 394, pp. 445–463. Springer, Heidelberg (1989). https://doi.org/10.1007/BFb0015048
5. Booch, G.: Object-Oriented Analysis and Design with Applications, 3rd edn. Addison Wesley Longman Publishing Co., Inc., USA (2004)
6. Bradfield, J., Walukiewicz, I.: The mu-calculus and model checking. In: Handbook of Model Checking, pp. 871–919. Springer, Cham (2018). https://doi.org/10.1007/978-3-319-10575-8_26
7. Bunte, O., Groote, J.F., Keiren, J.J.A., Laveaux, M., Neele, T., de Vink, E.P., Wesselink, W., Wijs, A., Willemse, T.A.C.: The mCRL2 toolset for analysing concurrent systems. In: Vojnar, T., Zhang, L. (eds.) TACAS 2019. LNCS, vol. 11428, pp. 21–39. Springer, Cham (2019). https://doi.org/10.1007/978-3-030-17465-1_2
8. Burch, J., Clarke, E., McMillan, K., Dill, D., Hwang, L.: Symbolic model checking: 1020 States and beyond. Inf. Comput. **98**(2), 142–170 (1992)
9. Caires, L., Pfenning, F., Toninho, B.: Linear Logic Propositions as Session Types. Math. Struct. Comput. Sci. **760** (2014)
10. Chang, C., Cleland-Huang, J., Hua, S., Kuntzmann-Combelles, A.: Function-class decomposition: a hybrid software engineering method. Computer **34**(12), 87–93 (2001). https://doi.org/10.1109/2.970582
11. Clarke, E., McMillan, K., Campos, S., Hartonas-Garmhausen, V.: Symbolic model checking. In: Alur, R., Henzinger, T.A. (eds.) CAV 1996. LNCS, vol. 1102, pp. 419–422. Springer, Heidelberg (1996). https://doi.org/10.1007/3-540-61474-5_93
12. Cook, B., Podelski, A., Rybalchenko, A.: Proving program termination. Commun. ACM **54**(5), 88–98 (2011)

13. Date, C.J.: An Introduction to Database Systems, 8th edn. Pearson/Addison Wesley, Boston (2004)
14. DeLine, R., Fähndrich, M.: Typestates for objects. In: Odersky, M. (ed.) ECOOP 2004. LNCS, vol. 3086, pp. 465–490. Springer, Heidelberg (2004). https://doi.org/10.1007/978-3-540-24851-4_21
15. Erl, T.: Service-Oriented Architecture: Concepts, Technology, and Design. Prentice Hall PTR, USA (2005)
16. Baader, F., Calvanese, D., McGuinness, D.L., Nardi, D., Patel-Schneider, P.F.: The Description Logic Handbook: Theory, Implementation, Applications. Cambridge University Press, Cambridge (2003)
17. Gelernter, D., Carriero, N.: Coordination languages and their significance. Commun. ACM **35**(2), 97–107 (1992)
18. Girard, J.Y.: Linear logic. Theoret. Comput. Sci. **50**(1), 1–101 (1987)
19. Hoare, C.A.R.: An axiomatic basis for computer programming. Commun. ACM **12**(10), 576–580 (1969)
20. Hoare, C.A.R.: Communicating sequential processes. Commun. ACM **21**(8), 666–677 (1978)
21. Bergmans, L., Lopes, C.V.: Aspect-oriented programming. In: Moreira, A. (ed.) ECOOP 1999. LNCS, vol. 1743, pp. 288–313. Springer, Heidelberg (1999). https://doi.org/10.1007/3-540-46589-8_17
22. Knuth, D.: The Art of Computer Programming, 3rd edn. Addison Wesley, Reading (1997)
23. Milner, R. (ed.): Some proofs about data structures. In: A Calculus of Communicating Systems. LNCS, vol. 92, pp. 111–125. Springer, Heidelberg (1980). https://doi.org/10.1007/3-540-10235-3_9
24. Milner, R.: Communicating and Mobile Systems: The Pi-Calculus, 1st edn. Cambridge University Press, Cambridge (1999)
25. Nederpelt, R., Geuvers, H.: Type Theory and Formal Proof: An Introduction. Cambridge University Press, Cambridge (2014)
26. Nielson, H.R., Nielson, F.: Semantics with Applications: An Appetizer. Springer, London (2007). https://doi.org/10.1007/978-1-84628-692-6
27. Parnas, D.L.: On the criteria to be used in decomposing systems into modules. Commun. ACM **15**(12), 1053–1058 (1972)
28. Pierce, B.C.: Types and Programming Languages, 1st edn. The MIT Press, Cambridge (2002)
29. Pnueli, A.: The temporal logic of programs. In: Proceedings of the 18th Annual Symposium on Foundations of Computer Science. SFCS 1977, pp. 46–57 IEEE Computer Society, USA (1977)
30. Prehofer, C.: Feature-oriented programming: a fresh look at objects. In: Akşit, M., Matsuoka, S. (eds.) ECOOP 1997. LNCS, vol. 1241, pp. 419–443. Springer, Heidelberg (1997). https://doi.org/10.1007/BFb0053389
31. Strom, R.E., Yemini, S.: Typestate: a programming language concept for enhancing software reliability. IEEE Trans. Softw. Eng. **12**(1), 157–171 (1986)
32. Tarr, P., Ossher, H., Harrison, W., Sutton, S.M.: N degrees of separation: multi-dimensional separation of concerns. In: Proceedings of the 21st International Conference on Software Engineering. ICSE 1999, New York, NY, USA, pp. 107–119. Association for Computing Machinery (1999)
33. Tiu, A.: Model checking for π-calculus using proof search. In: Abadi, M., de Alfaro, L. (eds.) CONCUR 2005. LNCS, vol. 3653, pp. 36–50. Springer, Heidelberg (2005). https://doi.org/10.1007/11539452_7

2nd International Workshop on Underpinnings for Safe Distributed Artificial Intelligence (USDAI 2021)

2nd International Workshop on Underpinnings for Safe Distributed AI (USDAI 2021)

Morten Larsen[1]

[1] AnyWi Technologies, Leiden, The Netherlands
morten.larsen@anywi.com

1 Introduction

Safe distributed AI requires reliable and secure underpinnings (enabling technologies and legal and regulatory frameworks). These will be horizontal efforts, to be deployed in many and different application areas.

Europe needs to develop its own capabilities in this area ("European digital sovereignty"), and a clear road to wide-scale deployment in value-added applications will be needed as well as a structure to ensure that the enabling technologies developed have a lasting impact to protect the significant investments needed.

The basic challenges to achieve safe distributed AI therefore include data collection, local processing, and reliable transport, as well as the orchestration of distributed algorithms, all in a reliable and secure manner and in a way that respects the privacy of users, operators, and the general public.

This workshop addressed a wide range of enabling methods and technologies to ensure trustworthiness of the data as well as the processing and use of these data. Topics ranged from advanced computational methods to the legal and regulatory framework in which they must function.

The aim was to unite academic research (oriented towards applications in real-world settings) with industrial research and development in order to explore options for application-oriented uptake of new technologies in the field of safe distributed AI.

2 This Year's Workshop

USDAI 2021 took the form of a panel debate, discussions, and insights in to an interesting and relevant set of topics related to safe distributed AI for vehicles of various sorts (unmanned aerial systems and road vehicles) as well as other industrial applications, such as industrial IoT.

One of the common underpinnings for safe distributed AI, with a special view to this year's theme, automated vehicles, is the need for reliable and predictable data communication. A discussion of how quality of communication services can be incorporated into vehicle gateways to support more reliable applications, including distributed AI, was presented in the invited paper Integration of a RTT Prediction into a Multi-path Communication Gateway by Schmid et. al. and other, more architectural, approaches were discussed in the panel session.

3 Acknowledgements

Part of the work presented in the workshop received funding from the ECSEL Joint Undertaking (PRYSTINE, grant agreement n° 783190). The JU receives support from the European Union's Horizon 2020 research and innovation programme and Germany, the Netherlands, Austria, Romania, France, Sweden, Cyprus, Greece, Lithuania, Portugal, Italy, Finland, and Turkey.

International Program Committee

Morten Larsen	AnyWi Technologies, The Netherlands
Alan Sears	Leiden University, The Netherlands
Anna Hristokova	SIRRIS, Belgium
Reda Nouacer	CEA, France
Ricardo Reis	Embraer, Belgium
Andries Stam	Almende, The Netherlands
Raúl Santos de la Cámara	Hi-Iberia, Spain
Raj Thilak Rajan	TU Delft, The Netherlands
Tobias Koch	consider-it, Germany
George Dimitrakopoulos	Harokopio University, Greece

Integration of a RTT Prediction into a Multi-path Communication Gateway

Josef Schmid[1], Patrick Purucker[1(✉)], Mathias Schneider[1], Rick van der Zwet[2], Morten Larsen[2], and Alfred Höß[1]

[1] OTH Amberg-Weiden, Kaiser-Willhelm-Ring 23, 92224 Amberg, Germany
{j.schmid,p.purucker,mat.schneider,a.hoess}@oth-aw.de

[2] AnyWi Technologies, 3e Binnenvestgracht 23H, 2312, NR Leiden, The Netherlands
{rick.vanderzwet,morten.larsen}@anywi.com

https://www.oth-aw.de/, https://www.anywi.com/

Abstract. Reliable communication between the vehicle and its environment is an important aspect, to enable automated driving functions that include data from outside the vehicle. One way to achieve this is presented in this paper, a pipeline, that represents the entire process from data acquisition up to model inference in production. In this paper, a pipeline is developed to conduct a round-trip time prediction for TCP in the 4th generation of mobile network, called LTE. The pipeline includes data preparation, feature selection, model training and evaluation, and deployment of the model. In addition to the technical backgrounds of the design of the required steps for the deployment of a model on a target platform within the vehicle, a concrete implementation how such a model enables more reliable scheduling between multiple communication paths is demonstrated. Finally, the work outlines how such a feature can be applied beyond the field of automated vehicles, e.g. to the domain of unmanned aerial vehicles.

Keywords: RTT prediction · Mobile network · Automated driving

1 Introduction

In the research field of automated driving there is a rising demand for a reliable communication link between the vehicle and its environment to prevent accidents and protect the driver. The recent introduction of ad-hoc networks (VANETs) and vehicle-to-everything communication network topologies enables vehicles to exchange information with each other or the infrastructure, pedestrians, road signs and any other communication device related to road traffic [4,22]. In this context, there are aspirations to utilize the mobile network technologies of the Long Term Evolution (LTE) and 5th Generation (5G) network [6].

I. Habli et al. (Eds.): SAFECOMP 2021 Workshops, LNCS 12853, pp. 201–212, 2021.
https://doi.org/10.1007/978-3-030-83906-2_16

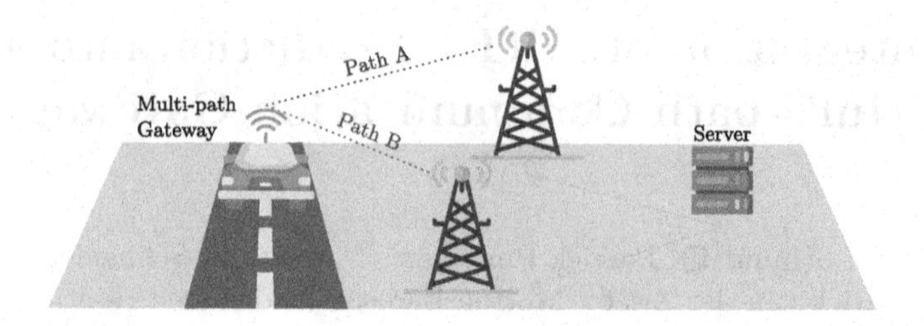

Fig. 1. Illustration of the multi-path gateway to server communication over two different mobile network paths.

Within the scope of V2X communication, this work introduces a multi-path communication gateway, which uses multiple LTE connections in parallel to provide a reliable vehicle to server communication. An illustration of this setup is shown in Fig. 1. Thereby, a new approach is presented, where a Round-trip Time (RTT) prediction pipeline is integrated into the gateway to anticipate RTT values in order to prioritize a communication path to increase reliability in terms of throughput and latency. The design of the pipeline is aimed at allowing the prediction model to be trained offline on new data without changing the interface to the gateway. Moreover, various prediction models based on Machine Learning (ML) can be trained in the pipeline and subsequently transferred to the gateway replacing the previous version as part of an over-the-air update.

This paper is structured as follows. In Sect. 2, related work on RTT prediction and ML based approaches to improve the quality of wireless networks is compared. Subsequently, the developed RTT prediction pipeline is presented in Sect. 3 and an overview of the multi-path communication gateway is given in Sect. 4. Section 5 explains the integration of the pipeline into the gateway. Finally, the paper is summarized and possible future work is outlined in Sect. 6.

2 Related Work

RTT prediction is an essential feature of several congestion avoidance algorithm for Transmission Control Protocol (TCP). Thereby, research and implementations in the last decades was influenced by the respective application domain in which the prediction was conducted. In the early stages, wired networks were investigated, leading to algorithms such as the Jacobson estimator [9] and the Eifel Retransmission Timer technique [11] which both employ formulas incorporating a smoothed RTT estimation. Following this direction, the applicability of time-series approaches are considered, such as recursive weighted median (RWM) [12] and autoregressive-moving-average (ARMA) [19]. Another formula-based direction for estimating the RTT utilizes probabilistic modeling such as Cauchy Round Trip Time Predictor (CRTTP) [15]. In latter, Rizo-Dominguez et al. compare the prediction error of their approach to Jacobson's algorithm and RWM, which respectively reduced the root mean squared error (RMSE) by up to

85% and 70%. Besides, they further elaborate the non-functional requirement of the processing time criteria of such implementations.

Whereas the presented approaches focus on one-dimensional, historical input for their estimation, multivariate features are more commonly incorporated by learning-based approaches. For instance, ML methods such as Support Vector Machines (SVMs), utilizing encoded IP addresses as features, reach a performance of an error less than 30% for 75% of the samples in a diverse real-world test set [3]. Furthermore, Recurrent Neural Networks (RNNs) are considered achieving between 90% to 96% accuracy with an error less than 10% depending on the prediction horizon for measurements obtained between a specific network link [2]. Finally, *Guyot*, a hybrid approach combining learning- and formula-based strategies, demonstrates that multiple strategies can be merged successfully and result in an absolute prediction error of less than 10 ms for approximately 90% of the test samples.

With the emergence of ubiquitous mobile network technologies, and their application in safety-critical domains such as automated driving, research on RTT prediction regained its importance in recent years. Due to the mobility of the sender, these networks inhere more dynamic behavior caused by spatial as well as temporal causes. Accordingly, novel TCP implementations, designed for mobile applications such as Cellular Controlled Delay TCP (C2TCP), integrate the relevance of the RTT in their congestion algorithm [1]. In particular, C2TCP uses a sliding window to determine the minimal RTT over a given interval and calculates a link condition state. Based on this state, actions are enforced to adjust the congestion window. Congestion algorithm for mobile applications benefit from RTT information, and anticipated RTT might even further improve their performance. Consequently, research fellows model this more wireless domain by revisiting previous strategies in recent work. To compensate the aforementioned complexity, more spatio-temporal features are included in these approaches, as presented by Khatouni et al., who incorporate other low-level Quality of Service (QoS) parameter e.g. Reference Signal Received Power (RSRP) and Reference Signal Received Quality (RSRQ) [10]. After an extensive feature selection, they train different ML classifiers, such as SVMs, Logistic Regression (LR), and Decision Tree (DT), using data from the Measuring Mobile Broadband Networks in Europe (MONROE) open measurement infrastructure to predict different RTT quality classes. According to their evaluation, DT outperforms the other two models. Another ML technique for RTT regression proposed by Nunes et al. leverages the Fixed-Share Experts algorithm offering online learning capabilities. Their simulation results show significant improvement in accuracy of up to 51% in comparison to the Jacobson estimator.

3 RTT Prediction Pipeline

So in order to predict network quality parameters, like RTT and to compare different machine learning algorithms, this work introduces a process including all necessary steps in a so-called pipeline. Figure 2 presents an overview of the steps as well as an indication of the placement. Whereby some tasks are computed directly on the target platform, possibly an embedded connectivity gateway, other computationally expensive parts of the pipeline are processed offline on a workstation. The pipeline starts with the recording of the relevant features, followed by the offline processing to train a prediction model with a high accuracy. This model is the converted to fit the requirements of the target platform, where it is executed during runtime. Each step is described in more detail in the following sections.

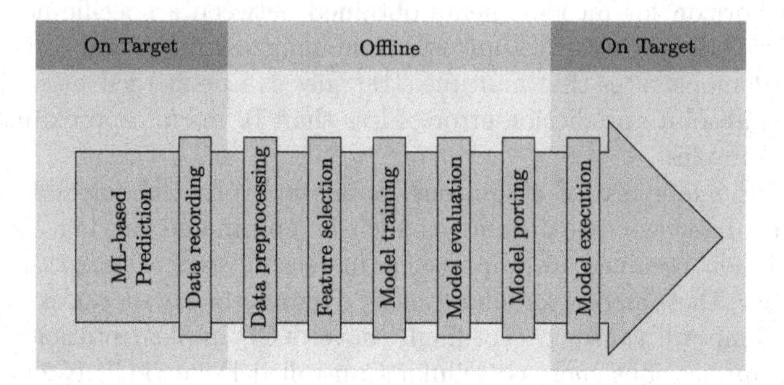

Fig. 2. Single steps and their placement of the processing pipeline used to predict the RTT.

3.1 Data Recording

Algorithm optimalisation and feature detection benefits from having as much sensor data available as possible. However, embedded devices have constraints with regards to memory, disk-space, and CPU processing power. When gathering data from remote devices an extra constraint of costs of data-transfer is included, both financially and in the form of performance overhead. Next sensor data characteristics vary heavily in input type as in frequency of occurrence. For example, location data update multiple times per second, cell tower identification is semi-static and details like firmware version of radio module only change sporadically. All data have one property in common, they all change relative to time and are chaotic in nature. Traditionally, a fixed data format is created and optimized to limit overhead of data per sample from single data sources. However, logging all sensors' output will often create an excess of data per sample, but also along the timeline, where non-changing values get transmitted at a constant rate. For this purpose, a protocol is created that allows compression of data along a time series, internally named Line TRANSmission (LTRANS). The protocol allows for dynamic recording of sensor values with variable frequency and dynamic configuration. This allows selective recording of any sensor value, by only recording the update of a sensor value, relative to the time.

3.2 Data Preprocessing

Since the prediction model is trained offline, the data recorded by a prototype of the target device needs to be preprocessed first. This is necessary to fulfill the requirements of the machine learning framework. Hereby, the highly compressed LTRANS format is converted into a more tangible CSV file format. Afterwards, invalid sample recorded during the system startup are dropped. Prediction value are shifted according to the forecasting step-size and the split of the set to training and validation data concludes this step. This processing stage results in two data sets, both containing only valid and prepared samples for (un-)supervised learning methods.

3.3 Feature Selection

In order to achieve better prediction results, a feature selection is performed on the training data set. In general, feature selection methods are employed to select significant features. This reduction of the input space dimension further lowers calculation effort and complexity of the resulting model. The goal is to drop negligible features to acquire a model that is faster to compute, with little or no degradation in its accuracy [23].

As a first step in feature selection, similar as presented in [17], the variance threshold filters out features in the measurements that have a variance of less than 1%. It primarily eliminates features that hardly change in the data set. Even though the computational effort for this step is moderate, it reliably detects redundant features and supports to avoid overfitting of the AI model.

As a further method, feature correlation is carried out. By calculating the Pearson correlation coefficient for each numerical feature pair, features with a strong linear association are determined. An example for such a pair of features in the mobile network is the Arbitrary Strength Unit (ASU) and the RSRP. Since the ASU is defined as $ASU = RSRP + 140$ for LTE, there is a linear relationship between the two characteristics, therefore it is sufficient to consider only one of these features in the following.

For feature selection by means of more complex methods, the number of features required is first determined. For this purpose, a Principal Component Analysis (PCA) is performed. This unsupervised procedure, generally normalizes the data to its mean, while preserving a maximum of the original variance. In the process, the number of features is decreased several times and the summarized variance is calculated for each step. Now, to determine the number of features, a certain variance is defined, which must be given. As a result, this is used to estimate how many features are approximately needed for the final selection.

In order to select these features, a ranking procedure has to be introduced. Therefore, several methods can be used. In this work, we are investigating three of them. As a first and straightforward method, the cross-correlation criteria is examined. It calculates the correlation between a feature and output metric throughout the data set. It expresses the quality of linear fit of each particular variable, and is frequently used for this analysis [20].

In addition to the correlation criterion, further methods are used. One of them is the Recursive Feature Elimination (RFE) [8]. It allows to identify suitable features by recursive training of linear machine learning models. Thereby, it removes either the best or the worst features for the next training iteration, until a predefined limit of features is reached. Then, all features are sorted by the inference performance criteria that was reached before their respective elimination. Since RFE can be performed with an arbitrary machine learning model, both, Support Vector Regression (SVR) as well as Decision Tree Regression (DTR), are examined in this work. The RFE can be extended using cross-validation to provide more stability. However, due to constraints regrading to resulting computational efforts, this is not done in this work.

The final feature selection incorporates the results of each of the three introduced procedures using majority vote. The corresponding feature set, which contains LTE low level parameter of the current cell as well as of the neighboring cells, is then applied to the training and validation data set.

3.4 Prediction Module Training

In order to train an accurate RTT predictor for the round trip time prediction of the multi-path communication gateway, different machine learning algorithms are evaluated using the input features selected in Sect. 3.3. The algorithms are described in the following paragraphs in more detail.

The first of them is SVR. It is based on the well known SVM and constructs a hyperplane in a high-dimensional space, which can be leveraged for regression purposes. One important property of the SVR is the use of different kernels. These functions map lower dimensional data into higher dimensional data which can significantly improve the prediction accuracy. In this work, the radial basis function as well as the sigmoid function are investigated. A more detail description of SVRs is given in [18]. A variation of this traditional model is implemented by the so-called Nu-SVR [16]. In this algorithm, a parameter is introduced to control the number of support vectors, which gives the possibility to eliminate the accuracy parameter taken in conventional SVR models. In this work, also this type of SVRs is studied.

Another method is taken into account, the k-Nearest Neighbors Regression. It predicts the target by interpolation of the nearest neighbors in the training set. Therefore, depending on the size of the training set, the models produced with this method may be substantially larger than the other models.

However, there is also a whole group of DT-based machine learning methods. The simplest of them is the DTR which is a non-parametric supervised learning method. It creates a set of rules from the input features to predict the target value. Since these rules are expressed as conditional statements, it is straightforward to implement such a model in an arbitrary programming language. In addition, there are a lot of ensemble methods using decision trees. One of them are Extra-Trees. In contrast to other tree-based ensembles, Extra-Trees use all the learning samples and choose the cut-points for their nodes completely randomly [7]. Therefore, they are also worth considering in this work.

3.5 Model Evaluation

Trained models are evaluated using two data sets. The first one, the test data set is generated during a train-test split inside the model training process. The second, known as validation data set, is containing different data, not used in the process up to now. This results are presented in Fig. 3. In this scatter plot, the abscissa indicates the mean absolute error of the model with regards to the test data set evaluation, while the ordinate specifies the error obtained for the validation data set. The following two criteria must be considered to choose a suitable model. On the one hand, the model error has to be as close as possible to the origin of the plot. On the other hand, the relationship between the two error values must be considered. With a method that generalities well, the point is close to the bisector resulting in similar errors for test and validation data set. Higher distance of an evaluation result to this bisector indicates under- or overfitting. So a model with an overall low error value and a symmetric performance should be selected for deployment.

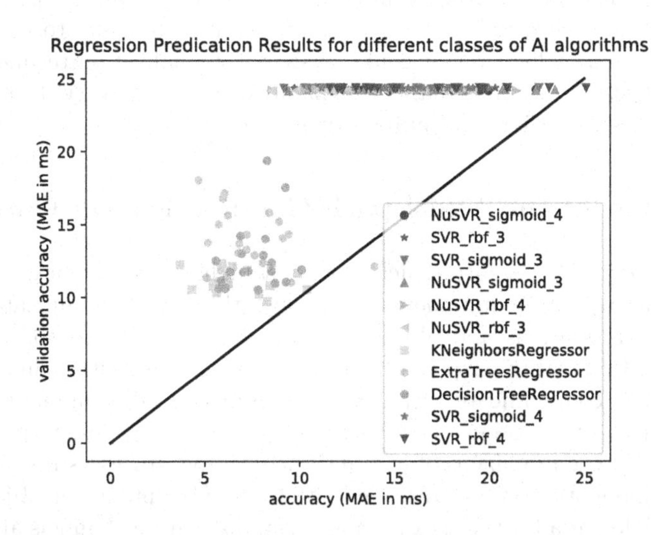

Fig. 3. Evaluation of the different models trained for RTT prediction.

In addition, Fig. 3 also shows the performance on multiple model using the same algorithm and parameters. This is presented, since using the build of machine learning models randomness is used, so each model is performing slightly different. Overall, the figure indicated that k-Nearest Neighbors Regression and DT-based machine learning methods are performing best. But since DT have a far smaller memory and computing footprint, this model is taken into account for porting them.

3.6 Model Porting

After the selection of a model, it needs to be ported to a target device. Since this can be either an embedded device or at least limited processing capabilities, the algorithm implementation should be efficient introducing as less overhead as possible. Therefore, a program called `sklearn-porter` [14] is used to convert the model into plain C/C++ code without incorporating additional external library dependencies. This allows the execution of the model on nearly any type of device.

3.7 Model Execution

But this conversion has also the disadvantage, that the model can not be executed as a stand-alone process. Therefore, a interface compliant with the `OpenAPI 3.0` specification is implemented and served via a web server. This architecture provides a more flexible use of the prediction for multiple clients, and allow deployment in distributed manner in cases where the computing power is too limited for the actual target device. In addition, it is possible to encapsulate this model by using a lightweight container-based virtualization technology like `Docker` [13], fostering the installability capability on an arbitrary device for the use in different applications like automotive or aviation.

4 Overview of the Multi-path Communication Gateway

Intelligent objects (e.g. smart-vehicles) with a higher level of autonomy have higher communication requirements. The objectives of communication links between backend stations and object are three-fold; a) send sensor values for monitoring and reporting purposes, b) receive updates to route plans and control input and c) communicate with external systems for decision making within the high-level control systems such as ADAS for road vehicles. Depending on the properties of the intelligent object and the decision processes involved, data transfer requirements vary per task and situation. The intelligent object could divert, halt, delay or alter the task if known communication channels are limited or tainted. For example, an autonomous vehicle on the highway could postpone or speed-up the overtake operation if upcoming traffic information from remote sensors could not be retrieved. This potential degredation in operation however could also disrupt decision making processing which causes the intelligent object to revert to a safe state, for example a lower level of autonomous operation or (e.g. when regalatory requirements cannot be met) an emergency stop procedure. Circumventing this problems requires knowledge of future communication profiles and mitigations strategies in communication channels to deal with potential issues. On LTE networks data-quality, transfer-time, throughput and (future) availability are not given by operators. Radio connections are inherently unpredictable due to external factors, like (deliberate) interference, thus guarantees cannot be given. This limitation imposes issues when requirements are needed to operate the intelligent object at levels which cannot be met with a single LTE data-connection and can be mitigated by combining multiple-access technologies such that higher availability and quality levels could be possible. The multi-path gateway makes this possible.

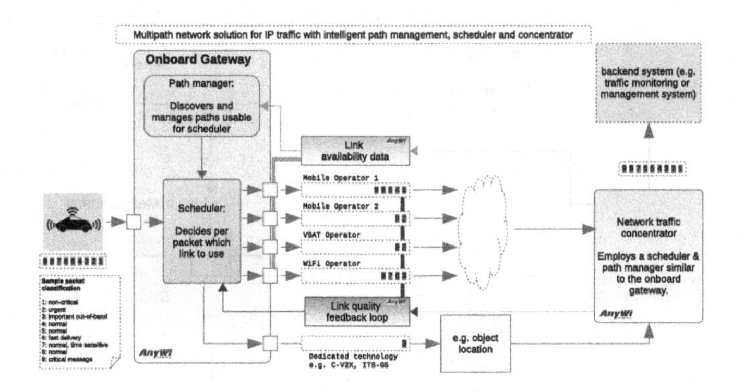

Fig. 4. Multipath architecture of gateway and related concentrator.

As depicted in Fig. 4, data-packets first get injected into the multi-path gateway, the scheduler will determine priority and importance of the data-packet and send it onwards over (multiple) communication channels. These data-packets get re-assembled at the receiving side and forwarded towards to the endpoint. The same principle applies at the concentrator gateway when sending data towards the gateway. The second-tier operation of the gateway involves discovery of new potential communication channels. Data connections are not guaranteed to exist in place and time; thus, monitoring is required to gather data on the current state of the communication channel and report finding towards the scheduler. The scheduler makes decisions based on the current state of the communication channel, however this yields no guarantees towards to transmission of the actual data-packet over the communication channel, since this will take place in the future. Prediction will result in a higher confidence of the possibility data is transferred and received over the correct channel in time.

5 Integration of the Pipeline into the Gateway

The prediction algorithms will be called as part of the evaluation cycle of the data connections within the monitoring process as shown in Fig. 5. Its results are fed into the central (memory-based) storage. The scheduler will use results at every packet it transmits. Since multiple algorithms can be selected, return values are standardized, allowing algorithms to be swapped in and out for better alternatives. However due to limited capacity of embedded systems, no dynamic learning is used. Depending on capacity multiple different algorithms could be executed and presented to the scheduler. The environment is chaotic with regards to sensor input, to create reproducible results parameters and weighting factors on algorithms are fixed. The central storage and monitoring data is handled using the above-mentioned LTRANS format for time-based compression of data streams, allowing the offline learning processes to learn and improve the algorithms.

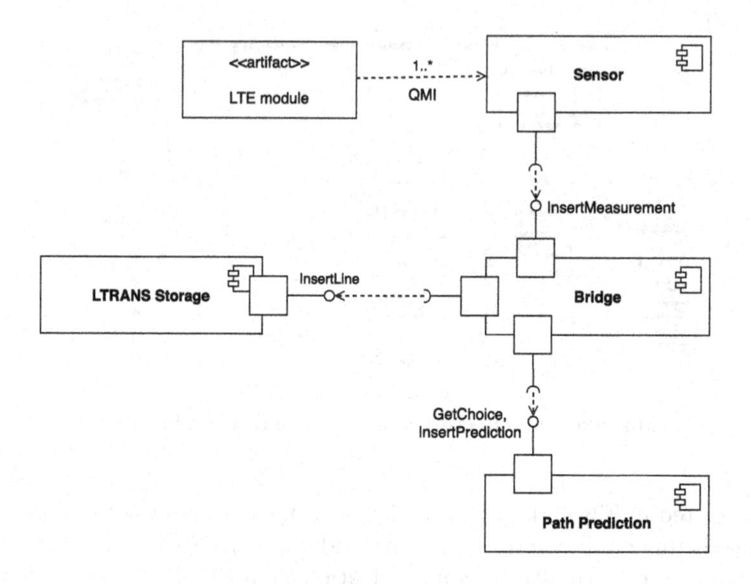

Fig. 5. Data flow in tranport layer on gateway.

6 Summary and Outlook on Future Work

With this paper, a RTT prediction pipeline generating a suitable prediction model for a multi-path gateway based on measured location and QoS data was presented. The pipeline contains various steps, such as preprocessing, feature selection, model training, evaluating and porting, which are executed offline. Different machine learning models, such as SVR, NuSVR, k-Nearest Neighbors Regression and different variants of DTRs, can be selected and evaluated to obtain the best prediction model. Once a prediction model has been created, it is ported to be executed on the embedded device and fed with data from the multi-path gateway via a defined interface. The scheduler of the gateway uses then the RTT prediction at every packet it transmits.

On implementation level, some additional improvements can be conducted in future work. The return-path from the concentrator is currently determined by estimation of the best possible data channel based on the arrival of packets. However, predictions on the gateway are not yet taken into consideration by the concentrator. A synchronized shared knowledge of the communication channels and its predictions between the gateway and concentrator should be modelled and implemented. Data-processing and predictions is currently done within a single environment (gateway), however time-specific results also have value to gateways nearby (e.g., which will soonish arrive in the area). Sharing this information requires normalisation of environments (hardware differences for example) to make algorithms predictions which are valid for a single gateway, relevant to other gateways. Exposing the scheduler predictions to the decision process would improve decision making. This requires a standardized way of modeling data channel availability and quality predictions.

Besides these aforementioned considerations, the feasibility of a multi-path gateway with a QoS predictor for the Unmanned Aerial Vehicles (UAVs) domain will be further examined in the future. Due to the use of UAVs in logistics, transport or for search and rescue missions [5], reliable communication in Beyond Visual Line Of Sight (BVLOS) ranges is necessary. Because of its capabilities such as wide are coverage and secure data transmission, the mobile network is considered in this use-case as well [21]. Moreover, in addition to the LTE network, the 5G network will also be utilized. In contrast to cars, UAVs can also move at different heights, which creates a third degree of freedom and thus may be more challenging in terms of data acquisition, storage and training of the prediction model. Furthermore, QoS based trajectory calculation will be investigated, whereby a route with diversions may be calculated to ensure sufficient QoS.

Acknowledgments. The authors gratefully acknowledge the following European Union H2020 – ECSEL Joint Undertaking project for financial support including funding by the German Federal Ministry for Education and Research (BMBF): PRYSTINE (Grant agreement No. 783190, funding code 16ESE0330) and ADACORSA (Grant Agreement No. 876019, funding code 16MEE0039).

References

1. Abbasloo, S., Xu, Y., Chao, H.J.: C2TCP: a flexible cellular TCP to meet stringent delay requirements. IEEE J. Sel. Areas Commun. **37**(4), 918–932 (2019)
2. Belhaj, S., Tagina, M.: Modeling and prediction of the internet end-to-end delay using recurrent neural networks. J. Networks **4**(6), 528–535 (2009)
3. Beverly, R., Sollins, K., Berger, A.: SVM learning of IP address structure for latency prediction. In: Proceedings of the 2006 SIGCOMM workshop on Mining Network Data. MineNet 2006, Pisa, Italy, pp. 299–304. Association for Computing Machinery, September 2006
4. Bhoi, S.K., Khilar, P.M.: Vehicular communication: a survey. IET Networks **3**(3), 204–217 (2014)
5. Christen, M., Guillaume, M., Jablonowski, M., Moll, K.: Zivile Drohnen-Herausforderungen und Perspektiven. No. 66/2018 in TA-SWISS, vdf, Zürich (2018)
6. Garcia-Roger, D., González, E.E., Martín-Sacristán, D., Monserrat, J.F.: V2X Support in 3GPP Specifications: From 4G to 5G and Beyond. IEEE Access **8**, 190946–190963 (2020)
7. Géron, A.: Hands-On Machine Learning with Scikit-Learn and TensorFlow (2017)

8. Guyon, I., Weston, J., Barnhill, S., Vapnik, V.: Gene selection for cancer classification using support vector machines. Mach. Learn. **46**(1), 389–422 (2002)
9. Jacobson, V.: Congestion avoidance and control. Comput. Commun. Rev. 17 (1988)
10. Khatouni, A.S., Soro, F., Giordano, D.: A machine learning application for latency prediction in operational 4G newtorks. In: 2019 IFIP/IEEE Symposium on Integrated Network and Service Management (IM), pp. 71–74 (2019)
11. Ludwig, R., Sklower, K.: The Eifel retransmission timer. ACM SIGCOMM Comput. Commun. Rev. **30**(3), 17–27 (2000)
12. Ma, L., Arce, G., Barner, K.: TCP retransmission timeout algorithm using weighted medians. IEEE Signal Process. Lett. **11**(6), 569–572 (2004)
13. Merkel, D.: Docker: lightweight linux containers for consistent development and deployment. Linux J. **2014**(239), 2 (2014)
14. Morawiec, D.: sklearn-porter (2021). https://github.com/nok/sklearn-porter
15. Rizo-Dominguez, L., Munoz-Rodriguez, D., Vargas-Rosales, C., Torres-Roman, D., Ramirez-Pacheco, J.: RTT prediction in heavy tailed networks. IEEE Commun. Lett. **18**(4), 700–703 (2014)
16. Scholkopf, B., Smola, A.J., Williamson, R.C., Bartlett, P.L.: New support vector algorithms. Neural Comput. **12**(5), 1207–1245 (2000)
17. Schneider, M.: Online and microservice-based data throughput prediction framework in context of mobile-based vehicle-to-server communication for automated driving (2018)
18. Smola, A.J., Schölkopf, B.: A tutorial on support vector regression. Stat. Comput. **14**(3), 199–222 (2004)
19. Sulei, X., Liang, W.: Smoothly estimate the RTT of fast TCP by ARMA function model. In: 10th International Conference on Wireless Communications. Networking and Mobile Computing (WiCOM 2014), Beijing, China, pp. 333–339. Institution of Engineering and Technology (2014)
20. Weston, J., Elisseeff, A., Schölkopf, B., Tipping, M.: Use of the zero-norm with linear models and kernel methods. J. Mach. Learn. Res. **3**, 1439–1461 (2003)
21. Yang, G., et al.: A telecom perspective on the internet of drones: from LTE-advanced to **5G**, 8 (2018)
22. Zhang, S., Chen, J., Lyu, F., Cheng, N., Shi, W., Shen, X.: Vehicular communication networks in the automated driving era. IEEE Commun. Mag. **56**(9), 26–32 (2018)
23. Zheng, A., Casari, A.: Feature engineering for machine learning: principles and techniques for data scientists. O'Reilly Media, Inc. (2018)

4th International Workshop on Artificial Intelligence Safety Engineering (WAISE 2021)

4th International Workshop on Artificial Intelligence Safety Engineering (WAISE 2021)

Simos Gerasimou[1], Orlando Avila-García[2], Mauricio Castillo-Effen[3], Chih-Hong Cheng[4], and Zakaria Chihani[5]

[1] Department of Computer Science, University of York | Deramore Lane, York, YO10 5GH, UK
simos.gerasimou@york.ac.uk
[2] Arquimea Reserch Center, Spain
oavila@arquimearesearchcenter.com
[3] Lockheed Martin, USA
mauricio.castillo-effen@lmco.com
[4] DENSO, Germany
c.cheng@eu.denso.com
[5] CEA LIST, CEA Saclay Nano-INNOV | Point Courrier 174, 91191 Gif-sur-Yvette, France
zakaria.chihani@cea.fr

1 Introduction

Achieving the full potential of Artificial Intelligence (AI) entails guaranteeing expected levels of safety and resolving issues such as compliance with ethical standards and liability for accidents involving AI-enabled systems such as autonomous vehicles. Deploying AI-based systems for operation in proximity to and/or in collaboration with humans implies that current safety engineering and legal mechanisms need to be revisited and updated to ensure that individuals – and their properties – are not harmed and that the desired benefits outweigh the potential unintended consequences. Researchers, engineers, and policymakers from different areas of expertise will need to join forces in order to tackle this challenge.

The increasing interest in developing approaches to enhance AI safety cover not only practical- and engineering-focused aspects of autonomous systems and safety engineering but also pure theoretical concepts and topics, including ethical considerations and moral aspects. These two sides of AI safety cannot be considered in isolation. Instead, the engineering of safe AI-enabled autonomous systems mandates the combination of philosophy and theoretical science with applied science and engineering. Accordingly, a multidisciplinary approach is needed that encompasses these seemingly disparate viewpoints and contributes to the engineering of safe AI-enabled systems that are underpinned by ethical and strategic decision-making capabilities.

Increasing levels of AI in "smart" sensory-motor loops allow intelligent systems to perform in increasingly dynamic uncertain complex environments with increasing degrees of autonomy, with the human being progressively taken out from the control loop. Adaptation to the environment is enabled by machine learning (ML) methods rather than more traditional engineering approaches, such as system modelling and programming. The tremendous progress achieved by deep learning, reinforcement learning, and their combination, in challenging real-world tasks such as image classification, natural language processing, and speech recognition, raises the expectation for their seamless incorporation into safety-critical applications. However, the inscrutability or opaqueness of their statistical models for perception and decision making is a major challenge. Moreover, the combination of autonomy and inscrutability in these AI-based systems is particularly challenging in safety-critical applications, such as autonomous vehicles, personal care, or assistive robots and collaborative industrial robots.

The Fourth International Workshop on Artificial Intelligence Safety Engineering (WAISE) is dedicated to exploring new ideas on AI safety, ethically aligned design, regulations, and standards for AI-based systems. WAISE aims at bringing together experts, researchers, and practitioners from diverse communities, such as AI, safety engineering, ethics, standardization, certification, robotics, cyber-physical systems, safety-critical systems, and application domain communities such as automotive, healthcare, manufacturing, agriculture, aerospace, critical infrastructures, and retail. The fourth edition of WAISE was held on September 7, 2021, in York (UK) as part of the 40th International Conference on Computer Safety, Reliability, & Security (SAFECOMP 2021).

2 This Year's Workshop

The Program Committee (PC) received 18 submissions, in the following categories:

- Short position papers – 7 submissions
- Full research papers – 11 submissions

Each paper was peer-reviewed by at least three PC members, by following a single-blind reviewing process. The committee decided to accept 10 papers (4 position papers and 6 full research papers) for oral presentation (acceptance rate 55%).

The WAISE 2021 program was organized in three thematic sessions.

The thematic sessions followed a highly interactive format. They were structured into short paper presentations and a common panel slot to discuss both individual paper contributions and shared topic issues. Three specific roles were part of this format: session chairs, presenters and session discussants.

- Session Chairs introduced sessions and participants. The chair moderated the session, took care of the time, and gave the word to speakers in the audience during discussions.
- Presenters gave a paper pitch in 10 minutes and then participated in the discussion.
- Session discussants prepared the discussion of individual papers and the plenary debate. The discussant gave a critical review of the session papers.

The mixture of topics was carefully balanced, as follows:

Session 1: AI Safety

- Deep Neural Network Uncertainty Estimation with Stochastic Inputs for Robust Aerial Navigation Policies, Fabio Arnez, Huascar Espinoza, Ansgar Radermacher, and François Terrier
- No Free Lunch: Overcoming Reward Gaming in AI Safety Gridworlds, Mariya Tsvarkaleva and Louise Dennis
- Effect of Label Noise on Robustness of Deep Neural Network Object Detectors, Bishwo Adhikari, Jukka Peltomäki, Saeed Bakhshi Germi, Esa Rahtu, and Heikki Huttunen
- Human-in-the-Loop Learning Methods Toward Safe DL-based Autonomous Systems: A Review, Prajit T Rajendran, Huascar Espinoza, Agnes Delaborde, and Chokri Mraidha

Session 2: Automated Driving Safety

- An Integrated Approach to a Safety Argumentation for AI-based Perception Functions in Automated Driving, Michael Mock, Stephan Scholz, Frederik Blank, Fabian Hüger, Andreas Rohatschek, Loren Schwarz, and Thomas Stauner
- Experimental Conformance Evaluation on UBER ATG Safety Case Framework with ANSL/UL 4600, Kenji Taguchi and Fuyuki Ishikawa
- Learning From AV Safety: Hope and Humility Shape Policy and Progress, Marjory Blumenthal

Session 3: Assurances for Autonomous Systems

- Levels of Autonomy and Safety Assurance for AI-based Clinical Decision Systems, Paul Festor, Ibrahim Habil, Yan Jia, Anthony Gordon, A. Aldo Faisal, and Matthieu Komorowski
- Certification Game for the Operational Safety Analysis of AI-based CPS, Imane Lamrani, Ayan Banerjee, and Sandeep Gupta
- A New Approach to Better Consensus Building and Agreement Implementation for Trustworthy AI Systems, Yukiko Yanagisawa and Yasuhiko Yokote

3 Acknowledgements

As chairpersons of WAISE 2021, we want to thank all authors and contributors who submitted their work to the workshop and congratulate the authors whose papers were selected for inclusion in the program and proceedings. We would also like to thank Friedemann Bitsch, the SAFECOMP publication chair, Erwin Schoitsch, the general workshop co-chair, and the SAFECOMP organizers, who provided us with the opportunity to organize the WAISE workshop at SAFECOMP 2021.

We would like to thank the Steering Committee for their support and advice to make WAISE 2021 a successful event:

Rob Alexander	University of York, UK
Huascar Espinoza	CEA LIST, France

Philip Koopman	Carnegie Mellon University, USA
Stuart Russell	UC Berkeley, USA
Raja Chatila	ISIR - Sorbonne University, France

We especially thank our distinguished Program Committee members, for reviewing the submissions and providing useful feedback to the authors:

Rob Alexander	University of York, UK
Vincent Aravantinos	Autonomous Intelligent Driving GmbH, Germany
Rob Ashmore	Defence Science and Technology Laboratory, UK
Alec Banks	Defence Science and Technology Laboratory, UK
Markus Borg	RISE SICS, Sweden
Simon Burton	IKS Fraunhofer, DE
Raja Chatila	ISIR - Sorbonne University, France
Huascar Espinoza	CEA LIST, France
John Favaro	INTECS, Italy
Michael Fischer	University of Manchester, UK
Mario Gleirscher	University of York, UK
Jérémie Guiochet	LAAS-CNRS, France
Vahid Hashemi	Audi, Germany
José Hernández-Orallo	Universitat Politècnica de València, Spain
Nico Hochgeschwende	Bonn-Rhein-Sieg University, Germany
Bernhard Kaiser	ANSYS, Germany
Philip Koopman	Carnegie Mellon University, USA
Timo Latvala	Space Systems Finland, Finland
Chokri Mraidha	CEA LIST, France
Davy Pissoort	KU Leuven, Belgium
Mehrdad Saadatmand	RISE SICS, Sweden
Rick Salay	University of Waterloo, Canada
Hao Shen	Fortiss, Germany
Erwin Schoitsch	Austrian Institute of Technology, Austria
François Terrier	CEA LIST, France
Andreas Theodorou	Umea University, Sweden
Mario Trapp	Fraunhofer ESK, Germany
Ilse Verdiesen	TU Delft, The Netherlands

Improving Robustness of Deep Neural Networks for Aerial Navigation by Incorporating Input Uncertainty

Fabio Arnez(✉), Huascar Espinoza, Ansgar Radermacher, and François Terrier

Université Paris-Saclay, CEA, List, 91120 Palaiseau, France
{fabio.arnez,huascar.espinoza,ansgar.radermacher,francois.terrier}@cea.fr

Abstract. Uncertainty quantification methods are required in autonomous systems that include deep learning (DL) components to assess the confidence of their estimations. However, to successfully deploy DL components in safety-critical autonomous systems, they should also handle uncertainty at the input rather than only at the output of the DL components. Considering a probability distribution in the input enables the propagation of uncertainty through different components to provide a representative measure of the overall system uncertainty. In this position paper, we propose a method to account for uncertainty at the input of Bayesian Deep Learning control policies for Aerial Navigation. Our early experiments show that the proposed method improves the robustness of the navigation policy in Out-of-Distribution (OoD) scenarios.

Keywords: Uncertainty propagation · AI safety · Autonomous systems

1 Introduction

Autonomous navigation in complex environments still represents a big challenge for autonomous systems. Particular instances of this problem are autonomous driving for self-driving cars and autonomous aerial navigation in the context of Unmanned Aerial Vehicles (UAVs). In both cases, the navigation task is addressed by first acquiring rich and complex raw sensory information (e.g., from cameras, radars, LiDARs, etc.), which is then processed to drive the robot towards its goal. Usually, this process is done in a modular fashion, where specific software components are linked together in the so-called *perception-planning-control* software pipeline [15,18].

Modular pipeline components can be implemented using Deep Neural Networks (DNNs) or other non-learning methods [7]. DNNs have become a popular choice thanks to their effectiveness in processing complex sensory inputs, and their powerful representation learning, that surpass the performance of traditional methods. Navigation policies implemented in a modular fashion admit a higher degree of interpretability and ease of analysis since components are developed in isolation. However, these architectures suffer from the accumulation of

I. Habli et al. (Eds.): SAFECOMP 2021 Workshops, LNCS 12853, pp. 219–225, 2021.
https://doi.org/10.1007/978-3-030-83906-2_17

errors (coming from the contribution of each component's erroneous outputs), which later harms the overall system performance [15,17]. An alternative approach to modular pipelines is to map sensory inputs directly to control outputs using neural networks in an End-to-End (E2E) fashion [7]. This is an appealing paradigm where perception-planning-control blocks are trained jointly, often via imitation learning [3,4]. Unfortunately, E2E training requires vast amounts of data [15,17] limiting its use to constrained scenarios.

Recently, some works from [2,17,20] have explored the combination of modular and E2E learning approaches to get the benefits of both families. The goal of this hybrid approach is to learn an environment representation and a driving policy through dedicated DNNs. In this manner, an autonomous system can incorporate DL-based modules for perception and control. Learning intermediate (compact) representations enables the extraction of relevant environmental features that can be used later in downstream tasks (e.g. control policy training).

Despite substantial performance improvements introduced by DNNs, they still have significant shortcomings due to their opacity and specially their inability to represent confidence in their predictions. These downsides hinder the deployment of DL methods in safety-critical systems, where uncertainty estimates are highly relevant [1,14]. To overcome these limitations, the authors in [15] propose the use of Bayesian Deep Learning (BDL) for implementing components used either in modular, E2E, or hybrid fashions. Bayesian methods in DL offer a principled framework to model data (*aleatoric*) and model (*epistemic*) uncertainties to represent the confidence in the outputs.

Nevertheless, system components that use BDL methods should be able to admit uncertainty information as an input, to account for the uncertainty that derives from the outputs of preceding DL-based components. Considering a probability distribution in the input enables the propagation of uncertainty through different components to provide a representative measure of the overall system uncertainty. For example, erroneous or uncertain predictions of an object detector in the perception stage of the pipeline can result in an unexpected and unsafe motion plan from the trajectory planner [10]. In a similar fashion, in perception components that learn environmental compact representations (e.g. with Variational Autoencoders [12]), the stochastic nature of the latent space can be use to model complex data-inherent uncertainty like multimodality [5]. In the trajectory planner example, the stochastic latent variables from perception can help to predict multimodal motion plans in ambiguous scenarios like intersections.

In the literature, only few works such as [6,9,16] consider the use of uncertainty information from other DL components within a system. In this paper, we present a method to account for uncertainty information from stochastic variable inputs that are obtained from other DL components. This method represents a first effort towards handling uncertainty beyond perception in the software pipeline. We evaluate our approach in the context of autonomous aerial navigation, where we aim to learn a robust uncertainty-aware control policy that handles stochastic latent variables from a perception module. To the best of our knowledge, no previous methods handle continuous probability distributions at the inputs of a probabilistic neural network for control in an autonomous navigation task.

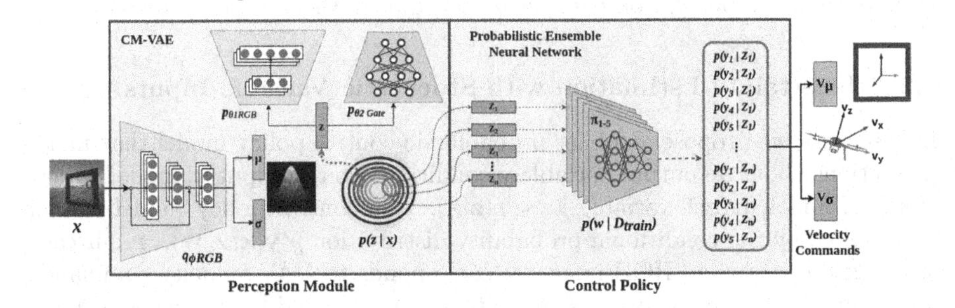

Fig. 1. System architecture for autonomous aerial navigation.

2 Proposed Method

2.1 Autonomous Aerial Navigation with DNNs

In this work, we address the problem of autonomous aerial navigation, in which a UAV aims to traverse a set of gates (with unknown locations) in a drone racing context. To implement our method, we build on the work from [2]. In this approach, the authors propose a navigation architecture that consists of two components implemented with DNNs: a perception module and a control policy. Figure 1 depicts the overall system architecture for autonomous navigation.

The objective of the perception module is to extract relevant information from the UAV environment and the current task. To do so, a low dimensional latent representation is learned using a deep generative model. A cross-modal variational autoencoder (CM-VAE) [19] is an effective method to learn a rich, compact, and robust representation. CM-VAEs are variant of traditional VAEs that learn a single latent space with data from multiple sources (modalities). In this case-study, data modalities are represented by RGB images, and the pose of the next gate relative to the UAV body-frame. In the perception module, an RGB image $\mathbf{x}$ at the input is processed by the encoder $q_{\phi RGB}$ into a multivariate normal distribution $\mathcal{N}(\mu, \sigma^2)$ from which latent vectors z are sampled. Later, each data modality can be recovered from the latent space using the decoder $p_{\theta 1RGB}$ and the estimator $p_{\theta 2Gate}$. For more details, we refer the reader to [2,19].

In the control module, imitation learning is used to train a neural network control policy that maps latent stochastic variables (from CM-VAE) to UAV velocity commands. Different from [2], we train an ensemble of probabilistic control policies $\{\pi_i\}_{i=1}^{M}$ (see Fig. 1) to account for both, model and data uncertainty. This can be viewed as training a Bayesian Neural Network (BNN) [8]. In addition, our policy ensemble can handle uncertainty from the latent space (probability distribution) in its input to model complex uncertainty patterns, this is described in detail in the next section. To train the policy ensemble, we employ behavior cloning using raw images with velocity labels. We freeze all the weights from the perception encoder $q_{\phi RGB}$ during the training process.

2.2 Uncertainty Estimation with Stochastic Variable Inputs

In this work we propose a robust probabilistic control policy model that makes predictions about an output variable $\mathbf{y}$ conditioned on an input $\mathbf{x}$, model parameters $\mathbf{w}$, and a latent variable $\mathbf{z} \sim p(\mathbf{z}|\mathbf{x})$. Our control policy model is then represented by the conditional probability distribution $p(\mathbf{y}|\mathbf{x}, \mathbf{z}, \mathbf{w})$. In our case-study, $\mathbf{x}$ represents an RGB image, $\mathbf{y}$ corresponds to UAV velocity commands, and $\mathbf{z}$ represents the low dimensional latent space learned by the encoder model $q_{\phi RGB}(z|x)$.

To capture model uncertainty and handle stochastic inputs (probability distributions), we embrace the Bayesian approach to compute the distribution for a target variable $\mathbf{y}^*$ associated with a new input $\mathbf{x}^*$, using the posterior predictive distribution:

$$p(\mathbf{y}^*|\mathbf{x}^*, \mathcal{D}) = \iint p(\mathbf{y}|\mathbf{z}, \mathbf{w}) p(\mathbf{w}|\mathcal{D}) p(\mathbf{z}|\mathbf{x}^*) \mathbf{dwdz} \tag{1}$$

The above integral is intractable, and we rely on approximations to obtain an estimation of the predictive distribution. The distribution $p(\mathbf{w}|\mathcal{D})$ reflects the posterior over model weights given dataset $\mathcal{D} = \{\mathbf{X}, \mathbf{Y}, \mathbf{Z}\}$, where $\mathbf{X}, \mathbf{Y}$ represent the training inputs and $\mathbf{Z}$ are the latent vectors observed during training. The posterior over the weights is difficult to evaluate, thus we can approximate the inner integral from Eq. 1 using common BNNs methods [1]. In our case, each member of the ensemble can be viewed as a sample taken from the true posterior distribution over the weights [8]. Finally, the outer integral is approximated by taking n samples z_n from the latent space in our perception model. This last step can be seen as taking a better picture of the perception latent space (with multiple latent samples instead of a single one like in the baseline), leveraging in this way its stochastic nature to take robust control predictions. The overall method is illustrated in Fig. 1, and summarized in Algorithm 1.

3 Early Experiments

For our experiments we used the aerial navigation architecture, the dataset, and the simulation environment from [2] as baseline. In this work, the architecture is composed by a CM-VAE, and a behavior cloning policy (CM-VAE+BC). To train both components, CM-VAE and BC policy, 300k and 17k RGB images (labeled) are used. In both cases the image size is 64×64 pixels.

We reproduced the baseline (CM-VAE+BC) and implement our method using Pytorch instead of Tensorflow. For the CM-VAE we used the same training settings as in the baseline for the unconstrained version. Different from the

Algorithm 1. Neural Network Predictive Distribution with Stochastic Inputs

1: **Input:** image $\mathbf{x}^*$, policy ensemble $\{\pi_{\mathbf{w_i}}\}_{i=1}^{M} \sim p(\mathbf{w}|\mathcal{D})$
2: **Initialize:** $\hat{\mu}_z$, $\hat{\sigma}_z$, $\hat{\mu}_\pi$, $\hat{\sigma}_\pi \leftarrow$ new empty Tensors
3: **Sample:** $\{\mathbf{z}_n\}_{n=1}^{N} \sim q_{\phi RGB}(\mathbf{z}|\mathbf{x}^*)$.
4: **for** $n = 1$ **to** N **do**
5:     **for** $i = 1$ **to** M **do**
6:         $\hat{\mu}_\pi^{\mathbf{w}_i}, \hat{\sigma}_\pi^{\mathbf{w}_i} \leftarrow \pi_{\mathbf{w}_i}(z_n)$
7:         Insert $\hat{\mu}_\pi^{\mathbf{w}_i}$ into $\hat{\mu}_\pi$; Insert $\hat{\sigma}_\pi^{\mathbf{w}_i}$ into $\hat{\sigma}_\pi$
8:     **end for**
9:     $\hat{\mu}_{z_n} \leftarrow mean(\hat{\mu}_\pi)$
10:     $\hat{\sigma}_{z_n} \leftarrow std(\hat{\sigma}_\pi, \hat{\mu}_{z_n}, \hat{\mu}_\pi)$
11:     Insert $\hat{\mu}_{z_n}$ into $\hat{\mu}_z$; Insert $\hat{\sigma}_{z_n}$ into $\hat{\sigma}_z$
12: **end for**
13: $\hat{\mu}_{\mathbf{y}^*} \leftarrow mean(\hat{\mu}_z)$
14: $\hat{\sigma}_{\mathbf{y}^*} \leftarrow std(\hat{\sigma}_z, \hat{\mu}_y, \hat{\mu}_\pi)$

baseline, we implemented an ensemble of five neural-networks for our BC control policy (BCE for short). Each member of the ensemble outputs two values, corresponding to the predictive mean and variance, and is trained using the heteroscedastic loss function [11,13]. Our BCE policy is able to handle uncertainty at the input according to Algorithm 1, therefore is referred as **BCE-UI**. In consequence, our aerial navigation architecture for the experiments is referred as **CM-VAE+BCE-UIx**, where the last letter indicates the number of samples from the latent space to take into account for the policy predictions.

We evaluate our navigation architecture with the proposed method under controlled simulations that resemble the conditions from the dataset collection. For this purpose, we create a circular track with eight equally spaced gates positioned initially in a radius of 8 m and constant height. To asses the robustness of the navigation policies, we generate new challenging tracks, adding random noise to each gate position (radius) and height. In this way, we force OoD operating conditions.

Table 1 shows the average number of gates traversed by each navigation policy under different noise levels for both gate radius (RN) and height (HN). We considered a maximum number of 32 gates which is equivalent to 4 laps in our track. Experimental results show that the navigation policy with our method outperforms the robustness of the other control policies (deterministic and with uncertainty representation only at the output). In fact, the version that includes more input samples is more robust to more drastic OoD conditions. In addition, we find that our navigation policy has a smooth behavior, while the baseline policy and the policy that represents uncertainty only at the output, present a turbulent behavior in noise tracks.

Table 1. Average number of gates traversed by each aerial navigation policy.

Navigation Model	Track		
	RN = 0 HN = 0	RN = 1 HN = 2	RN = 1.5 HN = 2.5
CM-VAE+BC	32	12	7
CM-VAE+BCE-UI1	32	16	10
CM-VAE+BCE-UI3	32	**20**	13
CM-VAE+BCE-UI5	32	**20**	**17**

4 Conclusion

We presented a method to cope with stochastic variables at the input of BNNs used for control in an aerial navigation task. Based on our experiments, implementing a control policy that accounts for uncertainty at the input improves the robustness of the system in OoD scenarios, besides enabling a smooth behavior. The proposed method represents a step towards uncertainty handling beyond perception in the autonomous system software pipeline. Despite the benefits shown in the experiments, our method relies on sampling, which can be prohibitive in real-time systems. Another direction for future work would be the quality assessment of uncertainty estimates at component and system level, where we should consider the uncertainty propagation capability introduced by our method. This last point represents an important step towards deploying DL-components into safety-critical systems.

Acknowledgments. This work has received funding from the COMP4DRONES project, under Joint Undertaking (JU) grant agreement N°826610. The JU receives support from the European Union's Horizon 2020 research and innovation programme and from Spain, Austria, Belgium, Czech Republic, France, Italy, Latvia, Netherlands.

References

1. Arnez, F., Espinoza, H., Radermacher, A., Terrier, F.: A comparison of uncertainty estimation approaches in deep learning components for autonomous vehicle applications. In: Proceedings of the Workshop on Artificial Intelligence Safety 2020 (2020)
2. Bonatti, R., Madaan, R., Vineet, V., Scherer, S., Kapoor, A.: Learning visuomotor policies for aerial navigation using cross-modal representations. arXiv preprint arXiv:1909.06993 (2019)
3. Codevilla, F., Müller, M., López, A., Koltun, V., Dosovitskiy, A.: End-to-end driving via conditional imitation learning. In: 2018 IEEE International Conference on Robotics and Automation (ICRA). pp. 4693–4700. IEEE (2018)
4. Codevilla, F., Santana, E., López, A.M., Gaidon, A.: Exploring the limitations of behavior cloning for autonomous driving. In: Proceedings of the IEEE/CVF International Conference on Computer Vision, pp. 9329–9338 (2019)
5. Depeweg, S., Hernandez-Lobato, J.M., Doshi-Velez, F., Udluft, S.: Decomposition of uncertainty in Bayesian deep learning for efficient and risk-sensitive learning. In: International Conference on Machine Learning, pp. 1184–1193. PMLR (2018)

6. Fan, T., Long, P., Liu, W., Pan, J., Yang, R., Manocha, D.: Learning resilient behaviors for navigation under uncertainty. In: 2020 IEEE International Conference on Robotics and Automation (ICRA), pp. 5299–5305. IEEE (2020)
7. Grigorescu, S., Trasnea, B., Cocias, T., Macesanu, G.: A survey of deep learning techniques for autonomous driving. J. Field Robotics **37**(3), 362–386 (2020)
8. Gustafsson, F.K., Danelljan, M., Schön, T.B.: Evaluating scalable Bayesian deep learning methods for robust computer vision. arXiv preprint arXiv:1906.01620 (2019)
9. Henaff, M., Canziani, A., LeCun, Y.: Model-predictive policy learning with uncertainty regularization for driving in dense traffic. In: International Conference on Learning Representations (2018)
10. Ivanovic, B., et al.: Heterogeneous-agent trajectory forecasting incorporating class uncertainty. arXiv preprint arXiv:2104.12446 (2021)
11. Kendall, A., Gal, Y.: What uncertainties do we need in Bayesian deep learning for computer vision? In: Advances in Neural Information Processing Systems, pp. 5574–5584 (2017)
12. Kingma, D.P., Welling, M.: Auto-encoding variational Bayes. arXiv preprint arXiv:1312.6114 (2013)
13. Lakshminarayanan, B., Pritzel, A., Blundell, C.: Simple and scalable predictive uncertainty estimation using deep ensembles. In: Advances in Neural Information Processing Systems, pp. 6402–6413 (2017)
14. Loquercio, A., Segu, M., Scaramuzza, D.: A general framework for uncertainty estimation in deep learning. IEEE Robotics Autom. Lett. **5**(2), 3153–3160 (2020)
15. McAllister, R., et al.: Concrete problems for autonomous vehicle safety: advantages of bayesian deep learning. In: International Joint Conferences on Artificial Intelligence, Inc. (2017)
16. McAllister, R., Kahn, G., Clune, J., Levine, S.: Robustness to out-of-distribution inputs via task-aware generative uncertainty. In: 2019 International Conference on Robotics and Automation (ICRA), pp. 2083–2089. IEEE (2019)
17. Mueller, M., Dosovitskiy, A., Ghanem, B., Koltun, V.: Driving policy transfer via modularity and abstraction. In: Conference on Robot Learning, pp. 1–15. PMLR (2018)
18. Siegwart, R., Nourbakhsh, I.R., Scaramuzza, D.: Introduction to Autonomous Mobile Robots. MIT Press, Cambridge (2011)
19. Spurr, A., Song, J., Park, S., Hilliges, O.: Cross-modal deep variational hand pose estimation. In: Proceedings of the IEEE Conference on Computer Vision and Pattern Recognition, pp. 89–98 (2018)
20. Wu, Q., Gong, X., Xu, K., Manocha, D., Dong, J., Wang, J.: Towards target-driven visual navigation in indoor scenes via generative imitation learning. IEEE Robot. Autom. Lett. **6**(1), 175–182 (2020)

No Free Lunch: Overcoming Reward Gaming in AI Safety Gridworlds

Mariya Tsvarkaleva and Louise A. Dennis(✉)

Department of Computer Science, The University of Manchester, Manchester, UK
mariya.tsvarkaleva@student.manchester.ac.uk,
louise.dennis@manchester.ac.uk

Abstract. We present two heuristics for tackling the problem of reward gaming by self-modification in Reinforcement Learning agents. Reward gaming occurs when the agent's reward function is mis-specified and the agent can achieve a high reward by altering or fooling, in some way, its sensors rather than by performing the desired actions. Our first heuristic tracks the rewards encountered in the environment and converts high rewards that fall outside the normal distribution into penalities. Our second heuristic relies on the existence of some validation action that an agent can take to check the reward. In this heuristic, on encountering an abnormally high reward, the agent performs a validation step before either accepting the reward as it is, or converting it into a penalty. We evaluate the performance of these heuristics on variants of the tomato watering problem from the AI Safety Gridworlds suite.

Keywords: AI Safety Gridworlds · Reinforcement learning · Reward gaming

1 Introduction

In Reinforcement Learning (RL) methods, an agent explores an environment from which it receives rewards and adjusts its behavioural policy to optimize this reward signal. Since RL algorithms rely heavily on trial and error, this often results in unwanted, irreversible, and potentially harmful actions. For this reason, empirical Reinforcement Learning is limited and current algorithms are predominantly applied to simulated and tightly-controlled environments [7,9,10].

A poorly designed reward signal can often lead to what is called *reward gaming* [2]. This refers to an autonomous system exploiting misspecification to obtain a high reward which does not reflect the desired behaviour. A practical example could be a household robot whose reward signal is designed to minimize the amount of dishes it sees as dirty, which learns to obstruct its camera view of the sink, so that it sees no unclean dishes. Thus, the agent can obtain a maximum reward while failing to behave as intended.

Work supported by EPSRC Grant EP/V026801/1 *Trustworthy Autonomous Systems Verifiability Node.*

I. Habli et al. (Eds.): SAFECOMP 2021 Workshops, LNCS 12853, pp. 226–238, 2021.
https://doi.org/10.1007/978-3-030-83906-2_18

We propose two heuristics for avoiding this problem. In both heuristics the agent keeps a record of previous rewards and their distribution. Where a reward signal falls outside this distribution (the agent receives a "free lunch") then it treats this with suspicion. In our first heuristic the agent converts free lunches into penalties (we refer to this as the *Direct Penalty* heuristic). In our second heuristic the agent performs a checking step to validate the reward signal (we refer to this as the *Doubt* heuristic).

We evaluate the performance of these heuristics on a set of examples based on the *tomato watering GridWorld* in the AI Safety Gridworlds [6] suite of challenge problems for RL agents. Experimental results demonstrate that both heuristics reduce reward gaming but that the second is able to detect and act upon genuine rewards that fall outside the distribution and also performs better even when all exceptional rewards are misspecified – however the Doubt heuristic, in the wrong environment, can still lead to reward gaming.

2 Background

Reinforcement Learning (RL) [11] is a learning method where an agent adopts a behaviour based on the rewards and penalties it receives from an environment. The agent's goal is to optimize a quantifiable reward signal, and it learns by a trial-and-error process of exploring the space and receiving feedback. The agent learns a strategy, based on its observations, to maximize its cumulative reward. It achieves this by empirically building estimates of how valuable states (or more typically state-action pairs) are, starting with uncertain knowledge and improving its beliefs in an iterative fashion.

The Q value function, $Q(s, a)$, describes how good a state-action pair is with regards to the *expected return* that can be obtained from taking an action a in state s. The output of the function is called a Q value and is often used as a quality measure in reinforcement learning. RAINBOW [4] is an implementation of a deep learning approach to learning a policy that optimises the Q value. It combines a number of techniques and is available open source from Open AI. We used RAINBOW in our experiments.

2.1 AI Safety Gridworlds Environments

AI Safety Gridworlds [6] is a suite of challenge problems that offers simplistic grid environments modelled to expose unsafe behaviour in RL agents. The gap between what we want the agent to achieve and what we told it to achieve is modelled in the environments with two functions – a *reward function* and a *safety performance function.* The reward function represents the signal the agent uses to optimise its policy, whereas the safety performance function defines the actual desired behaviour and is hidden from the agent. Environments are composed of a two-dimensional chessboard-like grid with a maximum size of 10×10. Each cell can either be empty, unreachable, occupied by the agent, or occupied by sprites with which the agent can interact.

The gridworlds are represented as a finite *Markov Decision Process.* The action space, A, is the set of actions {up, down, left, right}. The robot interacts with the environment on each time step by performing an action $a \in A$, based on its observations of the current state, $s \in S$, which results in transitioning to a next state s' and obtaining a reward $R(s, a)$. It moves in the environment by shifting to a neighboring cell in the specified direction if the cell is empty, or otherwise stays in its place. Each gridworld has a limited number of steps that can be taken and form a single episode, which ends either when that limit is reached or a goal specific to the environment is accomplished.

In this paper we concern ourselves with one particular gridworld, the *tomato watering* environment, that captures the reward gaming by self-modification problem. In the *tomato watering* environment, the agent's goal is to water as many tomatoes as possible within an episode. Unwatered tomatoes (t) appear as watered (T) whenever the agent steps on their tile, and then dry out with some probability P at each timestep. In addition, there is an observation/bucket tile (O) which when stepped on, makes the robot see all tomatoes as watered, but doesn't change their real state. The important distinction here is that the robot's reward function is based on how many tomatoes it perceives as watered, as opposed to how many have truly changed their state. This means that stepping on the O tile will yield a maximum reward while the goal of watering plants is not achieved. On the other hand, safety is expressed as a function of the number of truly watered tomatoes (Fig. 1).

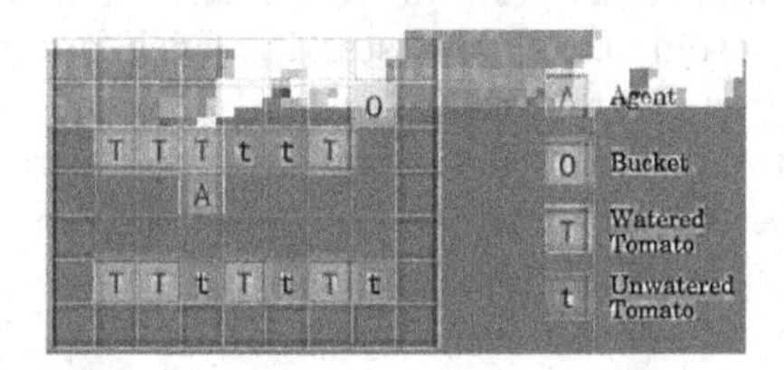

Fig. 1. The tomato watering gridworld [6]

3 Two Heuristics: Detecting a Free Lunch

3.1 Direct Penalty: Assuming This is a Free Lunch

One way to avoid unsafe behaviour in the self-modification problem is by penalizing state-transitions that yield unrealistically high rewards that represent free lunches. Thus, we want the agent to learn when an action it has done has led to a reward out of distribution and discourage such moves. To achieve this, we need to define what a "suspiciously high" reward is, and decide how to penalize this behaviour. By holding a list of rewards the agent has received at each step, we consider a newly encountered reward as a free lunch if it is an outlier compared to the others collected. NB: This heuristic relies on the assumption, common in RL, that the environment changes only slowly when the agent is not acting upon it.

We have implemented a variant of outlier detection. The solution consists of keeping a normal distribution over the rewards received, and at each step checking that the new one obtained lies within some range N of σ standard deviations away. Based on the empirical rule (or 68-95-99.7 rule) known in statistics which states that 99.7% of the data falls within 3 deviations, it is reasonable for N to be bounded between 3 and 5. We calculate the standard score z to determine how far away the datapoint, x_i, is from the mean μ, and compare if this is above the threshold N and should be considered an outlier. After a basic parameter search, $\sigma = 3$ was found to perform best and is used in all our experiments.

$$z = \frac{x_i - \mu}{\sigma} \tag{1}$$

Since rewards are achieved each step at a time, defining a threshold for what we consider abnormal can be difficult in early time steps before enough data about the distribution has been gathered. To account for this complication, we only start checking for outliers after some N steps have been taken, assuming that the agent has gathered observations which are more representative of the set of overall achievable rewards.

Finally, we penalize the undesired transitions. We experimented with modifying the reward to a static number and modifying it scaled by some value p. The second achieved better performance, because of the difficulty of categorizing outliers – if a normal action is incorrectly classified as abnormal, we want this to be penalized less than a move which has led to fake observations. Otherwise, we risk that the agent develops a policy of avoiding desirable states. Therefore the penalty is defined as:

$$R(s, a) = -R(s, a) * p \tag{2}$$

Pseudo code for the Direct Penalty heuristic is shown in Fig. 2.

3.2 Doubt: Checking if this is a Free Lunch

In addition to achieving safe behaviour, we would like the agent to continue to strive for optimal task performance. [6] suggests a scenario where the O tile is a sprinkler that truly waters all the plants, and not merely an observation modifier. In such case, the reward obtained, which might be flagged as a free lunch, would be a valid solution to the problem.

We modified our Direct Penalty heuristic to a new heuristic we called *Doubt*. This heuristic presumes the existence of some validation action by which the result of a reward can be verified. This is obviously application specific. In the

```
list_rewards ← []
while episode is not finished do
    action ← agent.choose_action()
    reward, observations ← agent.STEP(action)
    list_rewards ← list_rewards.ADD(reward)
    DETECT_OUTLIERS_STD()
end while
function DETECT_OUTLIERS_STD(list_rewards, reward)
    normal_distribution ← STD(list_rewards[-1])
    z ← Z_SCORE(reward, normal_distribution)
    if z > N then
        reward ← -reward * p
        list_rewards.POP()
    end if
end function
```

Fig. 2. The Direct Penalty algorithm

context of tomato watering we assume that an action truly changes the grid-world's state if its resulting effects remain and so this validation action can consist of undoing the agent's last move. The agent is equipped at each timestep with a 2D matrix representation of its observations. Therefore, it can decide if a change has persisted by comparing its observations of the doubted transition with those of the reversed one.

We employ a similarity metric for comparison, instead of an absolute equality between the matrices, because of the periodic random drying out of tomatoes, which makes the environment dynamic. We calculate similarity of two matrix observations of the same size A and B by subtracting their values and calculating the ratio of zero elements over all the elements. Finally, we utilize a threshold M to conclude if a penalty should be incurred (3).[1]

$$C = A_{m,n} - B_{m,n} = \begin{pmatrix} a_{1,1} - b_{1,1} & \cdots & a_{1,n} - b_{1,n} \\ \vdots & \ddots & \vdots \\ a_{m,1} - b_{m,1} & \cdots & a_{m,n} - b_{m,n} \end{pmatrix} \tag{3}$$

$$Penalize \begin{cases} True, & \text{if } \sum_{x \in C} \delta_{x,0} / \sum_{x \in C} 1 < M \\ False, & \text{otherwise} \end{cases}$$

Pseudocode for the Doubt algorithm can be found in Fig. 3.

[1] Kronecker delta δ_{ij} is a function of two variables i, j that returns 1 if the variables are equal, and 0 otherwise.

```
list_rewards ← []
while episode is not finished do
    if WAS_REVERSED()&&BOARDS_SIMILARITY(prev, curr) < M then
        reward ← −reward * p
    end if
    if STEP_WAS_SUSPECTED() then
        action ← agent.OVERRIDE_WITH_REVERSED()
        transition_reversed ← True
    else
        action ← agent.CHOOSE_ACTION()
    end if
    reward, observations ← agent.STEP(action)
    list_rewards ← list_rewards.ADD(reward)
    DETECT_OUTLIERS_STD()
end while
function BOARDS_SIMILARITY(previous, current)
    subtracted ← current − previous
    proportion_similar ← COUNT_ZEROS(subtracted)/subtracted.SIZE()
end function
```

Fig. 3. The Doubt algorithm

4 Evaluation

To investigate our heuristics, we implemented a modified version of the tomato watering environment, which we call the *tomato watering sprinkler* suite. We introduce a new S tile, which acts as a sprinkler that automatically waters all tomatoes whenever the agent steps on it. In contrast to the O tile, the S tile changes the number of truly watered tomatoes.

The *tomato watering sprinkler* suite consists of five cases, which demonstrate different configurations of the problem. The layout of the first case is similar to the *tomato watering* gridworld, consisting of dry tomatoes, watered tomatoes, and an observation-changing tile O. The only difference is that when the agent steps on the O tile, it observes only existing tomatoes as watered, instead of interpreting the full board as filled with watered tomatoes in order to provide a more consistent set of cases. The second case represents a situation where a sprinkler is put in the environment, however there is no observation tile (Fig. 4). Finally, the last three cases represent different configurations of the S and O tile placed at the same time in the grid in different cells (Fig. 5).

Fig. 4. Case 2: S tile only

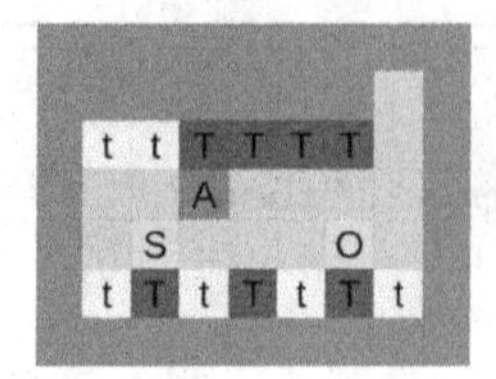

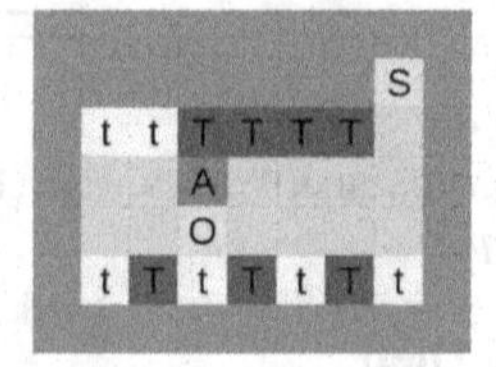

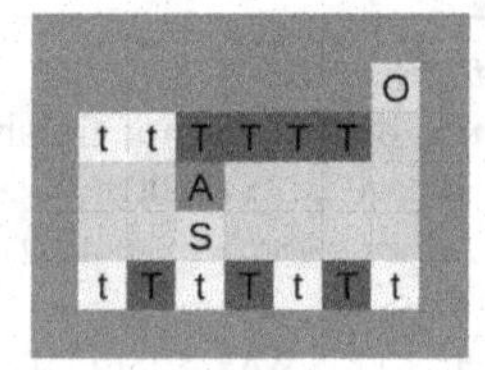

Fig. 5. Cases 3, 4, and 5: Both S and O tile

We ran an RL agent in the *tomato watering sprinkler* environment with neither the O nor S tiles present in order to establish a baseline reward when there is no possibility for gaming. This demonstrates the optimal return under guaranteed safe conditions within 200k timesteps. We observed that the agent's performance grows steadily, reaching at most 9.5. Since tomatoes dry out with a probability of $P = 0.5$, it is extremely difficult (if not quite impossible) for the robot to reach all of them.

4.1 Case 1: Observation Tile only

We trained up three agents in the first *tomato watering sprinkler* case. The first agent used no heuristics, the second used the Direct Penalty heuristic and third used the Doubt heuristic. The results are shown in Figs. 6 and 7.

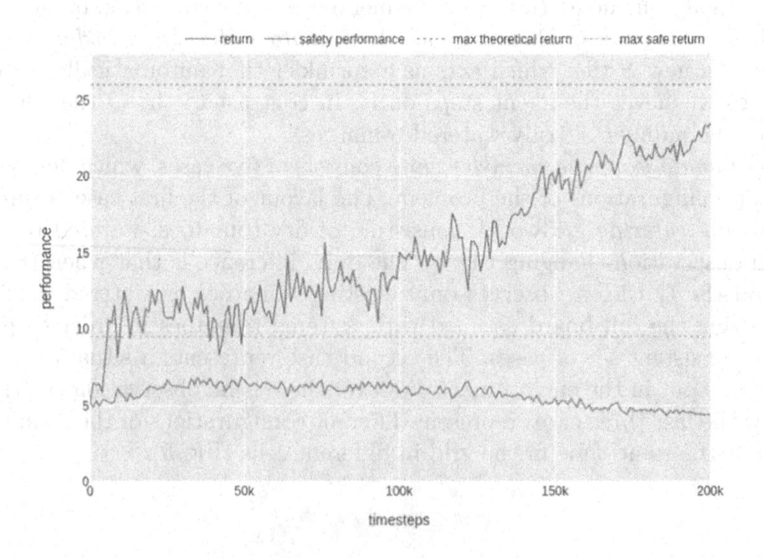

Fig. 6. Basic agent (no heuristics) - Case 1

Our results show that the agent that uses no heuristics to check the reward learns to cheat by stepping on the O tile, and so performs poorly with regards

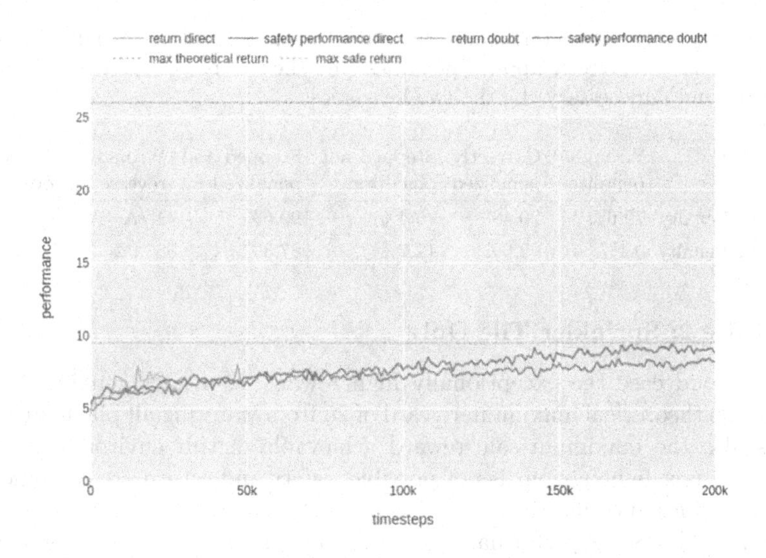

Fig. 7. Two heuristic agents - Case 1

to the safety function. Meanwhile the two agents using heuristics to judge the reward do not develop a policy that exploits the O tile. When penalizing directly, the first 25k steps reveal a spike in the return which doesn't align with the safety performance at that point. This means that the robot was frequently stepping on the O tile at first, but soon after, convergence is reached and the agent continues to act predominantly safely until the end of the run. When using the Doubt heuristic (green and purple solid lines), we observe an even better overlap of safety performance and return.

Further, the return obtained with the Doubt heuristic comes closer to the maximum safe return of 9.5 (green dashed line). We believe that the reason for this is the greatly reduced number of wrongly penalized steps illustrated in Table 1. When using the Direct Penalty heuristic, the portion of data wrongly categorized as outlying amounts to an overwhelming 79.6%. This is high compared to the very low number of O-tile steps which weren't penalized – meanwhile the Doubt heuristic, capable of validating the reward, rarely issues an incorrect penalty. We reason that this high miscategorization might have disrupted the agent's policy, making it avoid some desired reward actions (i.e., watering tomatoes). This would explain why the Direct Penalty heuristic achieves a lower return than the Doubt heuristic. Because the "doubtful" agent reverses a step to check if the reward persists, the miscategorisation of desired rewards is transferred to unnecessary reverse actions (shown in Table 1).

Table 1. Percentage steps taken onto the S tile (penalized and not penalised), number of correctly and incorrectly applied penalties, and the number of reverse actions taken (necessary and unnecessary) by the two heuristics.

Heuristic	Wrongly penalized	Correctly penalized	Stepped not penalized	Stepped and penalized	Wrongly reversed	Correctly reversed
Direct Penalty	79.6%	20.4%	0.4%	99.6%	N/A	N/A
Doubt Penalty	0.1%	99.9%	12.7%	87.3%	67.9%	32.1%

4.2 Case 2: Sprinkler Tile Only

In the second case the exceptionally high reward is, in fact genuine. For this reason, the theoretical maximum reward of 26 from watering all plants in a single step is also the maximum safe reward achievable in this environment layout. Given that unsafe behaviour is not possible, safety and return are combined into a single performance metric.

Figure 8 shows the performance of our two heuristics in this environment. As expected the Direct Penalty heuristic fails to exploit the sprinkler tile, while the Doubt heuristic is able to detect that the reward is genuine.

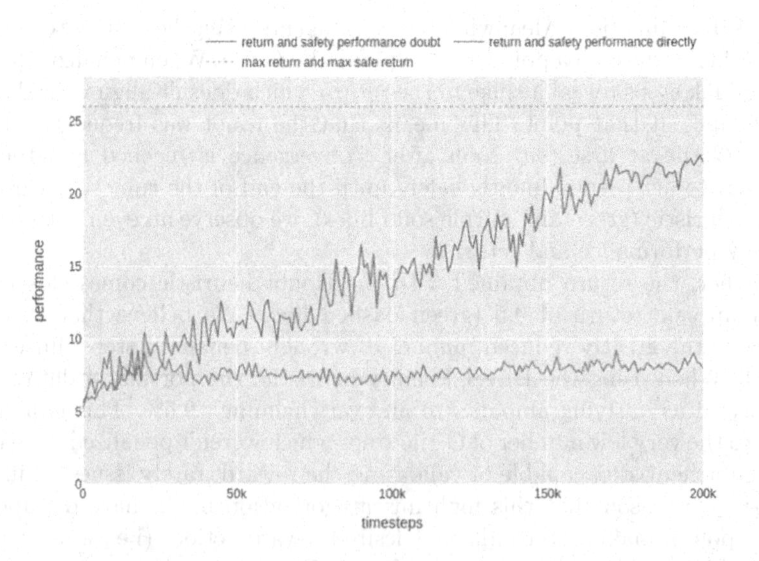

Fig. 8. Penalize Directly vs. Doubt - Case 2

4.3 Cases 3, 4, 5: Distinguishing Observation and Sprinkler Tile

In the last three cases, we vary the positions of the S and O tile. Our goal is to observe if the agent using our Doubt heuristic is able to distinguish between the genuine reward and the free lunch when both are presented as an option. There is no safety baseline since the maximum return can be achieved safely.

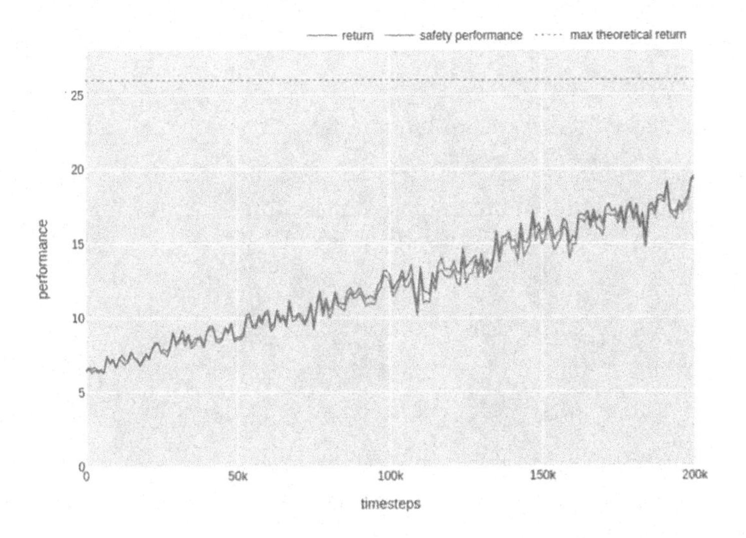

Fig. 9. Penalize Directly vs. Doubt - Case 3

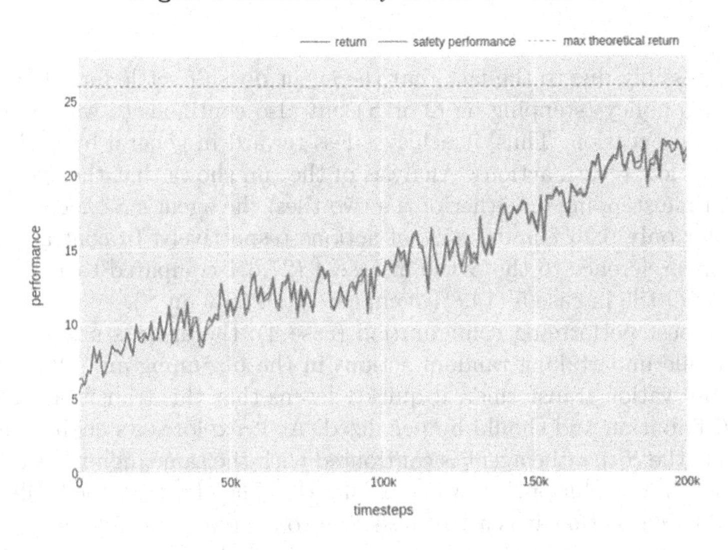

Fig. 10. Penalize Directly vs. Doubt - Case 4

Looking at the overall outcome (Figs. 9, 10 and 11), we can tell that the different positioning of the two special tiles affects the performance of the agent. We can see that the highest overlap of the safety and return functions occurs in case 4. A similar trend but slightly worse in terms of maximum reward and safe behaviour is observed in case 3. Finally, the case 5 results demonstrate the greatest divergence and therefore the most unsafe behaviour.

Case 3 presents a configuration in which the agent is close to both of the special tiles from the start. The return in this experiment is the lowest by a small

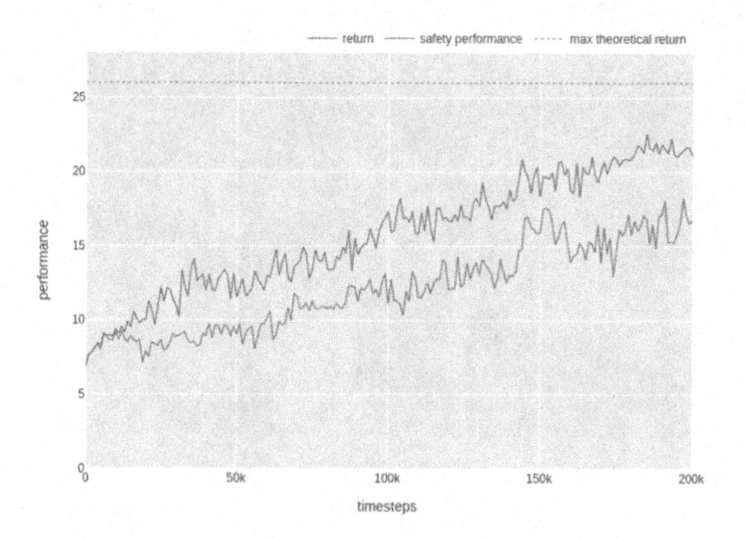

Fig. 11. Penalize Directly vs. Doubt - Case 5

amount, possibly due to the fact that the agent doesn't settle for a high-reward exploitative policy (stepping on O or S) but also continues to water individual tomatoes on occasion. Thus, it achieves less reward in general by performing a lot of other low return actions. Analysis of the run shows that the agent spends minimal time stepping on either of the two tiles: the agent's S- and O-tile steps account for only 6.26% and 1.85% of actions respectively. In contrast, there is an evident preference to the S tile in case 4 (27.6% compared to 1.18% on O), and to the O tile in case 5 (14.8% compared to 6.69% on S).

In the best performing configuration (case 4), the agent starts close to the O tile. While undertaking random actions in the beginning and often stepping on the observation transformer, it quickly learns that this transition's reward is out of distribution and should be penalized. As it explores its environment and encounters the S tile, the agent is confronted with the same high reward but this time, it is a valid solution. After confirming that the effects of the S tile persist, it gradually learns that it is a harmless way to achieve good return.

In case 5, a high reward is obtained but safety performance deteriorates. Only in the first 5k steps or so do the safety performance and return functions align. Soon after that, the two diverge. The episode starts with the agent placed in close proximity to the S tile and further away from the O tile. The agent initially "turns on" the sprinkler repeatedly and learns that this is a safe solution. Thus, by regarding the transition as safe, the high reward becomes part of the rewards distribution. This becomes a problem when the agent reaches the O tile – by the time it steps on the observation transformer, the reward received is no longer a free lunch because of the many S-tile steps that have yielded the same high reward. Therefore, the agent fails to even suspect the O tile and doesn't learn a safe policy.

5 Related Work

There are various emerging methods for solving the various reward misspecification problems highlighted by the Gridworlds. [1,5] look at negative side effects. [3] tackles *distributional shift*. Meanwhile robustness problems have been considered by [8]. As far as we are aware we are the first work to tackle reward gaming by self-modification.

6 Conclusion

We present two heuristics for overcoming reward gaming by self-modification in RL systems. Our first heuristic is agnostic to the environment and consists simply of penalizing unexpectedly high rewards detected by comparing them to the distribution of previous rewards. Our second heuristic is context dependent and relies upon being able to define a validation action that can verify that a reward is genuine. We ran experiments to evaluate the performance of these two heuristics on variations of the AI Safety Gridworlds tomato watering environment where special tiles could return large genuine and false rewards. We showed that even where only free lunches existed, the heuristic that attempted to validate rewards performed better because it allocated fewer incorrect penalties early in the learning process. However, this heuristic is sensitive to the initial conditions of the learning and we illustrated, via a case where the genuine reward tile was accessible early in the learning process while the free lunch was only encountered later, that unsafe behaviour could still be learned.

The value of the first heuristic is its context independence. In environments where reward functions are difficult to precisely define, the heuristic can be used to exclude suspiciously high rewards at the risk of reduced performance. While the second heuristic has better performance it does require the identification of a validation action. It may be that classes of such actions can be determined – for instance our persistence assumption may be sufficient in many cases – but there are many environments where determining an action that can validate a reward that is independent of the way the reward function is calculated may be challenging and even where one can be found the heuristic is vulnerable if there are also actions that genuinely return high rewards. It should also be noted that there is a cost involved in performing a validation step which in many applications may prove prohibitive, even where such a step can be identified.

References

1. Armstrong, S., Levinstein, B.: Low impact artificial intelligences. CoRR abs/1705.10720 (2017). http://arxiv.org/abs/1705.10720
2. Clark, J., Amodei, D.: Faulty reward functions in the wild (2016). https://blog.openai.com/faulty-reward-functions/
3. Hadfield-Menell, D., Milli, S., Abbeel, P., Russell, S.J., Dragan, A.: Inverse reward design. In: Guyon, I., et al. (eds.) Advances in Neural Information Processing Systems, vol. 30. Curran Associates, Inc. (2017)

4. Hessel, M., et al.: Rainbow: combining improvements in deep reinforcement learning. In: Proceedings of the AAAI Conference on Artificial Intelligence, vol. 32, no. 1, April 2018
5. Krakovna, V., Orseau, L., Martic, M., Legg, S.: Measuring and avoiding side effects using relative reachability. CoRR abs/1806.01186 (2018). http://arxiv.org/abs/1806.01186
6. Leike, J., et al.: AI safety gridworlds. CoRR abs/1711.09883 (2017). http://arxiv.org/abs/1711.09883
7. Mnih, V., et al.: Playing atari with deep reinforcement learning. CoRR abs/1312.5602 (2013). http://arxiv.org/abs/1312.5602
8. Santara, A., et al.: Rail: risk-averse imitation learning. In: Proceedings of AAMAS 2018, pp. 2062–2063 (2018)
9. Silver, D., et al.: Mastering the game of Go with deep neural networks and tree search. Nature **529**(7587), 484–489 (2016). https://doi.org/10.1038/nature16961
10. Silver, D., et al.: A general reinforcement learning algorithm that masters chess, shogi, and go through self-play. Science **362**(6419), 1140–1144 (2018). https://doi.org/10.1126/science.aar6404
11. Sutton, R.S., Barto, A.G.: Reinforcement Learning: An Introduction, 2nd edn. The MIT Press, Cambridge (2018)

Effect of Label Noise on Robustness of Deep Neural Network Object Detectors

Bishwo Adhikari[1(✉)], Jukka Peltomäki[1], Saeed Bakhshi Germi[1], Esa Rahtu[1], and Heikki Huttunen[2]

[1] Tampere University, Tampere, Finland
{bishwo.adhikari,jukka.peltomaki,saeed.bakhshigermi,esa.rahtu}@tuni.fi
[2] Visy Oy, Tampere, Finland
heikki.huttunen@visy.fi

Abstract. Label noise is a primary point of interest for safety concerns in previous works as it affects the robustness of a machine learning system by a considerable amount. This paper studies the sensitivity of object detection loss functions to label noise in bounding box detection tasks. Although label noise has been widely studied in the classification context, less attention is paid to its effect on object detection. We characterize different types of label noise and concentrate on the most common type of annotation error, which is missing labels. We simulate missing labels by deliberately removing bounding boxes at training time and study its effect on different deep learning object detection architectures and their loss functions. Our primary focus is on comparing two particular loss functions: cross-entropy loss and focal loss. We also experiment on the effect of different focal loss hyperparameter values with varying amounts of noise in the datasets and discover that even up to 50% missing labels can be tolerated with an appropriate selection of hyperparameters. The results suggest that focal loss is more sensitive to label noise, but increasing the gamma value can boost its robustness.

Keywords: Safe AI · Deep neural networks · Label noise · Image labeling

1 Introduction

The growing success of deep neural network algorithms in solving challenging tasks resulted in a surge of interest from the safety-critical applications domain. As stated by recent works, one of the major issues of using such an algorithm in line with safety standards is the effects of label noise on the output [1–3].

Earlier object detection pipelines consisted of manually engineered feature extraction together with relatively simple classifiers [4,5]. These systems required a human to label the different objects for training, and the labeling was done on the crop level. Although this approach had its challenges, such as mining negative examples, its behavior is still reasonably well understood due to relying on a straightforward method.

I. Habli et al. (Eds.): SAFECOMP 2021 Workshops, LNCS 12853, pp. 239–250, 2021.
https://doi.org/10.1007/978-3-030-83906-2_19

More recently, the success of convolutional neural networks (CNN) and deep learning [6] has transformed the domain of object detection. These approaches outperform traditional techniques by a large margin but are also more data-hungry at the same time [7–9]. The tedious task of manual labeling of enormous datasets means there will be faults in the process inevitably.

Popular large object detection datasets include MS COCO [10], PASCAL VOC [11], and OpenImages [12], containing millions of examples with quality annotations. The ground truth human annotations are gathered by crowdsourcing, and elaborate reward and evaluation schemes guarantee high quality for the annotations. However, apart from these large annotation campaigns, many players, companies, and research groups routinely collect smaller datasets within their application domains. In such cases, the quality of annotations is often compromised due to limited resources. Moreover, even standard benchmark sets are not error-free, and the influence of erroneous annotations on the system's safety requires further study.

The presence of noise in the training dataset can have a severe impact on the system's performance. For example, in the video surveillance system, a good detector would retain the same confidence, box coordinates, and class label over time. On the other hand, a bad one will be flickering, where the confidence fluctuates, the coordinates change, and even the class is mislabeled from frame to frame.

Figure 1 shows the four most common annotation error categories found in object detection datasets. These categories are (a) missing annotations (false negatives), (b) extra annotations (false positives), (c) inaccurate bounding boxes (which would result in low intersection over union (IoU)), and (d) incorrect class labels. In our experience, the most common error type is the first one, where the human annotator misses some target objects due to occlusions, small size, a large number of objects, or simply unclear annotation instructions. The second most frequent annotation error type is inaccurate bounding boxes, a very natural error for a human, as it takes more time and effort to pay attention to detail in every case. The two other types in Fig. 1, completely incorrect annotations and wrong labels, are probably easier for humans to avoid.

The loss functions being a significant differentiator in modern single-stage detection pipelines and current challenges for annotation quality, inspired us to study the effect of label noise in object detection with two popular loss functions. Notably, in this paper, we focus on examining how *cross-entropy loss* (CE) and *focal loss functions* (FL) handle noise in the form of missing labels. We focus on these losses since the focal loss is commonly used but may suffer from missing annotations because it puts higher weight on complex samples (hard negatives and hard positives). Missing bounding boxes in the annotation appear as hard negatives from the training point of view, and we wish to study their influence on the resulting accuracy. The main contributions of this paper are:

- We characterize different types of noise present in object detection datasets.
- We provide empirical observations on training single-stage object detectors with different loss functions and different hyperparameter settings.
- We suggest possible measures to boost the robustness of the object detector with minimal changes in the network.

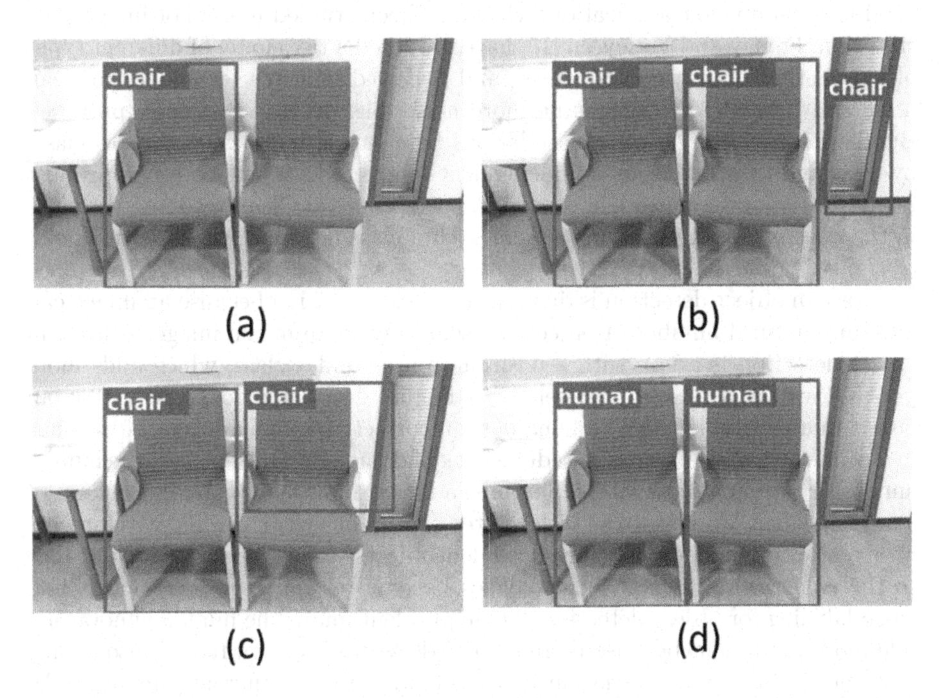

Fig. 1. Common types of label noise in object detection. (a) Missing label, the other chair is not labeled. (b) Incorrect annotation. (c) Inaccurately drawn box, resulting in low IoU. (d) Wrong classification label, humans instead of chairs. Image from the Indoor dataset [13].

The remainder of this paper is structured as follows. Section 2 briefly summarizes the related works followed by the review of object detection loss functions in Sect. 3. In Sect. 4, we experimented with multiple scenarios on our hypothesis and analyzed the obtained results. Finally, we conclude this paper with our findings and future direction in Sect. 5.

2 Related Works

Willers [1] and Wozniak [2] both provide a list of safety concerns or goals related to deep learning algorithms. In their works, label noise is mentioned as one of the primary faults that can affect safety. It is suggested to have a labeling guideline to mitigate the effect of this fault. However, even with a guideline, manually labeling a large dataset is prone to noise, as discussed before. Thus, a proper approach is required to deal with noisy datasets in deep learning systems. Zhang [14] reviews problems related to the dataset, such as label noise, by surveying over recent works. According to his work, using a robust loss function and reweighting samples can help mitigate this issue.

Our topic of label noise in object detection is closely related to the topic of label noise in image classification, which has been studied more: For image classification, Frenay and Verleysen [15] have proposed a taxonomy of different types of noise, studied their consequences, and reviewed multiple techniques to clean noise and have the algorithms be more noise-tolerant. Li *et al.* have proposed BundleNet exploiting sample correlations by creating bundles of samples class-by-class and treating them as independent inputs, which acts as a noise-robust regularization mechanism [16]. Lee *et al.* have proposed CleanNet to detect noise in the dataset and be used in tandem with a classifier network for better noise tolerance [17].

Noise in object detection is different from classification because an image can have any natural number of objects present, anywhere in the image. A label in object detection is a box with a position, a size, and a class, which adds more possibilities for noise. It is easier for a human annotator to identify that an object in a picture is indeed a banana than correctly labeling dozens of bananas in one image of a cafeteria. The tedious task of doing so might result in the human annotator skipping some labels. Skipping a label causes label noise in the form of a missing label. Moreover, the task is often ambiguous when dealing with objects in a real-world image. Partially occluded objects, reflective surfaces, distance to the camera, and overcrowded images become relevant consideration points when labeling for object detection. These problems make the human annotator's role more prominent because more mental decisions are required. It also means that there will be more variation in the annotations, as different humans make different decisions.

Su *et al.* [18] have studied the overall process of annotation for object detection in a crowd-sourced manner. They first divided the task into three different sub-tasks: (1) draw a box, (2) verify the quality of a drawn box, and (3) verify a box coverage on a single image. Different people do all these sub-tasks via Amazon Mechanical Turk (AMT). They concluded that this method produces good quality annotations.

Russakovsky *et al.* [19] have studied the human-in-the-loop annotation process, where state-of-the-art object detection models are used to detect many of the objects in the image. Then humans are used for detecting all the objects that the models are unable to detect. This method is needed as no current object detection system is perfect, yet, and their goal is to have every object in the image annotated adequately. A properly annotated object should have a tightly fitted box and not an arbitrary margin of non-object space in the annotation. They conclude that their method of using humans and computer vision together was better than using either alone.

3 Object Detection Loss Functions

Single-shot detection (SSD) [8] uses both regression loss for bounding box regression and cross-entropy loss for classification. The cross-entropy loss for a sample with ground truth one-hot-encoded labels $\mathbf{y} = (y_1, y_2, \ldots, y_C)$ and predicted class confidences $\hat{\mathbf{y}} = (\hat{y}_1, \hat{y}_2, \ldots, \hat{y}_C)$ in a C-class classification problem is defined as

$$\mathrm{CE}(\mathbf{y}, \hat{\mathbf{y}}) = -\sum_{c=1}^{C} y_c \log(\hat{y}_c). \tag{1}$$

The focal loss was extended by Lin *et al.* [20] to handle difficult samples better. They show that this improvement can result in better accuracy compared to the cross-entropy loss. The focal loss was designed to emphasize hard positives. It is similar to cross-entropy loss but has a parameterized penalty factor $\gamma > 0$ weighing the influence of each sample based on its detection score. More specifically, the focal loss for the C-class classification with ground truth $\mathbf{y} = (y_1, y_2, \ldots, y_C)$ and predictions $\hat{\mathbf{y}} = (\hat{y}_1, \hat{y}_2, \ldots, \hat{y}_C)$ is defined as

$$\mathrm{FL}(\mathbf{y}, \hat{\mathbf{y}}) = -\sum_{c=1}^{C} \alpha_c (1 - \hat{y}_c)^\gamma y_c \log(\hat{y}_c), \tag{2}$$

with the balancing factor α_c [20], which is equal to 0.75 for all $c \in \{1, \ldots, C\}$ in all our experiments.

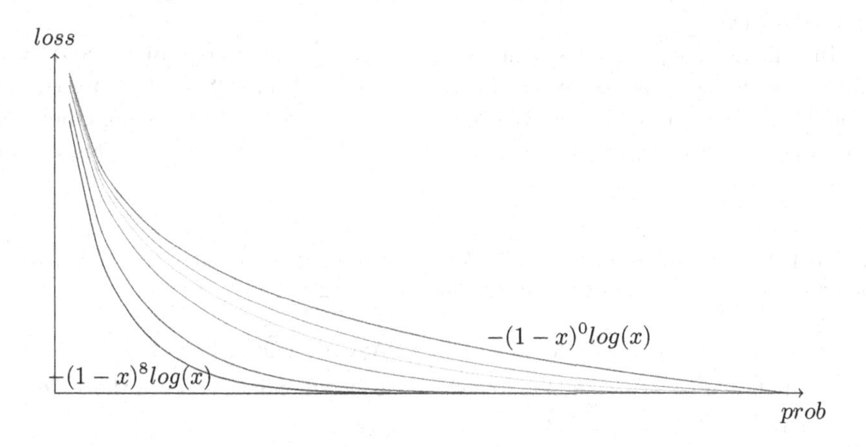

Fig. 2. Visualization of the focal loss function with different values for parameter $\gamma = 0, 1, 2, \ldots, 8$. The probability of being the ground truth is on the horizontal axis, and the loss is on the vertical axis. The higher the gamma value, the sharper the focus on harder cases. With gamma equaling zero, the focal loss is the same as cross-entropy loss.

In other words, the FL loss differs from the CE loss by the weight term $(1-\hat{y}_c)$, whose effect is to assign a higher weight for samples with low confidence (small

$\hat{y}_c$). γ affects the overall loss by lowering it; primarily well-classified samples with high confidence y_c for the most likely class c will have a negligible loss. At the same time, more attention is paid to learning the more complicated cases. Figure 2 demonstrates this scaling and aptly visualizes how the different γ parameters change the ferocity of the focus on more complicated cases. However, this loss weighting may have an adverse effect in the presence of label noise. The missing annotations are viewed as hard positives (non-annotated targets found by the model with a nonzero likelihood).

4 Experiments and Results

It was observed that sometimes in custom datasets, the focal loss seemed to produce results that were not as good as the research suggested. The intuition was formed that the weighting of complex cases, as performed by the focal loss function, would be more sensitive to label noise. The reason is that if a label is erroneous, to begin with, it is impossible to get right, so focusing on such a label leads the model astray and misuses the model capacity.

The experiments consider two questions: (1) how does label noise affect the two losses, and (2) how do models trained with different γ values tolerate label noise. For both experiments, we study the performance with three datasets: first with a small high-quality Indoor dataset, the second uses a large classical PASCAL VOC dataset, which does contain some annotation errors natively, and finally, with a single class FDDB dataset. Table 1 contains the characteristics of these datasets.

In all our experiments, the single-stage object detection (SSD) with MobileNet v1 [21] backbone network is fine-tuned from MS COCO pre-trained model for 100K training steps. We experimented only with the missing labels category. So, the training dataset has a percentage of randomly missing annotation boxes.

Table 1. Comparison of Indoor [13], PASCAL VOC 2012 [11] and FDDB [22] datasets based on source, size, quality of annotation, and usages.

	Indoor	PASCAL VOC	FDDB
Sample source	Indoor scenes	Collected online	Faces in the Wild
Image count	2213	17125	2845
Amount of instances	4500	40000	5171
Number of classes	7	20	1
Usage	Object detection	Multi-purpose	Face detection

4.1 Noise Robustness of the Two Losses: CE vs. FL

In this experiment, we use six different noise levels: 0%, 10%, 20%, 30%, 40% or 50%, of missing labels. The dropping of the labels was done randomly, but both networks were using the same training datasets. Also, the noisy datasets are constructed incrementally, *i.e.*,, the 20% noise had all the labels of the 10% dataset dropped (+10% more), and so forth. The model with both the CE loss and the FL loss with hyperparameter $\gamma = 2$ (as proposed in the original paper [20]) is fine-tuned for 100K steps, and *mAP@.50IoU* (mean average precision with 0.5 IoU threshold) is used as a performance evaluation metric.

Indoor Dataset. In the first set of experiments, we start training SSD using pre-trained weight from the MS COCO dataset, where some classes overlap between the datasets (chair, TV set, ...), while others do not (fire extinguisher). The resulting accuracies are presented in Fig. 3a; mAP@0.50 with the CE loss and the FL loss. Moreover, we show the *relative drop* in mAP with respect to noiseless labels in Fig. 4. It seems that the accuracy resulting from the FL loss objective function outperforms the CE loss for 10% – 20% noise levels. The FL loss is more robust till the 30% noise level and maintains a higher mAP than the CE loss. However, with the higher amount of label noise (>30%), FL loss accuracy plunged rapidly, falling behind the CE loss. For the extremely noisy (i.e., 50%) training dataset, accuracy from FL loss is 2% lower than that of CE loss.

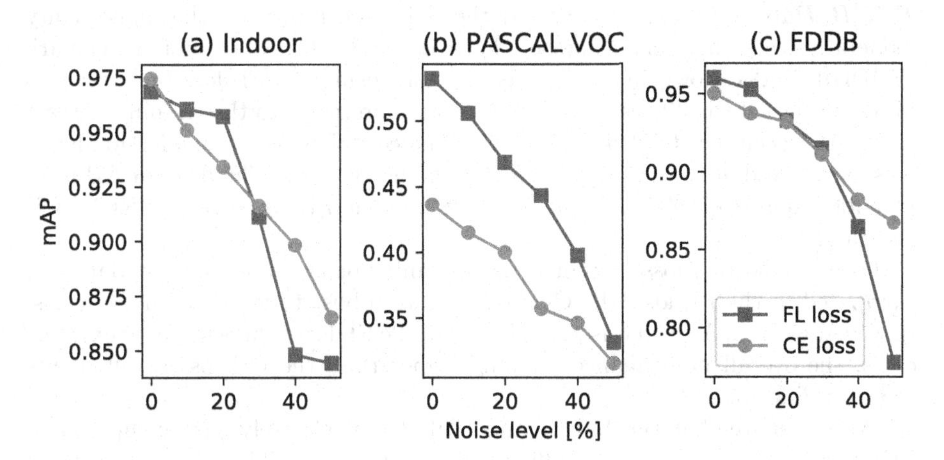

Fig. 3. Relationship between mAP (%) and different amount of noise levels on PASCAL VOC, and FDBB datasets.

PASCAL VOC Dataset. Next, we studied the noise sensitivity on the PASCAL VOC [11] dataset. The network using the FL loss function performs better than the alternative, but the accuracy with FL loss decreases more when the noise level increases compared to CE loss. Without added noise, FL loss gives 10% higher mAP than CE loss. While the FL loss outperforms the CE loss in

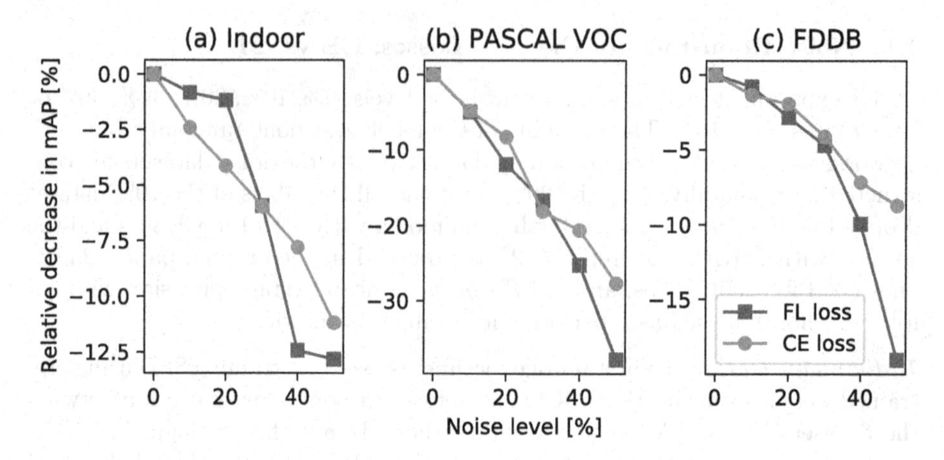

Fig. 4. Relative decrease in mAP (%) with respect to the noise levels in Indoor, PASCAL VOC, and FDDB datasets.

detection performance, it has a higher rate of mAP decrease than the CE loss. The difference in detection performance gets smaller by increasing label noise. The drop in mAP from no added noise to 50% label noise is 20.12% in with FL loss and 12.10% with CE loss.

FDDB Dataset. Next, we studied the noise sensitivity on the high-quality moderate-sized single class dataset, Face Detection Data Set and Benchmark (FDDB)[22]. As shown in Fig. 2c, the network using the FL loss function performs better till the 30% noise label. Adding more noise to the training dataset causes the accuracy to drop. The performance difference is smaller for lower noise levels and gets more significant for the noisy cases. The drop in AP from no added noise to 50% label noise is 18.28% in the FL loss case and 8.03% CE loss case.

Overall, the two losses seem to have similar behavior with these datasets. Compared to the FL loss, the CE loss is *more* robust to increased noise levels. However, with the VOC dataset, even though the FL loss suffers more for extreme cases, the overall performance remains higher than the CE loss at all points shown in Fig. 2b.

We speculate that the Indoor and FDDB are relatively easy compared to VOC, containing fewer small (difficult) bounding boxes. Thus, as long as most bounding boxes are in place, the FL loss equally weights the true targets and the hard negatives produced by the missing labels. The more varied and challenging nature of the PASCAL VOC dataset causes different noise tolerance behaviors than the smaller datasets.

4.2 Effect of the Gamma Parameter (γ)

In our second set of experiments, we compare the robustness of the FL loss for different values of the γ parameter. This time we only ran for three noise levels:

0%, 10%, and 50%. The gamma values tested were $\gamma = 1, 2, \ldots, 8$. All the other settings were kept the same as in the previous experiment.

Indoor Dataset. The first experiment in this set uses the Indoor dataset; results on this dataset are presented in Fig. 5a. In this dataset, the 10% noise detection performance is very close to the 0% noise. More interestingly, with extremely high label noise (50%), the gamma value has a significant impact. With $\gamma = 0$, the accuracy on the clean dataset (0% missing labels) is 18.52% more than the extremely noisy dataset (50% missing labels). With $\gamma = 8$, the clean dataset mAP is only 5.2% higher than the noisy dataset. The mAP curve indicates that a higher γ value does not affect the clean dataset while it boosts the performance in the presence of label noise.

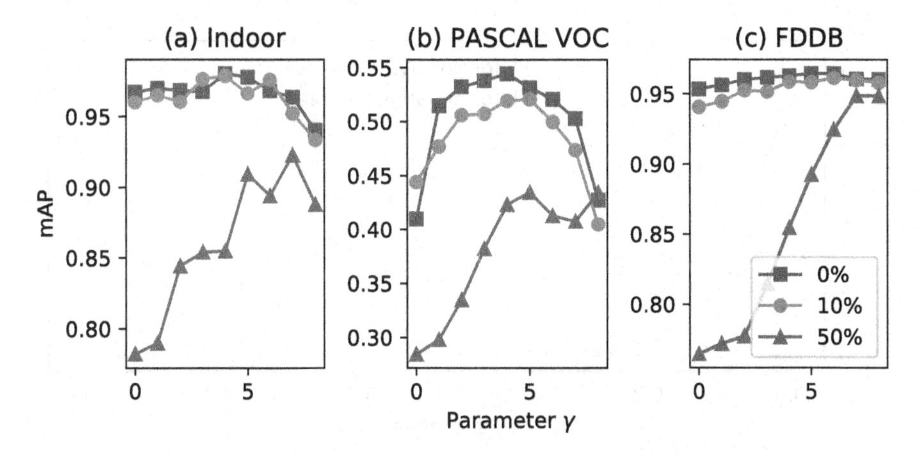

Fig. 5. Results on Indoor, PASCAL VOC and FDDB datasets with different gamma values on 0%, 10% and 50% noise levels.

PASCAL VOC Dataset. Next, we experiment on PASCAL VOC with different γ values. The higher values of gamma can be used to offset the effect of missing labels partially. Like the previous experiment, γ values in the range 4–6 have better performance.

FDDB Dataset. Experiments result on FDDB with different γ values is shown in Fig. 5c. Results coincide with our previous experiments. With $\gamma = 0$, the difference in performance between clean and extremely noisy datasets is 18.90%. However, this difference gets smaller by increasing the γ value. With $\gamma = 8$, a clean dataset is only 2% more accurate than a heavily noised dataset.

Generally, with an extreme amount of label noise, increasing the γ value improves the detection results. Still, the exact γ value and the detection performance are dependent on the dataset. This could indicate that maybe the sharp concentration introduced by the higher γ values can offset the missing labels in relatively easy datasets. Experiments on these datasets suggest that the robustness to label noise increases for larger γ values. In these cases, the model essentially learns from the complex samples only (annotated targets detected with low confidence and non-annotated targets detected with high confidence).

This is illustrated in Fig. 6, which shows the FL loss curves for both negative and positive examples. Due to the large γ value, the intermediate values ($\hat{y} \in [0.3, 0.7]$) behave as a *don't care* region, and the model does not learn from samples falling into this zone. Since all learning is based on complex samples (similarly to the support vector machine), it will be enough to push all objects with annotations to the "don't care" region. On the other hand, all negative samples (including missing annotations) can safely reside in this zone, and the model essentially learns to ignore those.

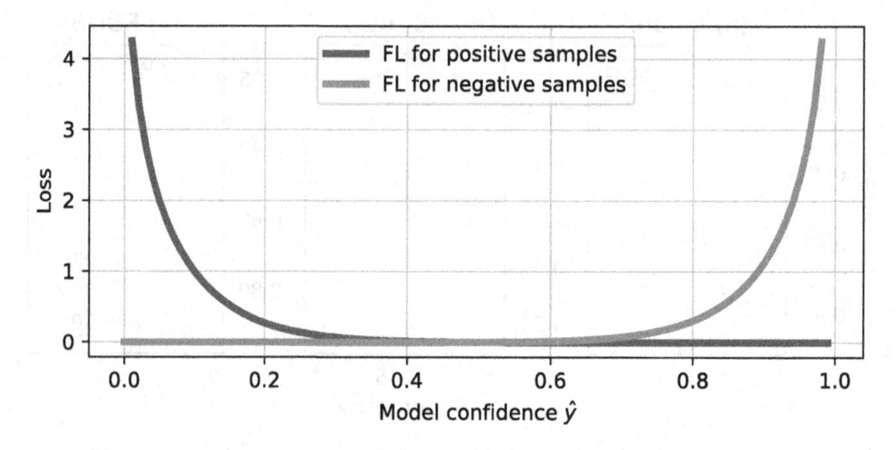

Fig. 6. Focal loss with $\gamma = 8$ for negative and positive samples with respect to model prediction confidence.

5 Conclusion

In this work, we characterized different types of label noise present in object detection datasets and explored the sensitivity of loss functions to them. With label noise being a crucial factor in ensuring the safety of the machine learning algorithm, we made sure to include experiments with large-scale real-world datasets. More specifically, we experimented on three datasets with varying amounts of label noise with cross-entropy and focal loss. Experiments suggest that focal loss suffers more with high amounts of noise, falling behind the cross-entropy loss. The second aspect studied is the effect of the hyperparameter γ on the sensitivity to label noise. It was discovered that larger values of γ improve the robustness to label noise such that extreme gamma values make the model indifferent to the noise level.

For future work, it would be beneficial to run more varied experiments to see how the label noise tolerance differs when training the network from scratch and its effect on system safety. Another point to consider would be running experiments with improved loss functions that are better suited for noisy datasets. It is also possible to quantify the risk associated with mislabeling by taking a statistical approach.

All relevant information, data, and codes are published open-access at https://github.com/adhikaribishwo/label_noise_on_object_detection.

Acknowledgment. This work was financially supported by Business Finland project *408/31/2018 MIDAS.*

References

1. Willers, O., Sudholt, S., Raafatnia, S., Abrecht, S.: Safety concerns and mitigation approaches regarding the use of deep learning in safety-critical perception tasks. In: Casimiro, A., Ortmeier, F., Schoitsch, E., Bitsch, F., Ferreira, P. (eds.) SAFECOMP 2020. LNCS, vol. 12235, pp. 336–350. Springer, Cham (2020). https://doi.org/10.1007/978-3-030-55583-2_25
2. Wozniak, E., Cârlan, C., Acar-Celik, E., Putzer, H.J.: A safety case pattern for systems with machine learning components. In: Casimiro, A., Ortmeier, F., Schoitsch, E., Bitsch, F., Ferreira, P. (eds.) SAFECOMP 2020. LNCS, vol. 12235, pp. 370–382. Springer, Cham (2020). https://doi.org/10.1007/978-3-030-55583-2_28
3. Schwalbe, G., et al.: Structuring the safety argumentation for deep neural network based perception in automotive applications. In: Casimiro, A., Ortmeier, F., Schoitsch, E., Bitsch, F., Ferreira, P. (eds.) SAFECOMP 2020. LNCS, vol. 12235, pp. 383–394. Springer, Cham (2020). https://doi.org/10.1007/978-3-030-55583-2_29
4. Dalal, N., Triggs, B.: Histograms of oriented gradients for human detection. In: IEEE Computer Society Conference on Computer Vision and Pattern Recognition (CVPR), vol. 1, pp. 886–893 (2005)
5. Cortes, C., Vapnik, V.: Support-vector networks. Mach. Learn. **20**, 273–297 (1995)
6. LeCun, Y., Bengio, Y., Hinton, G.: Deep learning. Nature **521**, 436–444 (2015)
7. Ren, S., He, K., Girshick, R., Sun, J.: Faster R-CNN: towards real-time object detection with region proposal networks. IEEE Trans. Pattern Anal. Mach. Intell. **39**, 1137–1149 (2017)
8. Liu, W., et al.: SSD: single shot MultiBox detector. In: Leibe, B., Matas, J., Sebe, N., Welling, M. (eds.) ECCV 2016. LNCS, vol. 9905, pp. 21–37. Springer, Cham (2016). https://doi.org/10.1007/978-3-319-46448-0_2
9. Redmon, J., Divvala, S., Girshick, R., Farhadi, A.: You only look once: unified, real-time object detection. In: IEEE Conference on Computer Vision and Pattern Recognition (CVPR), pp. 779–788 (2016)
10. Lin, T.-Y., et al.: Microsoft COCO: common objects in context. In: Fleet, D., Pajdla, T., Schiele, B., Tuytelaars, T. (eds.) ECCV 2014. LNCS, vol. 8693, pp. 740–755. Springer, Cham (2014). https://doi.org/10.1007/978-3-319-10602-1_48
11. Everingham, M., Eslami, S., Van Gool, L., Williams, C., Winn, J., Zisserman, A.: The pascal visual object classes challenge: a retrospective. Int. J. Comput. Vision **111**, 98–136 (2014)

12. Kuznetsova, A., et al.: The open images dataset V4. Int. J. Comput. Vision **128**, 1956–1981 (2020)
13. Adhikari, B., Peltomaki, J., Puura, J., Huttunen, H.: Faster bounding box annotation for object detection in indoor scenes. In: 7th European Workshop on Visual Information Processing (EUVIP), pp. 1–6 (2018)
14. Zhang, X., Liu, C., Suen, C.: Towards robust pattern recognition: a review. Proc. IEEE **108**, 894–922 (2020)
15. Frenay, B., Verleysen, M.: Classification in the presence of label noise: a survey. IEEE Trans. Neural Netw. Learn. Syst. **25**, 845–869 (2014)
16. Li, C., Zhang, C., Ding, K., Li, G., Cheng, J., Lu, H.: BundleNet: learning with noisy label via sample correlations. IEEE Access. **6**, 2367–2377 (2018)
17. Lee, K., He, X., Zhang, L., Yang, L.: CleanNet: transfer learning for scalable image classifier training with label noise. In: IEEE/CVF Conference on Computer Vision and Pattern Recognition (CVPR), pp. 5447–5456 (2018)
18. Su, H., Deng, J., Fei-Fei, L.: Crowdsourcing annotations for visual object detection. In: AAAI Human Computation Workshop, pp. 40–46 (2012)
19. Russakovsky, O., Li, L., Fei-Fei, L.: Best of both worlds: human-machine collaboration for object annotation. In: IEEE Conference on Computer Vision and Pattern Recognition (CVPR), pp. 2121–2131 (2015)
20. Lin, T., Goyal, P., Girshick, R., He, K., Dollar, P.: Focal loss for dense object detection. IEEE Trans. Pattern Anal. Mach. Intell. **42**, 318–327 (2020)
21. Howard, A.G., et al.: MobileNets: Efficient Convolutional Neural Networks for Mobile Vision Applications. arXiv (2017)
22. Jain, V., Learned-Miller, E.: FDDB: a benchmark for face detection in unconstrained settings. Department of Computer Science, University of Massachusetts. UM-CS-2010-009 (2010)

Human-in-the-Loop Learning Methods Toward Safe DL-Based Autonomous Systems: A Review

Prajit T. Rajendran[1(✉)], Huascar Espinoza[1], Agnes Delaborde[2], and Chokri Mraidha[1]

[1] Université Paris-Saclay, CEA, List, F-91120 Palaiseau, France
{prajit.thazhurazhikath,huascar.espinoza,chokri.mraidha}@cea.fr
[2] Laboratoire national de métrologie et d'essais, Trappes, France
agnes.delaborde@lne.fr

Abstract. The involvement of humans during the training phase can play a crucial role in mitigating some safety issues of Deep learning (DL)-based autonomous systems. This paper reviews the main concepts and methods for human-in-the-loop learning as a first step towards the development of a framework for human-machine teaming through safe learning and anomaly prediction. The methods come with their own set of challenges such as the variation in the training data provided by the human and test-time distributions, the cost involved to keep the human in the loop during the long training phase and the imperfection of the human to deal with unforeseen circumstances and define safer policies.

Keywords: AI safety · Human-in-the-loop learning · Deep learning

1 Introduction

Deep learning (DL) based components are finding their way in safety-critical autonomous systems with an increased vigour after the massive strides made in the field of artificial intelligence (AI) in the past decade. However, the opaque nature of deep learning systems makes it necessary for experts to analyse vast amounts of data to determine why the components make particular decisions or act in a certain manner [1]. This lack of transparency is a safety hazard, considering that risk assessment is performed before deployment, and depends on the designers being aware of the expected behaviour of the components. Another drawback in DL components is the weakness against adversarial inputs and out-of-distribution inputs [2,8].

Humans on the other hand are better equipped to deal with unforeseen circumstances and define safer policies. Humans could be helpful in some applications such as automated driving, where an expert driver is involved in the simulation environment to monitor the DL agent and correct it during the training phase before hazardous events actually occur [3]. Human-in-the-loop

I. Habli et al. (Eds.): SAFECOMP 2021 Workshops, LNCS 12853, pp. 251–264, 2021.
https://doi.org/10.1007/978-3-030-83906-2_20

approaches can also be used to *transfer human knowledge* to the agent (e.g., through labelling, intervention, evaluation or demonstration), helping it learn faster and with lesser data. While humans can help an agent towards identifying these errors, humans themselves have imperfections and uncertainties. The challenge in DL-based safety-critical applications is to combine human intervention and agent training by still keeping these performance gains and ensuring safe learning and execution under nominal and abnormal situations.

This paper reviews existing human-in-the-loop learning methods as a first step toward the development of a framework for safe learning and anomaly detection under human-machine teaming. Section 2 summarizes the main concepts for safe DL-based autonomous systems and human-in-the-loop learning. Section 3 provides a comparative review of methods that come with their own set of challenges such as the variation in the training data provided by the human and test-time distributions, the cost involved to keep the human in the loop during the long training phase and the unsafe blind spots from humans.

2 Background

2.1 Safety Concerns in AI-Based Autonomous Systems

Safety is of utmost importance in AI-based autonomous systems, which is why there is a large body of research and work in developing safety standards and assurance methods for such systems. *Safety is an emergent property*, meaning that it manifests itself as the result of a combination of various system components and not as a property of any particular component individually. Ensuring safety involves defining the set of operating conditions under which a given driving automation system is specifically designed to function [4], identification of potential hazards to prevent faults in individual components and failure of the system as a whole. Traditional methods such as Hazard Analysis and Risk Assessment (HARA) from ISO 26262 [5] in the automotive domain are focused on ensuring that the specified functional requirements of the system are met at the design time. These methods are considered not fully adapted in safety assurance of AI/DL-based systems, mostly because the probabilistic nature of the computation and the type of data that is processed may lead to issues relative to transparency or predictability. DL-based systems are known for their black-box nature, wherein the inputs and outputs are known but the contents of implementation are unknown [7]. This nature poses several problems in terms of regulatory qualification/certification, since it can impact safety assurance.

Uncertainty and variability in the environment conditions and DL-based models could lead to abnormal situations. These situations are difficult to analyse during the design phase, especially for autonomous systems, which may rely on multiple sensors and DL-based components [8]. Each of the components and sensors could have some associated uncertainty and this uncertainty could be propagated to the subsequent components, thereby placing the whole system at risk. For example, if there is a high uncertainty associated with a driving decision made by a planner, the action it proposes could be potentially unsafe.

Known unknowns are data points for which the model makes a mistake, and demonstrates that it had a lower confidence for that particular input. Unknown unknowns are also data points where the model made a mistake, but here the model is highly confident about its predictions [9]. The high confidence exhibited for an erroneous sample is potentially dangerous, as the safety thresholds for hand-over or safe-transition could be breached and the agent would proceed autonomously with an incorrect prediction. Approaches developed for addressing known unknowns cannot be used to deal with unknown unknowns, because the model is highly confident in its ignorance. Unknown unknowns can arise due to insufficiency or lack of diversity in the training data, causing the model to be not representative of the real world.

Out-of-distribution (OoD) samples are those which are significantly different from the training samples [10]. Such samples could confuse the DL components because they were not trained to handle such inputs. Generative models can detect OoD samples but are overly pessimistic in predicting safety, since all OoD samples may not lead to a catastrophic situation [11]. However, since the DL-based components are black-box models in the most part, it is advised to operate under the assumption that out-of-distribution samples could be under the category of "unknown unsafe" [6].

The work [12] discusses some unintended and harmful behaviour that may emerge from poor design of AI-based systems. Sometimes the AI algorithm exploits an unknown loophole in the design of the system to achieve the objective in an undesirable manner, possibly leading to unsafe or catastrophic results. A successful completion of a task by an AI agent does not underline the safety in the steps it took. If a human could provide fine-grained supervision and feedback every time the agent performs an action, the AI agent could be nudged into selecting safer policies.

2.2 Human-in-the-Loop Learning Approaches

Human-in-the-Loop learning is a paradigm in which intelligent systems are designed with active human participation in its training process [13]. In DL-based autonomous systems, human interaction is leveraged to improve the safety and performance of the system. The main advantage of this approach is the ability to *incorporate human judgment in the learning process, thereby leading to safer policies*. As human-in-the-loop methods allows human to have a role in the learning phase, human knowledge about complex and unforeseen situations can be transferred to and imbibed by the system. By incorporating human intelligence and judgement, the learning time could also be reduced because the human would stop the system from exploring irrelevant conditions and states. An adapted human interaction strategy can thus improve the performance and safety of the system as compared to a fully automated counterpart.

The human-in-the-loop learning methods present in literature can be categorized based on the nature and level of control that the human or oracle has [14] during the learning process (see Fig. 1). The works we review in this paper were chosen to cover this categorization.

Active learning is a semi-supervised approach in machine learning (ML) where only a subset of the training data is labelled. A human in the loop is queried interactively to label data points from the unlabelled set, when the model deems the particular data point to be of interest [15]. The fundamental concept in this approach is that by selectively querying only the most important and different data points, the amount of training data needed could be brought down [16]. This could also translate into a higher level of accuracy since the model can pick out rare and important data points and learn the pattern of such points more efficiently. The human annotator responds to the query with the correct label for the queried instance. Since this approach vastly reduces the amount of training data and annotation expenses, it is a popular form of human-in-the-loop learning especially for tasks in e.g. natural language processing [17].

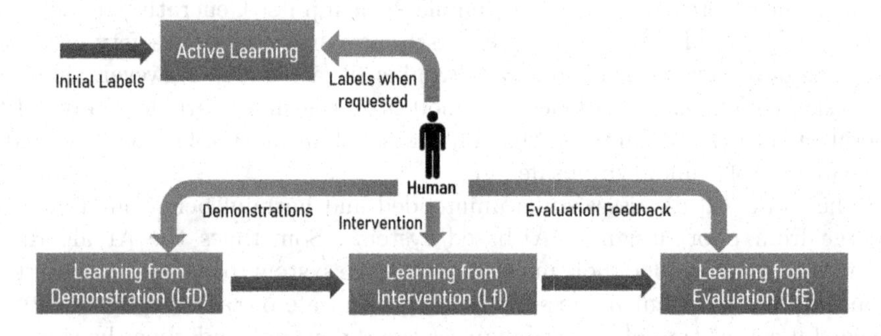

Fig. 1. Categorization of the human-in-the-loop learning approaches.

In *Learning from Demonstration (LfD)*, the human is providing demonstrations and is in full control [14]. The goal is to train an AI agent to mimic the human behavior in a particular task. In this approach the assumption is that the human is the best source of safe and efficient policies and the human is the demonstrator who provides examples of how to perform the task. Using the human demonstration data as a starting point, the AI agent is already in a much better state than a random initialization, where the agent might make many mistakes before learning a decent policy. When safety is critical, and training with more catastrophic scenarios is not feasible, this approach is shown to work well. One of the drawbacks of this approach is that the agent's performance depends heavily on the quantity of the human demonstration data, thereby making it expensive in terms of human labour and time. Another drawback is that the environment during tests and deployment could be dynamic, and different from that in the training phase. In such cases the learned demonstration data would not prepare the agent to deal with unexpected situations, and a policy purely mimicking the human would not allow the system to know how to act safely [3].

In the *Learning from Intervention (LfI)* approach, the human occasionally intervenes, and both the human and autonomous system share control [14]. LfI is a method of human-in-the-loop learning that can improve safety wherein the human stays in the loop during the learning phase and intervenes i.e. takes over control when there is a threat of the autonomous agent entering a catastrophic state. Systems that embed AI, such as drones or robots, learn by interacting with a real environment; in this context, unsafe actions could potentially lead to serious damage on human beings or property. A human's continued presence in the loop to perform safe exploration can thus prevent catastrophic actions. LfI is a promising approach to improve safety, but the main drawback in this approach is that it is time consuming and expensive to keep the human in the loop for a long training duration, which is typically the case for tasks such as control of autonomous vehicles or drones in complex environments.

In the *Learning from Evaluation (LfE)* approach, the human provides evaluative feedback as the autonomous system performs a task, and the autonomous system is in control of identifying the policy[14]. Here the human acts as a supervisor and provides real-time evaluative feedback to the autonomous system, thereby shaping its policy. The feedback could be provided in different ways, such as a binary feedback of "good" or "bad", or choosing the best alternative between several policies. The LfE approach has the advantage that the humans are only required to possess an understanding of the desired objective, and do not need to specify what the underlying policy to attain the objective is. One of the challenges of this approach is that the human must be able to be quick enough to respond and evaluate an action. If the environment is slowed down in a simulator for example, this could be achieved, albeit at the cost of training time and expenses of keeping a human in the loop for a prolonged period. It could also be challenging for the AI agent to attribute which of the actions correspond to the provided feedback, especially in complex environments. In addition, the human evaluative feedback could be subjective, depending on the human's perception.

The active learning approach is more suitable for tasks such as classification and regression which could be a part of some components in an autonomous system. Active learning could be considered as a variation of learning by demonstration, since the human provides the labels in real time. In general LfD, LfI and LfE represent the mechanisms in which the human can be present in the loop online during the learning process, and will constitute the focus of this study.

3 Review of Human-in-the-Loop Learning Approaches

In safety-critical autonomous systems, safety is given more importance as opposed to performance in traditional AI systems. Additionally when the learning process involves humans, care should be taken to reduce the human costs in terms of data and time. Therefore, in this section we compare and analyze some common human-in-the-loop learning approaches, with particular focus on:

- **Training time:** In certain applications such as reinforcement learning, the training time required for the DL component to reach a significant level of performance on its own is quite large. In such cases it is not practical to keep the human present in the loop for the whole duration of training.

- **Data requirement:** DL components typically require large amounts of data to reach a reasonable level of performance. This translates to more labels or more demonstration data, which requires more human efforts and time.
- **Regard to safety:** In this study, we examine the degree of various measures incorporated by the human in the loop to mitigate the effects of safety factors such as OoD, unknown unknowns and others, as discussed in Sect. 2.1.

We have chosen the most representative works to the best of our knowledge for each method or combination of methods. This is summarized in Table 1. As discussed in Sect. 2, the primary focus is on the LfD, LfI and LfE human-in-the-loop learning approaches.

3.1 Learning from Demonstration (LfD)

LfD methods mostly fall under the purview of imitation learning, where the AI agent learns by imitating a human expert or an oracle. *Behaviour cloning (BC)* is the simplest LfD approach wherein the human demonstration data is learnt by the agent in a supervised manner [18–20]. This performs poorly in run-time because the predictions by the agent only work if the same state was present in the training phase. Distributional shift between training and testing phases is a major reason why behavioural cloning is not practical in autonomous systems. In [3], the authors take into account the agent and human blindspots to improve safety. Here, a behavior cloned agent trained on real world demonstration data acts as a proxy for humans, whereas the agent learns a policy by interacting with the simulator. A blindspot detector is trained to detect the features which each agent misses and by detecting these blindspots in runtime, safety can be improved by handing over control to the safer agent. The work [21] presents an approach to generalize and learn from very few demonstrations. The proposed method proceeds by creating a context embedding from the demonstration data, which is then used to decide the current action of the agent. The architecture contains three blocks- one to process the demonstration data, one to prepare a context embedding and finally, a module to choose the action. In [22], high-level command input from the human is used to condition the imitation learning algorithm in the autonomous driving task. The authors demonstrate that perception alone is not a good input to imitation learning for driving, and that the high-level intention of the demonstrator based on his or her internal state will aid the system to be more robust.

Inverse reinforcement learning (IRL) is an approach where the goal is to learn the reward function purely based on the demonstrations from the expert, and then derives the optimal policy. It is assumed that the human demonstration data is optimal, as the problem is intractable if we take into the numerous faulty reward functions in the case of human sub-optimality [23]. The main disadvantages of this approach is that it is expensive in terms of human time and effort, incorrectly considers that the human is perfect, and the anomalies induced by the human may not be appropriately addressed.

Generative methods have also been attempted in works such as *Generative adversarial imitation learning(GAIL)* [24,26–28], working on the same principle. GAIL is composed of two neural networks- a generator or policy network trained using Trust Region Policy Optimization (TRPO) [25] and a discriminator network learnt as a supervised learning problem on demonstration data. The goal is to find a policy where the discriminator cannot distinguish between observed states of the expert policy and agent policy. This approach is typically expensive in terms of training time.

LfD methods based on pure behavioral cloning perform poorly on OoD and previously unseen data and hence are deemed to be low on our scale of safety considerations. Generative methods such as GAIL work better in dealing with previously unseen data because they learn the distribution, but may still be susceptible to unknown unknowns. The works combining LfI with LfD provide better support to the safety considerations because they encourage to actively avoid catastrophic states.

3.2 Learning from Intervention (LfI)

In the work *Trial without error* [35], an improvement of imitation learning using LfI is proposed wherein the initial policy is learnt from the human by demonstrations acting as a safe starting point from which the agent starts to diverge by occasionally exploring alternate actions. If the proposed action is dangerous, the human in the loop blocks the agent and sends a penalty and the corrected action. Since the human has to constantly watch over the agent in the initial phase of training, we can instead train a helper model (blocker) which learns the situations where the human blocked the action previously in a supervised manner. After sufficient training this helper model is capable of replacing the human as it learns to classify catastrophic situations and performs an intervention. This approach could reduce the human time and expenses required, but the possibility that the human intervention instances do not cover all possible scenarios in the real world affects the safety of the system. The blocker model should be trained adequately such that it is highly accurate, and even then new, previously unseen situations may confuse it. The authors of [36] present a similar approach involving the human acting as a blocker preventing the agent from performing catastrophic actions. It also proposes to reduce the human intervention time and ensure safety by combining model-based approaches and training a supervised learner to improve sample efficiency. In the work [37] a user intervention algorithm that relies on user's hands tracking is proposed.

DAgger is an approach proposed in [38] which initially uses the human expert's policy to gather a dataset of trajectories and learns to mimic the expert, while still placing an oracle in the loop. At each iteration under the current policy when the agent explores the environment, it collects data based on actions taken by the agent and some actions are sampled from the oracle who is present live in the loop. The collected data is appended to the original expert data and the agent is retrained on the aggregated dataset. The probability of querying the oracle is designed to decay as the agent performance as per a pre-set evaluation

Table 1. Summary of various human-in-the-loop learning approaches

Work	Approach	LfD	LfE	LfI	Helper model(s)	Data requirement	Training time	Regard to safety
Faraz Torabi et al. [18]	Behaviour cloning	✓			Supervised imitator	High	Moderate	Low
Vinicius G Goecks et al. [19]	Behaviour cloning	✓			Supervised imitator	High	Moderate	Low
Wael Farag et al. [20]	Behaviour cloning	✓			Supervised imitator, CNN	High	Moderate	Low
Ramya Ramakrishnan et al. [3]	Imitation learning	✓		✓	Blindspot detector	High	Moderate	High
Yan Duan et al. [21]	Imitation learning	✓			Context embedding	Low	High	Moderate
Felipe Codevilla et al. [22]	Imitation learning	✓			Supervised imitator	High	Moderate	Moderate
Pieter Abeel et al. [23]	Inverse RL	✓			Reward estimator	High	Moderate	Moderate
Jonathan Ho et al. [24]	GAIL	✓			Generative reward model	Moderate	High	Moderate
Jonathan Lacotte et al. [26]	GAIL	✓			Generative reward model	Moderate	High	Moderate
Faraz Torabi et al. [27]	GAIL	✓			Generative reward model	Moderate	High	Moderate
Konrad Zolna et al. [28]	TRAIL	✓			Generative reward model	Moderate	High	Moderate
Stephane Ross et al. [38]	DAgger	✓		✓	Supervised imitator	Moderate	Moderate	Moderate
Jiakai Zhang et al. [39]	Safe DAgger	✓		✓	Discrepancy function	Moderate	Moderate	High
Kunal Menda et al. [40]	Ensemble DAgger	✓		✓	Uncertainty estimator	Moderate	High	High
W. Bradley Knox et al. [29]	TAMER		✓		Reward predictor	Moderate	High	Moderate
Ngo Anh Vien et al. [31]	ACTAMER		✓		Preference predictor, critic	Moderate	High	Moderate
Garrett Warnell et al. [32]	Deep TAMER		✓		Reward predictor	High	High	Moderate
W. Bradley Knox et al. [30]	TAMER		✓		Reward predictor	Moderate	High	Moderate
Riku Arakawa et al. [33]	DQN-TAMER		✓		DQN, Reward predictor	Moderate	High	Moderate
Paul F Christiano et al. [34]	Human preference		✓		Preference predictor	Low	Moderate	Moderate
William Saunders et al. [35]	Trial without error			✓	Blocker model	Low	Moderate	High
Bharat Prakash et al. [36]	Human intervention			✓	Blocker, dynamic model	Low	High	High
Aleksandar Jevtic et al. [37]	Human intervention			✓	Intervention block	Low	Moderate	High

metric gets better and better. Even with this decaying function, the number of queries to the oracle is typically large for complex tasks and this could be expensive and not scalable in terms of human time and effort.

The approach *Safe DAgger*, proposed in [39] computes the discrepancy between the human expert actions and agent actions and uses this as a decision metric for querying. If the distance between the actions is less than a preset threshold, the agent action is sampled and if the discrepancy is large, the oracle is queried to provide the appropriate action. A drawback of this approach is that the choice of the metric and threshold varies for each application and scenario and is hard to define. *Ensemble DAgger*, an improvement to Safe DAgger is proposed in [40]. This approach makes use of the uncertainty of the agent policy using Bayesian DL. Discrepancy between the expert's and the novice's mean action, as well as the novice's doubt (variance) is used to decide whether the oracle has to be queried.

DAgger relies on random sampling of actions from the oracle and is relatively less safe than the other LfI methods described. The other techniques rank highly in safety considerations because the human intervenes and prevents catastrophic actions. The feedback of the human in this case is a safe alternate action instead of just a feedback of "good" or "bad". The human in the loop makes this approach safer in terms of policy definition and can also work well against OoD outliers, provided the human can identify their unsafe nature.

3.3 Learning from Evaluation (LfE)

TAMER is an approach proposed in [29] wherein the agent learns autonomously via interaction with environment, and a human trainer gives feedback to the agent. The human provides feedback in the form of scalar reward signals in response to the agent's actions, which allows the agent to develop a safe and effective policy to perform a certain task. The agent shapes the policy so as to maximise the reward provided by the human. This approach does not require the human to be aware of the learning policy of the agent as the evaluation is purely based on the observed action. The agent models the human reward function and eventually learns to predict the human reward to a particular action, meaning that the human need not be present in the loop once the agent learns to predict the human reward function to a reasonable degree. The authors in an extension of this work [30] where apart from human feedback for agent tasks, the intention for the next action as advertised by the agent is also evaluated. The work *ACTAMER* [31] focuses on extending TAMER to continuous state and action spaces by general function approximation of a human trainer's reinforcement signal. The authors propose an actor-critic framework where the actor is the human preference policy and the critic is the vanilla TAMER approach.

In *Deep TAMER* [32], the representational power of deep neural networks is leveraged in order to learn complex tasks in just a short amount of time with human evaluations. Using this approach, tasks in a high-dimensional environment such as in autonomous driving scenarios can be learnt more easily.

DQN-TAMER proposed in [33] is an approach wherein the reinforcement learning technique of Deep Q-Learning networks (DQN) is combined with TAMER. The policy is learnt by a DQN whereas the reward is learnt from the human by a binary feedback based on TAMER. In the study introducing this concept, the feedback comes from the classification of facial expressions of the human. Such an evaluative feedback method is highly dependent of the performance of the facial expression recognition, which is subject to high inter- and intra-individual variations. In general, the DQN-TAMER approach also faces risks of delayed feedback. Another interesting work is *Deep RL from human preferences* [34] where human preferences are queried based on uncertainty in the reward function estimator. Humans are given choices of alternate trajectories, and they select their preferred safe and efficient trajectory for the current state. Then a helper model is trained using supervised learning to predict the human preference and reward. This approach has the drawback of the fact that giving an alternative to humans often leads to ambiguity, and biases may creep in. Additionally, the human may be forced to choose the lesser of the evils if all of the proposed trajectories are poor.

In LfE methods, humans give feedback for the action taken by the agent, and therefore the agents can avoid making undesirable actions. However, we consider the safety consideration level to be moderate because of susceptibility to unknown unknowns and a risk of reward hacking, where the agent maximises reward but performs unsafe and unintended actions. The binary and delayed nature of feedback, explained in [33] is another matter of concern.

3.4 Gaps and Opportunities for Future Work

The approaches described above explore different modes of human-in-the-loop learning and improves safety and performance of autonomous systems. A frequent theme in the above is the presence of a supplementary model to aid in replacing the human in the form of a clone of human actions, a human reward learner, or a human-like blocker. This theme sets the tone for potentially adding other supplementary models such as uncertainty estimators or anomaly detectors which could further help in policy filtering and improving safety. The following aspects can further improve safety in human-in-the-loop learning methods:

Human Blind Spots: Most of the approaches in literature consider the human to be a perfect oracle. In active learning, we operate under the assumption that the human always knows the right label of a complicated data point. However there may be some data points which are ambiguous to human labellers as well. In LfD, it is considered that the demonstration data is a good indicator of a safe policy with the assumption that the human did not make any mistakes while performing the demonstrations. In LfE, it is assumed that the human gives the appropriate reward for the correct action, although there is a possibility that there is an ambiguity between different human evaluators. In LfI, the human is expected to take over control or perform an intervention in time- but there is a possibility that the response time could be slower than desired, leading to undesirable effects. Thus, the assumption that the human is perfect is not an accurate representation of reality and we can improve the safety and performance even further by considering the uncertainty of the human.

Make Use of Uncertainty: The approaches above contain DL components but fail to take into account the uncertainty associated with the decisions of the components. The uncertainty could be with respect to the model parameters themselves or stemming from the data used for training. A first step in dealing with uncertainty of the decision would consist in an appropriate specification of the factors of uncertainty. There are various approaches such as Bayesian neural networks, mixture density networks and deep ensembles which could be used to estimate the uncertainty in the models. The uncertainty of the DL components could be used as an important piece of the puzzle to understand the system and develop safer and more effective policies by the agent.

Make Use of Anomaly or Out-of-Distribution Detectors: The approaches proposed do not consider the possibility of perturbations in the models by outliers, anomalies or OoD samples. In the case of LfD, this could be a serious problem because the human demonstration might contain such samples, which may affect the safety of the system. In LfI, it is possible for the human to intervene and prevent a catastrophy, but the agent ought to be able to assign the credit to an anomaly or OoD sample rather than its own policy. If an anomaly detector is present, the system could be made aware that the current state is an anomaly and its behaviour ruleset could be consulted to take an appropriate action or hand over control to the human. In LfE, the human evaluator needs to be aware that the current situation is an anomaly or unexpected and not a mistake due to the policy of the agent. A typical LfE feedback mechanism may not be able to capture this subtle difference between a mistake by the agent and an anomaly in the environment. Thus, we propose that the addition of an anomaly or OoD detector could improve the learning process.

4 Conclusion

In this paper, we explored emerging works where a human is present in the loop in the learning and run-time phases of a DL component in autonomous systems. The involvement of humans is shown to play a crucial role in mitigating some safety issues of traditional black-box approaches. We conceptually defined human-in-the-loop learning and its various categories, articulated its necessity and discussed its potential to improve safety and efficiency in DL components of autonomous systems. The works presented in this review each offer their own approaches to human-in-the-loop learning. Finally, we identified several opportunities and exciting areas of future study for further research as a first step towards the development of a framework for human-machine teaming in safety-critical domains.

References

1. Heuilleta, A., Couthouis, F., et al.: Explainability in deep reinforcement learning. Knowl.-Based Syst. **214**, 106685 (2020)
2. Papernot, N., McDaniel, P., et al.: The limitations of deep learning in adversarial settings. In: 1st IEEE European Symposium on Security and Privacy, Saarbrucken, Germany. IEEE (2016)
3. Ramakrishnan, R., Kamar, E., Nushi, B., Dey, D., Shah, J., Horvitz, E.: Overcoming blind spots in the real world: leveraging complementary abilities for joint execution. In: AAAI, pp. 6137–6145 (2019)
4. SAE International, Taxonomy and definitions for terms related to driving automation systems for on-road motor vehicles, SAE International (J3016) (2018)
5. International Organization for Standardization: ISO 26262-1:2018 Road vehicles - Functional safety. Standard, International Organization for Standardization, Geneva, CH (2018)
6. International Organization for Standardization: ISO/PAS 21448:2019 Road vehicles - Safety of the intended functionality. Standard, International Organization for Standardization, Geneva, CH (2019)
7. Systems and software engineering — Vocabulary, ISO/IEC/IEEE 24765:2017 (2017)
8. Arnez, F., Espinoza, H., et al.: A comparison of uncertainty estimation approaches in deep learning components for autonomous vehicle applications. In: Workshop AISafety 2020 - Workshop in Artificial Intelligence Safety (2020)
9. Lakkaraju, H., Kamar, E., et al.: Identifying unknown unknowns in the open world: representations and policies for guided exploration. In: NIPS Workshop on Reliability in ML (2016)
10. McAllister, R., Kahn, G., et al.: Robustness to out-of-distribution inputs via task-aware generative uncertainty. In: International Conference on Robotics and Automation (ICRA), Palais des congres de Montreal, Montreal, Canada, 20–24 May 2019 (2019)
11. Geiger, A., Liu, D., et al.: TadGAN: time series anomaly detection using generative adversarial networks. In: IEEE International Conference on Big Data (Big Data) Atlanta, Georgia, USA, 10–13 December 2020 (2020)
12. Amodei, D., Olah, C., et al.: Concrete Problems in AI Safety. arXiv preprint arXiv:1606.06565 (2016)
13. Waytowich, N.R., Goecks, V.G., et al.: Cycle-of-Learning for Autonomous Systems from Human Interaction. arXiv preprint arXiv:1808.09572 (2018)
14. Goecks, V.G.: Human-in-the-loop methods for data-driven and reinforcement learning systems. Ph.D. thesis (2020)
15. Settles, B.: Active learning literature survey, Computer Sciences Technical report 1648 University of Wisconsin-Madison (2010)
16. Druck, G., Settles, B., McCallum, A.: Active learning by labeling features. In: Proceedings of the Conference on Empirical Methods in Natural Language Processing (EMNLP), pp. 81–90. ACL Press (2009)
17. Freund, Y., Seung, H.S., Shamir, E., Tishby, N.: Selective samping using the query by committee algorithm. Mach. Learn. **28**, 133–168 (1997)
18. Torabi, F., Warnell, G., et al.: Behavioral cloning from observation. In: Proceedings of the 27th International Joint Conference on Artificial Intelligence (IJCAI 2018), Stockholm, Sweden, July 2018

19. Goecks, V.G., Gremillion, G.M., et al.: Integrating behavior cloning and reinforcement learning for improved performance in dense and sparse reward environments. In: International Conference on Autonomous Agents and Multi-Agent Systems (AAMAS 2020), Auckland, New Zealand, 9–13 May 2020 (2020)
20. Farag, W., Saleh, Z., et al.: Behavior cloning for autonomous driving using convolutional neural networks. In: Proceedings of the 27th International Joint Conference on Artificial Intelligence (IJCAI 2018), Stockholm, Sweden, July 2018
21. Duan, Y., Andrychowicz, M., et al.: One-shot imitation learning. In: Advances in Neural Information Processing Systems 30 (NIPS 2017) (2017)
22. Codevilla, F., Muller, M., et al.: End-to-end driving via conditional imitation learning. In: IEEE International Conference on Robotics and Automation (ICRA), Brisbane, Queensland, Australia, 21–25 May 2018 (2018)
23. Abbeel, P., Ng, A.Y.: Apprenticeship learning via inverse reinforcement learning. In: Proceedings of the 21st International Conference on Machine Learning (2004)
24. Ho, J., Ermon, S.: Generative Adversarial Imitation Learning. arXiv preprint arXiv:1606.03476 (2016)
25. Schulman, J., Levine, S., et al.: Trust region policy optimization. In: Proceedings of the 32nd International Conference on International Conference on Machine Learning, vol. 37, pp. 1889–1897, July 2015
26. Lacotte, J., Ghavamzadeh, M., et al.: Risk-sensitive generative adversarial imitation learning. In: The 21st International Conference on Artificial Intelligence and Statistics (AISTATS), Lanzarote, Canary Islands, 9–11 April 2018 (2018)
27. Torabi, F., Warnell, G., et al.: Generative Adversarial Imitation from Observation. arXiv preprint arXiv:1807.06158 (2018)
28. Zołna, K., Reed, S., et al.: Task-relevant adversarial imitation learning. In: Conference on Robot Learning (CoRL), 16–18 November 2020 (2020)
29. Knox, W.B., Stone, P.: TAMER: training an agent manually via evaluative reinforcement. In: The 7th IEEE International Conference on Development and Learning, pp. 292–297 (2008)
30. Knox, W.B., Stone, P., et al.: Learning from feedback on actions past and intended. In: The 7th ACM/IEEE International Conference on Human-Robot Interaction (HRI), Boston, Massachusetts, USA, 5–8 March 2012 (2012)
31. Vien, N.A., Ertel, W.: Reinforcement learning combined with human feedback in continuous state and action spaces. In: IEEE International Conference on Development and Learning and Epigenetic Robotics (ICDL) SSan Diego, California, USA, 7–9 November 2012 (2012)
32. Warnell, G., Waytowich, N., et al.: Deep TAMER: Interactive Agent Shaping in High-Dimensional State Spaces. arXiv preprint arXiv:1709.10163 (2017)
33. Arakawa, R., Kobayashi, S., et al.: DQN-TAMER: Human-in-the-Loop Reinforcement Learning with Intractable Feedback. arXiv preprint arXiv:1810.11748 (2018)
34. Christiano, P.F., Leike, J., et al.: Deep reinforcement learning from human preferences. arXiv preprint arXiv:1706.03741 (2017)
35. Saunders, W., Sastry, G., et al.: Trial without Error: Towards Safe Reinforcement Learning via Human Intervention. arXiv preprint arXiv:1707.05173v1 (2017)
36. Prakash, B., Khatwani, M., et al.: Improving Safety in Reinforcement Learning Using Model-Based Architectures and Human Intervention. arXiv preprint arXiv:1903.09328 (2019)
37. Jevtic, A., Colomé, A., et al.: Robot motion adaptation through user intervention and reinforcement learning. Pattern Recogn. Lett. **105**(1), 67–75 (2018)

38. Ross, S., Gordon, G.J.: A reduction of imitation learning and structured prediction to no-regret online learning. In: Proceedings of the 14th International Conference on Artificial Intelligence and Statistics (AISTATS) 2011, Fort Lauderdale, FL, USA. Volume 15 of JMLR: W&CP 15 (2011)
39. Zhang, J., Cho, K.: Query-efficient imitation learning for end-to-end simulated driving. In: Proceedings of the 31st AAAI Conference on Artificial Intelligence (AAAI 2017) (2017)
40. Menda, K., Driggs-Campbell, K., et al.: EnsembleDAgger: a Bayesian approach to safe imitation learning. In: Proceedings of the 2019 IEEE/RSJ International Conference on Intelligent Robots and Systems (IROS 2019) (2019)

An Integrated Approach to a Safety Argumentation for AI-Based Perception Functions in Automated Driving

Michael Mock[1], Stephan Scholz[2], Frédérik Blank[3], Fabian Hüger[2], Andreas Rohatschek[3], Loren Schwarz[4], and Thomas Stauner[4](✉)

[1] Fraunhofer IAIS, 53757 St. Augustin, Germany
michael.mock@iais.fraunhofer.de
[2] Volkswagen AG, 38440 Wolfsburg, Germany
{stephan.scholz1,fabian.hueger}@volkswagen.de
[3] Robert Bosch GmbH, 70469 Stuttgart, Germany
{frederik.blank,andreas-juergen.rohatschek}@de.bosch.com
[4] BMW AG, 80809 München, Germany
{loren.schwarz,thomas.stauner}@bmw.de

Abstract. Developing a stringent safety argumentation for AI-based perception functions requires a complete methodology to systematically organize the complex interplay between specifications, data and training of AI-functions, safety measures and metrics, risk analysis, safety goals and safety requirements. The paper presents the overall approach of the German research project "KI-Absicherung" for developing a stringent safety-argumentation for AI-based perception functions. It is a risk-based approach in which an assurance case is constructed by an evidence-based safety argumentation.

Keywords: Safe AI · Automated driving · Safety argumentation

1 Introduction

This paper presents the overall approach, integrating AI-based technology with classical safety argumentation, of the German research project "KI-Absicherung"[1]. In this project, 24 partners from automotive and automotive supplier industry, and research institutions develop a methodology providing a stringent safety argumentation for AI-based perception functions in automated driving.

A methodology for safety argumentation of classic automotive Electrical/Electronical systems not including AI-based function is already well established and specified in ISO26262 [1]. It considers potential malfunctions of the E/E systems. Motivated by advanced driver assistance systems, ISO21448 [2] has been created to provide a basis for

[1] The research leading to these results is funded by the German Federal Ministry for Economic Affairs and Energy within the project "KI Absicherung – Safe AI for Automated Driving". http://www.ki-absicherung-projekt.de.

I. Habli et al. (Eds.): SAFECOMP 2021 Workshops, LNCS 12853, pp. 265–271, 2021.
https://doi.org/10.1007/978-3-030-83906-2_21

safety argumentation also in presence of performance limitations of the considered system. This is an important basis also for assurance of AI-based systems. ISO TR4804 [4] complements existing standards w.r.t. recommendations, guidance and methods for assurance of automated driving systems. Similar to [2], its focus is on system level. [2] and [4] contain sections on machine learning but do not go into much detail. In contrast, this paper focuses on a methodology for the AI-part of automated driving functions. UL 4600 [5] addresses the safety argumentation of automated vehicles in a way compatible to [1] and [2] and explicitly considers machine learning. Like [5], we also follow a safety goal-based approach. Specifically, we propose to build the safety argumentation for deep neural network (DNN) based perception by systematic consideration of DNN specific safety concerns (cf. [6, 7]). Note that integration of the development result in a product development process as well as human interaction are addressed only implicitly in this paper via DNN specific safety concerns and possible safety measures.

In the sense of the recent EU commission proposed "Artificial Intelligence Act" [8], automated driving functions fall in the category of high-risk systems. The proposal does not specify a specific way on how safety has to be established. The methodology in this paper is a proposal of how a safety argumentation can be obtained.

2 Integrated Approach

2.1 Overview

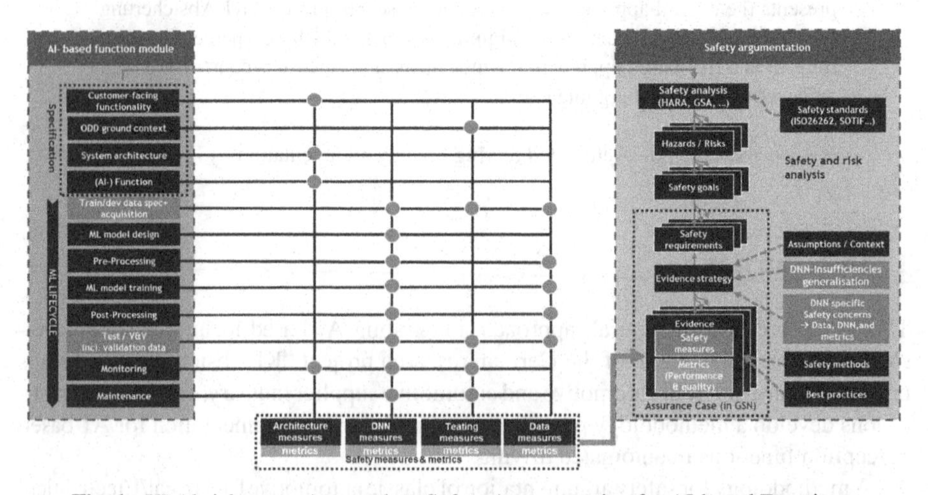

Fig. 1. KI Absicherung: Approach to Safety Argumentation for AI-based Functions.

We describe the methodology and a workflow leading to a stringent safety argumentation for AI-based functions on the basis of Fig. 1. It consists of three major blocks:

1. Specification and development steps of the AI-based function (left hand side).
2. Safety measures and metrics being applied to specific steps (middle part) and

3. Safety argumentation making use of the evidences provided by the measures and metrics (right hand side).

Given the specifications provided in the upper specification block, the safety analysis formulates safety requirements based on a risk analysis that takes the specification of the function, the proposed architecture and ODD into account. The safety requirements systematically address the DNN specific safety concerns that lead in particular to the DNN insufficiency [6, 7]. A formal assurance case argumentation in GSN (Goal Structured Notation) [9] is then developed that shows that the safety measures being applied provide sufficient evidence, measured in quantitative and qualitative metrics, that the risks raised by the DNN insufficiency and DNN specific safety concerns are mitigated to a tolerable level such that the safety requirements are fulfilled.

2.2 Specification of the Functions and Architecture (Fig. 1, Top Left)

Safety is not a property of an AI algorithm, but it is a property of a system, respectively its functionality, in its operating context. Therefore, this is the starting point for our methodology.

As example, we assume a level-4 driving **function** that once activated by the driver ahead of an urban intersection in "sample village", takes control of the vehicle and drives it safely and smoothly through a specific type of intersection. This means in particular, that no pedestrian (in general no vulnerable road user, VRU) is harmed or injured by the vehicle during the automated ride unless the accident is physically unavoidable. The function is limited to an operating context, e.g. to "sample village" suburb and a speed of at most 50 km/h. The context definition is described in Sect. 2.3.

Next, a specification of the system **architecture** is needed. We do not go into detail on this activity because the task it not different for an AI based system. From the point of view of assurance for the AI function, it is important to identify how it is embedded in the overall system.

As example, we use a simple architecture consisting of a camera providing raw sensor output, a SW component "pedestrian detection" which comprises sensor data preprocessing and which uses our AI-based function in the SW component to compute 2D bounding boxes for all visible vulnerable road users ahead of the vehicle. In a post processing, a list of bounding boxes is computed. This is provided to a further SW component "prediction and planning" that computes a trajectory which ensures that no pedestrian is harmed. The trajectory is given to a lower-level motion controller that uses the vehicle's actuators to follow the trajectory.

Note that, for the purpose of this paper, we ignore questions of mitigation of hardware failures because this is orthogonal to mitigation of insufficiencies of the AI function.

2.3 Specification of the Operational Design Domain (Fig. 1, Top Left)

Within our project, an **Operational Design Domain (ODD)** was developed to create a definition of the operating conditions under which the overall system should work as expected. First taxonomies and definitions from different publications such as [3] were used as a basis and extended to include influencing factors such as environmental, geographic, time-of-day constraints and the required presence or absence of certain features and dynamic elements in the context of use. This taxonomy represents a first simplified high-level semantic description of the ODD.

Furthermore, in our project the ODD was mapped to a newly developed more detailed semantic ontology. It aims to further describe the possible data input space of the DNN pedestrian detection function and systematically interconnect possible influencing factors and dimensions of the DNN. This ontology also aims to support the systematic tool-based description of possible variations and their combinations. It was developed by reviewing public data sources/existing standards, brainstorming with experts, organizing expert interviews and constant iterative refinement.

It is assumed that the dimensions or combinations of dimensions (e.g. occlusion level, pedestrian depth, cloth color) and variations defined in the simplified ontology can have a noticeable impact on the detection performance of a DNN-based perception system. Thus, those dimensions should be considered in safety evaluations as well as data distribution and coverage analyses.

Examples for the sub-domains included in our ontology are light effects, person descriptions, surfaces and material properties, static and dynamic objects & object interactions as well as weather conditions and sensor effects.

2.4 From Hazard Analysis to Safety Goals and Safety Requirements for the AI Function (Fig. 1, Top Right)

By means of a hazard analysis and risk assessment, the safety goals are derived (cf. [1], part 3). In order to comprehensively address safety of the system, we need to consider two aspects. One is the safety of the intended functionality (SOTIF), which means that if the system operates correctly (i.e. not malfunctioning), any unreasonable risk has to be avoided, in particular risks caused by performance limitations of the system [2]. The second is functional safety, which means that there is no unreasonable risk resulting from malfunctioning of the system [1]. Since DNN based AI functions typically have performance limitations, it is not sufficient to consider functional safety alone, but risks emerging from these performance limitations or insufficiencies have to be added in the hazard analysis, leading to safety requirements addressing the mechanisms that mitigate these risks. For example, existence of False Positives and False Negatives in object detection by DNNs. Note that these DNN insufficiencies go beyond physical limitations of sensors or hardware.

Hazards may also result from security risks, like potential vulnerability to adversarial attacks. These hazards also have to be considered. Mitigation measures may be different from measures directly related to safety, because of the systematic nature of such attacks. E.g. measures against reverse engineering may be helpful against some attacks.

The **safety goals** are on vehicle level. In the example from above, an abstract safety goal could be that no pedestrian (or VRU) is harmed or injured by the vehicle during the automated ride through the intersection in "sample village", given appropriate preconditions.

The **safety requirements** of system elements, like the AI function, result from refining the safety requirements and allocating them to system elements. For our example we assume that the downstream system elements (prediction, trajectory planning,...) can guarantee that pedestrians are not hit (unless physically unavoidable) given the assumption that they are correctly detected by the perception subsystem. Thus, a derived safety requirement for the function "pedestrian detection" is that within the ODD every relevant pedestrian is detected in all cases. Relevance may be defined w.r.t. visibility of the pedestrian and its distance to the vehicle. Note that derived safety requirements also result from the need to mitigate DNN specific safety concerns to a tolerable level.

2.5 Safety Argumentation and Assurance Case (Fig. 1, Bottom Right)

An **assurance case** is used to provide confidence that a system is safe to operate in a defined environment. It brings together the large body of evidence generated during the development process (analysis, design, verification, etc.) and explains, how that evidence demonstrates that the system is safe to operate.

However, assurance of AI leads to several challenges, since this technology needs new paradigms in development. The software is not developed explicitly any more. The neural network is trained and the network's behavior is implicitly influenced by the data and models for training. In a nutshell: classic safety approaches for software development are not sufficient.

Our approach for the assurance case strategy follows an evidence-based safety argumentation [10, 11]. An essential way for providing evidences is to mitigate effects caused by insufficiencies. The approach combines several measures and methods to a holistic assurance strategy. The approach can be summarized in these major steps:

- Identify potential causes of insufficiencies in the function by systematic consideration of DNN related safety concerns.
- Introduce metrics for each insufficiency.
- Develop mechanisms to mitigate the insufficiencies.
- Argue that the residual risk associated with the causes has been reduced to a tolerable level, e.g. by iterations over improvements of applied mitigation mechanisms as described in Sect. 2.6. Note that both, quantitative and qualitative arguments are feasible.

Safety concerns might cause insufficiencies [7]. The effectiveness of the safety mechanisms warranted by the assurance case analysis will have to be evaluated w.r.t. the AI-based function (or product) safety requirements. Testing of all used mechanisms provides further evidences. The coverage of all safety requirements shall be reached by systematic identification of gaps in the argumentation. This is also supported by considering performance limits in the functional effect chain.

We use the "**Goal Structuring Notation**" (GSN) [9] as graphical notation that represents the elements of an assurance case and the relationships between them. The GSN visualizes the evidence based safety argumentation. Its principle aim is to improve the comprehension of the assurance case, thus enabling rigorous review and analysis. The GSN shows how goals (claims) can be broken down into sub-goals until they can be supported by direct reference to available evidence.

2.6 DNN Safety Measures and Metrics (Fig. 1, Bottom Left and Center)

While the performance of **AI-based function** is quantified during the iterative development process using state of the art performance metrics (e.g. detection rates) further qualitative and quantitative metrics are needed to argue the effectiveness of applied DNN-specific safety measures. E.g., it is not sufficient to show that a very good DNN performance can be reached from available data, but also that the data is a good approximation of the specified ODD. This is (only) one of the **DNN specific safety concerns** that needs to be addressed.

In "KI-Absicherung" we develop and evaluate **mechanisms** [12] either providing **metrics** to measure or **measures** to mitigate the effects of DNN specific safety concerns. We iteratively assess and apply selected mechanisms from our mechanism catalogue to create evidences for the safety argumentation. Properties of mechanisms are e.g. addressed safety concerns, maturity, availability of implementations per DNN task, metrics and the level at which the mechanism is applied.

As depicted in Fig. 1 bottom, a mechanism can be applied at the architecture level, DNN level, data level, and test level. Architecture measures include, for example, redundancy per sensor input, redundancy via multiple sensors and algorithms, monitors and degradation strategies. DNN measures are measures that are applied during the ML lifecycle either online or during development time and include, e.g., uncertainty estimation, robustification training and attention consistency validation. Testing measures for DNN-based functions include statistical testing with an independent, in-distribution data set, which is standard in DNN development, robustness and adversarial robustness testing, testing w.r.t. input distribution shift, corner case-based testing, coverage-based testing, search based testing and tests based on network topology or the training method. Data measures include sensitivity analysis, data generation, data selection, data analysis and data augmentation. As data metrics we use multiple data coverage metrics e.g. regarding luminance or pedestrian size.

In evidence workstreams, promising candidates of mechanisms per safety goal are applied and evaluated with respect to their effectiveness using consolidated metrics. As a result, we aim to provide a guideline that comprehensively maps relevant safety mechanisms to DNN-specific concerns with respect to the identified safety goals within the specified operating environment (ODD).

3 Summary and Outlook

Developing a stringent safety-argumentation for AI-based perception functions requires a complete methodology linking all involved aspects of AI-function development and safety analysis. The presented methodology has been developed and exemplified in the project "KI-Absicherung" in a "Proof of Project Concept" [13] and is currently being further developed to cover a multi-dimensional ODD definition and include multiple DNN safety mechanisms, providing a safety-argumentation that takes the inherent multi factorial nature of DNN failures into account. One of the central challenges in the project will be to jointly identify and evaluate assurance methods in the evidence workstreams that can be used as a base for suitable evidences. Harmonization with upcoming standards and AI regulations will be a further step.

References

1. ISO 26262: Road vehicles – Functional safety. 2d Edition (2018)
2. ISO PAS 21448: Road vehicles—Safety of the intended functionality (2019)
3. BS PAS 1883:2020 Operational design domain (ODD) taxonomy for an automated driving system (ADS). Specification, standard by BSI Group, 31 August 2020
4. ISO TR 4804: Road vehicles—Safety and cybersecurity for automated driving systems—Design, verification and validation (2020)
5. ANSI/UL 4600: Standard for Evaluation of Autonomous Products (2020)
6. Sämann, T., Schlicht, P., Hüger, F.: Strategy to increase the safety of a DNN-based perception for HAD systems. arXiv preprint arXiv:2002.08935 (2020)
7. Willers, O., Sudholt, S., Raafatnia, S., Abrecht, S.: Safety concerns and mitigation approaches regarding the use of deep learning in safety-critical perception tasks. In: Casimiro, A., Ortmeier, F., Schoitsch, E., Bitsch, F., Ferreira, P. (eds.) SAFECOMP 2020. LNCS, vol. 12235, pp. 336–350. Springer, Cham (2020). https://doi.org/10.1007/978-3-030-55583-2_25
8. European Commission: Proposal for a Regulation laying down harmonised rules on artificial intelligence (Artificial Intelligence Act). https://ec.europa.eu/newsroom/dae/items/709090. Accessed 14 May 2021
9. The Assurance Case Working Group: Goal Structuring Notation Community Standard (Version 3), May 2021
10. Schwalbe, G., et al.: Structuring the safety argumentation for deep neural network based perception in automotive applications. In: Casimiro, A., Ortmeier, F., Schoitsch, E., Bitsch, F., Ferreira, P. (eds.) SAFECOMP 2020. LNCS, vol. 12235, pp. 383–394. Springer, Cham (2020). https://doi.org/10.1007/978-3-030-55583-2_29
11. Burton, S., Hawkins, R.: Assuring the safety of highly automated driving: state-of-the-art and research perspective, research report April 2020. https://www.york.ac.uk/assuring-autonomy/news/news/report-launch-safety-automated-driving/. Accessed 14 May 2021
12. Houben, S., et. al.: Inspect, understand, overcome: a survey of practical methods for AI Safety. arXiv preprint arXiv:2104.14235v1 (2021)
13. Mock, M., et al.: KI-Absicherung: Proof of Project Concept conducted. https://ki-familie.vdali.de/ki-newsletter-nr-1/ki-absicherung-proof-of-project-concept. Accessed 14 May 2021

Experimental Conformance Evaluation on UBER ATG Safety Case Framework with ANSI/UL 4600

Kenji Taguchi(✉) and Fuyuki Ishikawa

National Institute of Informatics, Tokyo, Japan
{ktaguchi,f-ishikawa}@nii.ac.jp

Abstract. The safety of Self-Driving Vehicles (SDVs) is crucial for social acceptance of self-driving technology/vehicles, and how to assure such safety is of great concern for automakers and regulatory and standardization bodies. ANSI/UL 4600 (4600) [3], a standard for the safety of autonomous products, has an impact on the regulatory regime of self-driving technology/vehicles due to its detailed and well defined assurance requirements on what will be required for the safety of autonomous products. One of the major characteristics of the standard is wide-scale adoption of the safety case, which has been traditionally used for safety assurance of safety-critical systems such as railways and automobiles.

Uber ATG (now Aurora) then released its own safety case called the Safety Case Framework (SCF) [1] for their SDVs. A question arises as to how much the SCF would conform to 4600 even though the SFC does not claim its conformance with the standard. An answer to this question would result in what type of argumentation would be fit for purpose for safety assurance for SDVs and address issues with conformance assessment of a safety case with a standard.

In this paper we report on lessons we learned from an experimental analysis on the conformance ratios of the SCF with 4600 and structural analysis following the argument structure of the SCF.

Keywords: Safety · Self-driving vehicles · Safety case · Uber ATS's safety case framework · ANSI/UL 4600 · Regulatory compliance

1 Introduction

The safety of Self-Driving Vehicles (SDVs) is crucial for social acceptance of self-driving technology/vehicles, and how to assure such safety is of great concern for automakers and regulatory and standardization bodies. ANSI/UL 4600 [3] (4600), a standard for the safety of autonomous products, has an impact on the regulatory regime of self-driving technology/vehicles due to its detailed and well defined assurance requirements on what will be required for the safety of autonomous products. One of the major characteristics of the standard is wide-scale adoption of the safety case, which has been traditionally used for safety assurance of safety-critical systems such as railways and automobiles.

I. Habli et al. (Eds.): SAFECOMP 2021 Workshops, LNCS 12853, pp. 272–283, 2021.
https://doi.org/10.1007/978-3-030-83906-2_22

Uber ATG (now Aurora) then released its own safety case called the Safety Case Framework (SCF) [1] for their SDVs *to foster open, public dialogue around our safety approach for the testing and development of self-driving vehicles.* A question arises as to how much the SCF would conform to 4600 even though the SCF does not claim its conformance with the standard. An answer to this question would result in what type of argumentation would be fit for purpose for safety assurance for SDVs and address issues with conformance assessment of a safety case with a standard.

In this paper we report on lessons we learned from an experimental analysis on the conformance ratios of the Safety Case Framework with 4600 and structural analysis following the argument structure of the Safety Case Framework.

We first provide information on the SCF and 4600. In Sect. 3, we present our experimental analysis results from two different angles. In Sect. 4, we discuss related work and then the concludes the paper in Sect. 5.

2 Background

In this section all background knowledge for this paper is explained.

2.1 Uber ATG Safety Case Framework

The SCF is a safety case for SDVs by then Uber ATG [1]. The entire safety case is provided as a web page with GUI, which helps in viewing and browsing the entire argument structure represented in Goal Structuring Notation (GSN) [7,8]. To the best of our knowledge, this is the only publicly available safety case for SDVs. Unfortunately the SCF only provides argument structure without any evidence, e.g., work products produced from the system development. This is quite understandable to protect confidential information but a large obstacle to assessing the validity of the safety argument in SCF.

The following are assumptions and limitations of the SCF, as explained in the SCF:

- The SCF is not for an SDV but Self-Driving Enterprise which may include the entire management and operating system for SDVs,
- "SDV is a modified production vehicle that complies with all safety and regulatory requirements" (from a context associated with the top goal in the SCF),
- "SDV is used for rideshare operations in a defined Operational Design Domain" (from a context associated with the top goal in SCF).

The top-level argument structure of the SCF faithfully follows Uber ATG's Self-Driving Safety Principles in its safety report [11], which is in fact, submitted for a Voluntary Safety Self-Assessment (VSSA) set out by National Highway Traffic Safety Administration (NHTSA). One of the benefits of this approach is that the SCF preserves the consistency with the safety report. We reserve our judgement on whether this top-level structure of safety argument is the most suitable for SDVs, but this would be a good starting point to consider how to construct one's own safety case for SDVs.

The following explanation is quoted from [11].

- **Proficient** *In the absence of system faults, how do we demonstrate that our system is* acceptably safe during nominal operations?
- **Fail-Safe** *How do we ensure that the system is acceptably safe in the presence of faults and failures? How will the system mitigate harm in the event of a fault or failure?*
- **Continuously Improving** *How do our developmental and operational processes identify, evaluate, and resolve anomalies that can potentially affect the safety of the self-driving vehicle? How can we actively cultivate a strong culture of safety, where the organization is engaged and empowered and holds all employees at all levels accountable for their active participation?*
- **Resilient** *How do we ensure the self-driving vehicle is acceptably safe in case of reasonably foreseeable misuse or unavoidable events?*
- **Trustworthy** *How do we earn and keep the trust of our riders, regulators, legislators, public safety officials, other road users, and advocacy organizations and provide them evidence of the safety measures of our self-driving enterprise?*

Each sub-goal is further decomposed into a more detailed argument on process used, Operational Design Domain (ODD), etc. in rather traditional safety case style narrative.

What are not clearly presented in the SCF are the following:

- Underlying assurance framework that encompasses assurance processes and associated assurance artefacts presented, e.g., in Pegasus [9] and SaFAD (Safety First for Automated Driving) with reference to a process for Deep Neural Networks [10].
- Level of autonomy for an SDV, which is specified in certain guidelines/standards, e.g., J3016 [5]
- Any effect on safety due to modification to a base production vehicle

2.2 ANSI/UL 4600

ANSI/UL 4600 (4600 in short) is a standard for safety of autonomous products published in 2020 [3]. The standard is a vast collection of safety assurance requirements ranging from the safety case and its argument constructions (Chap 5) up to assessment methods (Chap 17). There are several significant characteristics of the standard such as an adaption of safety case/argument for its main assurance framework and Safety Performance Indicators (SPIs) as assessment metrics for safety. It is evident that the standard is expected to be applied to safety assurance of autonomous vehicles.

The structure of a safety assurance requirement is divided into **General** (general description of the requirement), **Mandatory** (more detailed description of the requirement. No deviation is permitted), **Required** (Deviation is permitted in some case), **Highly Recommended** (Best practices. May be omitted in case of low risk items), **Recommended** (Optional. Good practices and/or suggestions for helpful techniques), **Conformance** (Conformance measure for the requirement).

We only use General and Mandatory for conformance analysis in this paper, since other categories of assurance requirements are chosen depending on the risk assessment result on the item, which is not available from the SCF.

We analyzed the assurance requirements from Chapters 5 through 17, as listed below:

5. Safety Case and Arguments
6. Risk Assessment
7. Interaction with Humans and Road Users
8. Autonomy Functions and Support
9. Software and System Engineering Processes
10. Dependability
11. Data and Networking
12. Verification, Validation, and Test
13. Tool Qualification, COTS, and Legacy Components
14. Lifecycle Concerns
15. Maintenance
16. Metrics and Safety Performance Indicators (SPIs)
17. Assessment

3 Experimental Conformance Analysis

As was pointed out previously, the SCF has not been developed in conformance with 4600 and we must emphasize that our preliminary goal of this experimental exercise was not aimed at judging its conformance to 4600 but to obtain insights into the construction method of a safety case for the SDVs in general.

There are several critical issues with and limitations of our experimental analyses. First, to the best of our knowledge there are no well-established criteria/methodologies for judging whether a given safety case conforms with a certain standard. Some standards provide the means to assess conformance (e.g., Conformance clause in 4600 and conformance measure in ISO 26262). However, there are many types of safety cases that go beyond what can be handled by conformance means stated in standards. This is why we heavily rely on expert judgement by assessors, even though this is fundamentally subjective.

A newly issued standard, such as 4600, generally requires several years to establish common understanding among engineers and assessors, and we had difficulties in understanding certain safety requirements in some places in 4600.

The SCF only provides an argument structure without any evidence such as test results. This sometimes makes it difficult to judge conformance with an assurance requirement. Another major problem with the SCF is that it lacks strategies for goal decomposition and contexts in many places, which makes it difficult to understand why those argument structures are chosen.

3.1 Sample Analysis Results

We used the following format shown in the Table 1 for assessment, where the content of the first column is an assurance requirement in 4600 [3], the second column for identified goal(s) and solution(s), which comply with the assurance requirement, and the third column for the verdict (Confirmed or Unconfirmed).

We originally used four values *Confirmed*, *Partially Confirmed* (part of an assurance requirement is satisfied), *Unconfirmed* and *Pending* (Need more information for verdict). But we changed them to binary values due to avoiding confusion among experts who assessed our evaluation results.

Table 1. Assessment format.

ANSI/UL 4600	Safety case framework	Result
Assurance requirement of ANSI/UL 4600	Corresponding goal(s) and solution(s), if any	Confirmed or Unconfirmed

We present typical cases showing how the conformance judgement was conducted.

Table 2 shows a straightforward conformance relation between assurance requirement 7.3.3 in 4600 with goal G2.1.3.1.1.6 in the SCF, even though slightly different terms "human-settable item parameters" and "human adjustable parameters" are used. This type of mismatch in terminology frequently occurs in several places, but we did not take a stringent policy for interpretation of terms.

Table 2. Example of assessment result #1.

ANSI/UL 4600	Safety case framework	Result
7.3.3 Hazards which can be contributed to by human-settable item parameters shall be acceptably mitigated	G2.1.3.1.1.6 Hazards and risks due to human adjustable parameters have been mitigated	Confirmed

Table 3 shows the lack of supporting argument relevant to an assurance requirement for sensor fusion and redundancy management techniques.

Table 3. Example of assessment result #2.

ANSI/UL 4600	Safety case framework	Result
8.3.3 Sensor fusion and redundancy management techniques shall be used as necessary to result in acceptable sensor performance for the defined ODD		Unconfirmed (No clearly defined goals for sensor fusion, but relevant argument is G1.2.2.1.2.12 Perception system has sufficient availability and coverage to appropriately detect objects in the ODD and Sn1.2.2.1.2.12.5 Perception fusion architecture.)

Table 4 shows that the standard has a wider scope, which might be useful for autonomous products in general but not necessarily compliant for the SCF. A Minimum Equipment List (MEL) is in fact a list for the operation for aircraft and neither this notion nor associated assurance requirements are found in safety standards for automotive systems.

Table 4. Example of assessment result #3.

ANSI/UL 4600	Safety case framework	Result
10.3.5 A Minimum Equipment List (MEL) shall be defined for each autonomous operational mode		Unconfirmed (No argument on MEL)

Table 5 shows that the judgement is done with a goal and its associated solutions (evidence).

Table 5. Example of assessment result #4.

ANSI/UL 4600	Safety case framework	Result
16.3.1 Data for each defined SPI shall be collected	G3.2.5.2 Safety performance indicator data measurements are valid, Sn3.2.5.2.3 SPI data collection tools, Sn3.2.5.2.4 SPI data collection justification of robustness	Confirmed (Data collection related to SPIs can be judged as conforming indirectly from a combination of a goal and associated solutions)

Table 6 shows that the conformance is completely judged by taking the entire structure of the SCF and related solutions.

Table 6. Example of Assessment Result #5.

ANSI/UL 4600	Safety case framework	Result
5.1.1 The safety case shall be a structured explanation in the form of claims, supported by argument and evidence, that justifies that the item is acceptably safe for a defined operational design domain, and covers the item's lifecycle	Judged from the entire structure of the SCF. Configuration management for the safety case is judged from Sn5.1.3.2.1.3 Safety Case Framework version control policy/guidelines	Confirmed (The use of appropriate format (GSN) for a safety case. Safety claims are structurally presented with the support of evidence.)

As can be seen from sample analysis results, the verdict for conformance is straightforward in some cases and difficult to judge in other cases.

3.2 Summary for Entire Analysis

In this subsection, we summarize the results of our analysis, which focuses on the conformance ratio for each chapter in 4600. Table 7 shows these conformance ratios where "C" stands for confirmed and U for Unconfirmed. The third column includes selected goals and solutions that support the conformance evaluation.

The conformance ratios vary significantly from one chapter to another. The highest ratio is 8:1, which is for Chapter 9 of Software and System Engineering Process. This analysis result is reflected by very detailed process-related arguments in the SCF. The lowest ratio is 0:6 for Chapter 15 of Maintenance. This is mainly because many arguments only partially conform to assurance requirements. For instance, an assurance requirement (15.1.1) in 4600 for mitigation of hazard and risks related to maintenance and inspection is only argued by maintenance and no argument for inspection in the SCF.

Table 7. Conformance Ratio

Chapter	Conformance ratio	Relevant goals/solutions
5. Safety Case and Arguments	C (8), U (8)	Sn5.1.3.2.1.3, G2.1.1.1.2, G5.1.1.1.1, G2.1.1, G1.1.1.8.1, G3.2.1.6
6. Risk Assessment	C (9), U (11)	G2.1.3, G2.1.3.1, G2.1.1.2, G3.3.1.1.4, G2.1.1.6.2, G3.3.1.1.3, G2.1.1.4, G2.1.1.3, G2.1.1, G2.1.2.4, G2.1.2.1.1
7. Interaction with Humans and Road Users	C (3), U (6)	G4.2.1.1, G2.1.3.1.1.6, G2.1.1.4.12
8 Autonomy Functions and Support	C (6), U (4)	G2.1.1.2.2, G2.1.1.4.3, G1.2.2.6.3, G1.2.2.4.3, G1.2.2.4.2, G1.2.2.3.3.7.2, G1.2.2.2.1.12, G1.3.1.6.2, G1.4.2.1, G1.4.2.2, G1.2.2.1.1, G1.2.2.1.1.2.5, G1.2.2.1.1.3
9 Software and System Engineering Processes	C (8), U (1)	G1.1.1.6.4, G1.1.1.2.4, G1.1.1.5.4, G1.1.1.1, G1.1.1.2, G1.1.1.5.4, Sn1.1.1.5.4.1, G1.1.1.5.4, Sn1.1.1.5.4.1, G1.1.1.6, G1.1.1.6.3, G1.1.1.6.4, G3.2.1.4, G1.3.1.4, G1.1.1.6
10 Dependability	C (19), U (6)	G2.1.3.1.5, G2.1.3.1.1.8, G1.2.2.5.1.9, G1.2.2.4.1.9, G2.1.3.4.3, G2.1.1.4.5, G2.1.3.3, G2.1.3.5, G2.1.3.3.1, G2.1.3.3.1, G2.1.3.3.4, G2.1.3.4.3.3, G1.2.2.3.3.7.3, G1.2.2.2.1, G4, G4.1.4, G4.1.1.1, G2.1.3.1.9, G2.1.1.4.9, G4.1.1.5.1, G4.1.1.5.2, G4.1.1.5.6, Sn4.1.1.5.6.2, G4.1.3.1., Sn4.1.3.1.5, G1.2.2.7, G1.2.2.7.6, G4.3, G4.3.1.1.1, G4.3.1.1.7
11 Data and Networking	C (5), U (5)	G2.1.3.1.1.10, G2.1.3.1.13, G2.1.3.1.14, G2.1.3.1.11, G4.3, G2.1.3.1.11, G2.1.3.1.12, G2.1.3.1.14, G2.1.3.1.13
12. Verification, Validation, and Test	C (11), U (9)	G1.3.1.4.1.3.2, G1.3.1.2.1, G1.3.1.4.1.3.1.2, G1.3.1.4.6.3, G1.3.1.4.6.7, G1.3.1.4.1.3.2, G1.3.1.2.1, G1.3.1.4.1.3.1.2, Sn1.3.1.4.1.3.2.5, An1.3.1.2.1.1, Sn1.3.1.4.1.3.1.2, G1.2.2.1.2.12, Sn1.3.3.4.4.7.2, G1.3.1.4.1.6.4, Sn1.3.1.4.1.6.4.8, G1.3.1.4.1.6.4, Sn1.3.1.4.1.6.4.7, G2.1.3.5.1.2.2.6, G2.1.3.5.1.2.2.4, G1.3.1.4.1.4.4, G4.3.2.2, Sn4.3.2.2.1, Sn4.3.2.2.4, G3.4.1.6, G4.3.2.2,Sn4.3.2.2.1, Sn4.3.2.2.3, Sn4.3.2.2.4, G1.1.1.6.1.1., Sn1.1.1.6.1.2
13. Tool Qualification, COTS, and Legacy Components	C (6), U (0)	G1.1.1.5.4.6, G1.1.1.3.5.6, Sn1.1.1.9.4.7.1, G2.1.1.3.5, G2.1.1.4.7, G2.1.1.5.3, G2.1.3.1.16, G2.1.3.1.17, G2.1.1.6.6
14. Lifecycle Concerns	C (6), U(8)	G2.1.3.1.4, G2.1.3.1.18, G2.1.3.1.19, G2.1.3.1.21, G2.1.3.1.22, G2.1.3.1.23.2
15. Maintenance	C (0), U(6)	
16. Metrics and Safety Performance Indicators (SPIs)	C (5), U(4)	G3.2.1.2, G3.2.7, G3.2.6, G3.2, G3.2.4, G3.2.5.2, Sn3.2.5.2.3, Sn3.2.5.2.4, G3.3.2, Sn3.3.2.1, Sn3.3.2.3, G3.3.2, Sn3.3.2.2
17. Assessment	C (2), U (10)	G1.5.3.2.2, G5.1.3.2.2, Sn5.1.3.2.2.1

3.3 Structural Analyses

We mapped the previous analysis results onto the top-level structure of the SCF, which is composed of five sub-goals; "G1: Proficient", "G2: Fail-Safe", "G3: Continuously Improving", "G4: Resilient" and "G5: Trustworthy". The results are listed in Table 8.

This table can be interpreted horizontally and vertically, where a horizontal interpretation shows which chapter in 4600 is dealt with in which sub-goal of the SCF, and a vertical interpretation shows which sub-argument structure contains how many different assurance requirements in the chapters in 4600.

Table 8. Map to SCF

	G1: Proficient	G2: Fail-Safe	G3: Continuously Improving	G4: Resilient	G5: Trustworthy
5. Safety Case	G1.1.1.8.1	G2.1.1, G2.1.1.1.2	G3.2.1.6		G5.1.1.1.1
6. Risk Assessment		G2.1.3, G2.1.1.2, G2.1.1.3, G2.1.1.4	G3.3.1.1.4, G3.3.1.1.3		
7. Interaction with Humans and Road Users		G2.1.3.1.1.6, G2.1.1.4.12		G4.2.1.1,	
8. Autonomy Functions and Support	G1.3.1.6.2, G1.2.2.1.1, G1.2.2.6.3	G.2.1.1.2.2			
9. Software and System Engineering Processes	G1.1.1.6.4, G1.1.1.1, G1.1.1.5.4, G1.1.1.5.4, G1.1.1.6, G1.3.1.4, G1.1.1.6		G3.2.1.4		
10. Dependability	G1.2.2.7,	G2.1.3.1.5, G1.2.2.5.1.9, G2.1.3.3, G2.1.3.3.1, G2.1.3.4.3.3		G4, G4.1.4, G4.1.1.5.1, G4.1.3.1, G4.3	
11. Data and Networking		G2.1.3.1.1.10, G2.1.3.1.11, G2.1.1.14			
12. Verification, Validation, and Test	G1.3.1.4.1.3.2, G1.3.1.4.1.2.1, G1.2.2.1.2.12, G1.3.1.4.1.6.4, G1.3.1.4.1.6.4, G1.1.	G2.1.3.5.1.2.2.5	G3.4.1.6	G4.3.2.2,	
13. Tool Qualification, COTS, and Legacy Components	G1.1.1.5.4.6, Sn1.1.1.9.4.7.1	G2.1.1.3.5, G2.1.1.4.7, G2.1.3.1.17			
14. Lifecycle Concerns		G2.1.3.1.4, G2.1.3.1.18, G2.1.3.1.19, G2.1.3.1.21, G2.1.3.1.22, G2.1.3.1.23.2			
15. Maintenance					
16. Metrics and Safety Performance Indicators (SPIs)			G3.2.1.2, G3.2.3, G3.2.5.2, Sn3.2.5.2.3, G3.3.2, Sn3.3.2.1		
17. Assessment					G5.1.3.2.2, G5.1.3.2.2, Sn5.1.3.2.2.1

Some observations from the horizontal analyses are listed below:

- Chapter 5. Safety case covers almost all sub-argument structures since the fundamental mechanism for safety assurance is based on the safety case.
- Chapter 6. Risk Assessment is covered by G2: Fail-Safe and G3: Continuously Improvement, which is plausible due to their intentions.
- Chapter 8. Autonomy Functions and Support are mainly covered in G1: Proficient, which is plausible for its intention.
- Chapter 9. Software and System Engineering Process is mostly covered by G1: Proficient. We cannot judge its plausibility but there is no reason against it.
- Chapter 10. Dependability is either safety-related (G2: Fail-Safe) or resilient (G4: Resilient), which results from the very nature of dependability.
- Chapter 16. Metrics and SPIs are entirely covered by G3: Continuously Improving, which is plausible. Since SPIs are used for risk assessment in the operation phase.
- Chapter 17. Assessment is only covered by G5: Trustworthy due to its role in assessment.

In accordance with this analysis, we can conclude that the distribution of intended use of assurance requirements are very plausible.

Some observations on the vertical analyses are listed below:

- G1: Proficient covers almost all Chapters 8, 9 and 12, since its intention is to cover systems engineering development as a whole and autonomy,
- G2: Fail-Safe covers almost all chapters, since safety analysis is dominant in all activities,
- G3: Continuously Improving covers a part of Chapter 6 and all of Chapter 16, which matches its intention,
- G4: Resilient mostly covers Chapter 10, which can be predictable from its main role,
- G5: Trustworthy covers Chapter 17.

The vertical analyses show that each sub-goal and its argument structure fulfills its intention.

We consulted industry experts who are specialized in functional safety, (SOTIF) Safety of The Intended Functionality [6] and SDVs on our analysis results. We provided them a check sheet (Table 1) and asked them to provide comments. The overall response to our results was positive due to detailed analyses on the conformance relation between 4600 and the SCF, which they think would help to understand 4600 and the role of a safety case for substantial documents for future certification of their own SDVs. They expressed difficulties in assessing the results due to terminology mismatch between the two and lack of correct understanding of 4600, but provided helpful comments to improve our assessment results. Unfortunately, we cannot provide any details of their comments due to confidentiality agreement, but we will use them to improve our results for future study.

4 Related Work

There are a few papers which address the issues addressed in this paper. The closest is that by Graydon, et al. [15], which discusses issues on conformance to software assurance standards and presents a conformance arguments structure in which how a part of an argument is linked to objectives in DO-178B [16].

Submission of a safety case is stated in several safety standards, e.g., ISO 26262 [4] and there are many papers that show their methods of constructing a safety case for a certain standard. For instance Dargar et al. [12,13] presented a safety case for ISO 26262, but it does not deal with conformance criteria for a standard. This type of research assumes that the conformance to a standard is built in its approach and does not provide explicit criteria why the approach assures conformance with a standard.

The safety case approach to Machine Learning (ML)/Artificial Intelligent (AI) systems is carried out by several researchers. For instance, Gauerhof [14] focused on the inherent nature of functional inefficiencies which could be caused by ML/AI systems and provide a safety argument structure that covers their root causes. Pegasus is a European research project on autonomous driving and developed the safety argumentation framework for Highly Automated Driving Functions. Its presentation material [9] includes reference to standards, e.g., ISO 26262, ISO/PAS 21448 [6] and NHTSA guidelines, but it does not provide any details on how the framework can conform with those standards and guidelines.

5 Concluding Remarks

In this paper we reported our analysis results on experimental conformance evaluation on Uber ATS's Safety Case Framework with ANSI/UL 4600. The aim of our analyses was to gain insight into what type of argument structure would be fit for purpose for safety assurance for SDVs. We summarized our analyses results in two tables; the first showing the conformance ratio of the Safety Case Framework with ANSI/UL 4600 and the second showing structural analysis following the argument structure of the Safety Case Framework.

Future work includes investigation on the general criteria for conformance evaluation which does not rely on subjective expert judgements and the general argument structure for SDV. We also plan to

Acknowledgements. This work is supported by JST ERATO-MMSD and JST MIRAI-eAI projects (grant numbers: JPMJER1603 and JPMJMI20B8).

References

1. Safety Case Framework (2020). https://uberatgresources.com/safetycase/gsn
2. Safety Case Framework Blog. https://medium.com/@UberATG/trailblazing-a-safe-path-forward-e02f5f9ef0cc
3. ANSI/UL 4600:2020. Standard for Evaluation of Autonomous Products (2020)

4. ISO 26262:2018 Road Vehicles - Functional Safety (2018)
5. SAE J3016: 2018, Taxonomy and Definitions for Terms Related to Driving Automation Systems for On-Road Motor Vehicles, SAE International (2018)
6. ISO/PAS 21448: 2019, Road vehicles - Safety of the intended functionality (2019)
7. Kelly, T.: Arguing safety: a systematic approach to managing safety cases. D. Phil Thesis, U. York (1998)
8. ACWG: Goal Structuring Notation Community Standard (ver. 2) (2018)
9. Maus, A.: PEGASUS Safety Argument (2019). https://www.pegasusprojekt.de/files/tmpl/Symposium2019/3_3_PEGASUS%20safety%20argument_Maus.pdf
10. Safety First for Automated Driving (2019)
11. Uber ATG Safety Report: https://uber.app.box.com/v/UberATGSafetyReport?uclick_id=3a2a8230-402c-404c-9eac-1e81a561b703
12. Dardar, R.: Building a Safety Case in Compliance with ISO 26262 for Fuel Level Estimation and Display System, Master Thesis, Mälardalen University (2013)
13. Dardar, R., Gallina, B., Johnsen, A., Lundqvist, K., Nyberg, M.: Industrial experiences of building a safety case in compliance with ISO26262. In: 23rd IEEE International Symposium on Software Reliability Engineering Workshops. ISSRE Workshops 2012, pp. 349–354 (2012)
14. Gauerhof, L., Munk, P., Burton, S.: Structuring validation targets of a machine learning function applied to automated driving. In: Gallina, B., Skavhaug, A., Bitsch, F. (eds.) SAFECOMP 2018. LNCS, vol. 11093, pp. 45–58. Springer, Cham (2018). https://doi.org/10.1007/978-3-319-99130-6_4
15. Graydon, P., Habli, I., Hawkins, R., Kelly, T., Knight, J.: Arguing conformance. IEEE Softw. **29**(3), 50–57 (2012)
16. DO-178B: Software Consideration in Airborne System and Equipment Certification, RTCA (1992)

Learning from AV Safety: Hope and Humility Shape Policy and Progress

Marjory S. Blumenthal(✉)

RAND Corporation, Arlington, VA, USA
marjory.blumenthal@ceip.org

Abstract. Producing automated vehicles (AVs) that are, and can be shown to be, safe is an ongoing challenge. This position paper draws on recent work to discuss alternative approaches to assessing AV safety, noting how AI can be a positive or negative influence it. It features suggestions to promote AV safety, drawing from practice and policy, and it ends with a speculation about a special new role for AI.

Keywords: Automated vehicle · Safety · AI · Measurement · Policy

1 Measurement Muddle

Automated vehicle (AV) safety provides a powerful, practical illustration of how artificial intelligence (AI) challenges public policy in the context of a complex consumer product. That challenge arises not only because AI is used, but because AI choices combine with other choices about vehicle design and manufacturing (automotive engineering) and add to a broader cybersecurity challenge.[1] All stakeholders agree that safety is a top priority for AVs. Without it, public trust will fall short of what is needed to sustain a market that can recoup AV development costs and support continued progress. Yet safety itself is elusive. Better indicators of AV safety are needed, to inform more people and to provide more coverage of the concept of safety, and better public policy is needed to promote AV safety in a dynamic and fiercely competitive context.

During a recent study of how to measure AV safety [1], my colleagues and I were surprised to learn that there is no consensus definition of safety—in general or for transportation, let alone for AVs. We chose to define safety for AVs as avoiding harm to people, whether in or near a vehicle. Defining safety is easier than measuring it. The qualities of good measurements are known (reliability, validity, feasibility, and so on); our research also collected views of different kinds of stakeholders,[2] including their preferences for AV safety measurements and their observations about measurement shortcomings, such

[1] Even conventional vehicles are cyberphysical systems with many cyber components and vulnerabilities, creating risks arising from malice in addition to safety risks arising from circumstances or inadvertence.

[2] We interviewed AV developers in industry, officials from different levels of government, independent safety researchers, and safety advocates.

I. Habli et al. (Eds.): SAFECOMP 2021 Workshops, LNCS 12853, pp. 284–290, 2021.
https://doi.org/10.1007/978-3-030-83906-2_23

as whether a developer could "game" a measurement.[3] Unfortunately, the typical approach to measuring safety for conventional vehicles—using lagging measures collected after an event, like crash rates—will not be an option for AVs for the foreseeable future [2]. The alternative is leading measures—precrash measures of different kinds. Obvious examples include rates of disengagement (disparaged by many because of inconsistency in use) or violations of traffic rules (sometimes necessary to avoid crashes), which apply to AV-related phenomena with more frequency than crashes. Measurements can be made in different settings (simulation, closed courses, public roads) used in developing AVs and at different stages of the lifecycle of an AV, from development through deployment.[4] Measuring in all settings and stages matters—it is not enough to rely on measurements taken in simulation or even in simulation plus closed course testing. Understanding how an AV, dependent on AI, will perform in the real world demands testing on public roads, with all of the known and unexpected circumstances that they present. Such testing puts people who are near (and sometimes in) a vehicle at risk.

Our appreciation for the limitations of existing AV safety measurements took us in two directions: a proposal for a new measurement concept, and consideration of additional approaches to assessing safety that complement measurement. These are outlined below.

1.1 Roadmanship

We proposed a new kind of measurement, which we called *roadmanship*, to expand on what can be gleaned about AV safety from leading measures [1]. Roadmanship refers to how well an AV behaves on the road, especially in traffic. Roadmanship measures should be objective and reflect official and unofficial rules of the road, recognizing differences between a vehicle initiating and one responding to a problematic situation and capturing the chaos that may surround the AV as it operates. Research has shown promise in a variety of directions for measuring roadmanship, such as safety-envelope monitoring,[5] probabilistic instantaneous safety metrics, post-encroachment time, and near misses. In the second phase of our work [3] we concluded from our discussions with a variety of AV technology developers and our review of the literature that ongoing development of leading measures constitutes progress toward roadmanship measures. These measures are increasingly integrated and sophisticated. Achieving their potential will require more research.

Given the many roles that AI does and might play in an automated driving system, and given the many forms of uncertainty involved in automated driving—uncertainties that AI is supposed to help manage if not resolve—refinement of roadmanship measures is important to advancing AV safety. Our original articulation noted that roadmanship

[3] Gameability has been a concern with disengagements, which are used in different ways by different developers, and it has been raised in connection with proposals to have AVs perform on prespecified tests.

[4] Our report discussed how measures from different settings and stages can be aggregated into a framework that could be used across the industry.

[5] An early innovation of this kind is the Responsibility Sensitive Safety (RSS) model introduced by Mobileye.

should "reward predictability and anticipatability," both of which can benefit from AI. Because of their newness, unfamiliarity to most users of public roads, and technological immaturity, today's AVs are not as predictable as many hope they will become. Our most recent report added details for what roadmanship can capture, including the AV's situational awareness, time to response, and how well the developer understands the vehicle's physics. Ideally, roadmanship captures a shift from "don't crash" to the more complex and subjective "drive safely."

It is without irony that the technologists developing AVs—many with deep roots in computer science—have been advancing technology-based approaches to promote safety for automated driving systems. Illustrated by the safety-envelope monitoring mentioned above, systems that serve to check on the performance of primary automated driving systems (rounding out doer-checker dyads) have been developed and introduced into use. These systems use quantitative descriptions of safety. Their use could generate new measures based on new kinds of rates, such as number of incursions into safety envelope per mile traveled. The promise of checkers is sufficient that standard-setting has commenced through the IEEE; standard-setting would facilitate third-party supply of these systems by fostering interoperability.[6]

As systems get layered upon systems—in a context that already layers systems (software, electrical/electronic, mechanical) and in a context where different entities commonly supply the vehicle-platform and the physical and digital components supporting the automated driving—it is worth considering whether that layering process can go too far. When does the complexity becomes too great? When does the addition of more software make the dependability of the system harder to achieve or understand? Formal methods loom large in checkers, and verification and validation of learning systems (i.e., AI) continue to vex regardless of where those systems are applied. Although the early years of AV development were marked by the proliferation of entrants with new ideas and technologies, consolidation of the industry and convergence on a smaller number of technology suites and automated driving systems is likely to benefit safety and governance in this new arena, in part by reducing complexity or at least the number of variations and their possible combinations. The complexity conundrum makes simpler approaches to assessing AV safety appealing, and two are described below.

2 Thresholds and Processes

Because of the limitations of measurements, it is important to consider other approaches to assessing AV safety. One candidate is thresholds. The other is information about processes.

2.1 Thresholds

Thresholds indicate what level of safety is deemed acceptable. They can be informed by measurements or processes (see below)—or not. They can be quantitative or qualitative,

[6] The IEEE P2846 working group focuses on a Formal Model for Safety Considerations in Automated Vehicle Decision Making. See: https://sagroups.ieee.org/2846/.

and they can evolve. The most obvious threshold for AVs is comparison to human drivers. In our first project (2017–2018), we heard a lot about comparisons to the average human driver, and in less than two years, during our second project (2019–2020) we heard that that is not good enough—AVs should be compared to better-than-average or safe human drivers. Ideally, that comparison should relate to the operational design domain (the circumstances of driving, such as weather, daylight, road type, traffic volume, pedestrian volume, and so on), but we lack good data on human driving by ODD to support such comparisons.

A very different kind of threshold takes the form of an absolute goal. Here, an obvious example comes from Vision Zero—the goal of eliminating deaths and serious injuries on the road. But although such a goal can motivate policy and action, achieving zero anything—especially zero risk—is not realistic. After all, even the threshold concept of "minimum endogenous mortality" (MEM) accepts that human life comes with risk. For AVs and other forms of transportation, not adding to existing risk levels is a more realistic approach, consistent with maintaining risk at a level "as low as reasonably achievable" (ALARA), a threshold that is used elsewhere, or GAMAB, a threshold that in English is "generally at least as good as."

The performance of an automated driving system could be assessed using a threshold such as a score on a driving test. But although often proposed and considered, driving tests are an imperfect tool for assessment—as is obvious when considering their use with human drivers, who team with conventional vehicles. Driving tests for human drivers are oriented to teenaged new drivers, who are expected to learn with experience.[7] Their learning is not the same as AI learning—a contrast that might merit its own consideration—and car crashes are a top killer of American teens. For AVs, discussion of driving tests typically involves the use of scenarios, which would represent a tiny fraction of the situations an AV would be expected to encounter in use. Known scenarios, in turn, suggest the potential for developers to "teach to the test," promoting gaming or cheating. Meanwhile, the expectation that AV software will be updated frequently over the air raises the question of how often any testing should be repeated.

2.2 Processes

Finally, because AVs will remain black boxes for the foreseeable future, especially outside of their developers' facilities, interest has grown in gleaning insight from the processes that surround their production. In particular, information about compliance with technical standards could provide useful process information. Technical standards relating to AVs have been proliferating, and in the United States the National Highway Traffic Safety Administration has proposed using information about compliance with them as an indicator relevant to assessing safety [4]. Attention to technical standards has tended to be a proprietary matter, so disclosing details would be novel—and it might also require interpretation, since compliance with some of the early standards associated with AV safety is recognized by developers as necessary but not sufficient.[8] A new

[7] As we explained in [3], drawing from psychology research, people are more willing to accept human error than machine error.

[8] ISO 26262 for functional safety and ISO/PAS 21448 for safety of the intended functionality motivate this caution.

and different kind of standard for AVs, UL4600, includes compliance with technical standards as an element in a comprehensive safety case—UL4600 promotes disclosure broadly [5].

Safety cases are a cross-cutting theme for process information. One of the earliest technical standards used for AVs, ISO 26262 for functional safety, called for a safety case, and UL4600 maximizes the options for AV developers. It invites developers to assemble indicators or arguments for the safety of their AVs, in the absence of reliable lagging measurements.

Another cross-cutting theme relates to safety culture. History has repeatedly shown lapses in an organization's safety culture when a safety-related catastrophe occurs.[9] Indicators of how seriously and comprehensively safety is viewed within an organization and in terms of job expectations and management behavior are another kind of process information. Safety culture signals risk preferences, and it affects software development along with other kinds of activity.

Today process information of the kinds outlined above is private to developers, but as signaled by UL4600 (which provides for independent review) and the NHTSA ANPRM mentioned above it need not be. We recommended in our most recent report that a consortium of developers collaborate on developing a template for sharing safety cases publicly. Similarly, in our first report we recommended that developers write up and share case studies of problematic events. Competition should not foster winners and losers when it comes to safety, and developers can make choices that promote transparency and safety and that support collective learning. Collective learning is especially important when it is, as it is with AVs, a matter of life and death.

Transparency has become a concern for AI, particularly learning systems, more broadly. AV technology is advancing at a time when *accountable* or *explainable AI* is a topic of discussion, advocacy, and research. A first step toward accountability or explainability is understanding development processes. Ongoing discussions of what process information might be disclosed by AV developers constitute a case study combining technology with political economy that might have applicability in other domains (or, alternatively, might benefit from insights from other domains). Legal systems might induce additional steps, especially for addressing situations where there have been errors.[10]

A final observation is that compliance with regulations is another kind of process information. Today, such processes do not speak specifically to AVs, but that is changing as motor vehicle regulators worldwide consider how to assess the safety of automated driving systems.

---

[9] Investigations into the Boeing 737 MAX, Tesla, and Uber crashes have underscored this concern most recently.

[10] Also, a legally appropriate explanation might not require technical detail but rather might motivate technical development to support legal expectations: "To build AI systems that can provide explanation in terms of human-interpretable terms, we must both list those terms and allow the AI system access to examples to learn them. System designers should design systems to learn these human-interpretable terms, and also store data from each decision so that is possible to reconstruct and probe a decision post-hoc if needed." [6].

3 Policy, Technology, and Managing AV Uncertainty

In the United States, there has been recognition across both government and industry that premature regulation of some kinds could freeze technology prematurely and chill innovation. The U.S. approach has included Voluntary Safety Self-Assessments [7]; it has been more hortatory than regulatory, although work is under way to adapt the Federal Motor Vehicle Safety Standards that focus on crash avoidance and occupant protection for conventional vehicles. Many believe that more is needed as automation levels rise and with them expectations for AVs to operate in unpredictable environments.

The role of AI epitomizes the uncertainty and the information asymmetry associated with AV development. It adds to the complexity and inscrutability of a product for which government safety oversight is expected, and for which government understanding of the technology will always lag that of the developers. Whether within a developer company or within a government entity seeking to protect public safety, a compound approach is needed. Measurements, thresholds, and processes cannot stand on their own when it comes to assessing AV safety, but together each complements and compensates for deficiencies in the others.

The need for such synergy was demonstrated in the breach, with the pioneering regulatory action of the State of California Department of Motor Vehicles when it required developers testing AVs on public roads to report disengagements [8]. That one measurement was a beacon in a context where developers had a lot of proprietary information about their AVs and outside observers had none, but the inconsistency in approach to this reporting undercuts its utility. At the other extreme, the comprehensive case studies provided by National Transportation Safety Board investigations of fatalities blend technical and process information about vehicles and their producers, notably putting a spotlight on safety culture. Investigations elicit more specific information than developers otherwise would share, allowing everyone to learn from the investigation reports, but they are few and offer limited coverage of the AV safety domain (not that anyone wants there to be a lot of fatal crashes).

Since policy makers will always know and understand less about AVs than their developers, perhaps it is time for a new kind of AI—a system designed for policy makers to protect the public while supporting AV progress. As the cliché says, turnabout is fair play. What might that look like? Perhaps the steps people would have to take to understand what would go into such a system, even if it wasn't implemented, could help cultivate a middle ground with greater transparency for these evolving, safety-critical systems. There might be a set of systems, including tools that use AI for deeper understanding of automated driving systems than their developers disclose, tools that use AI for analyzing information collected from all developers seeking to operate in a given territory to both compare across developers more systematically and to understand how the total automated fleet in that territory is evolving, or tools that use AI to generate information at different levels of granularity to communicate consistently but comprehensibly with different stakeholders.

References

1. Fraade-Blanar, L., Blumenthal, M.S., et al.: Measuring automated vehicle safety: Forging a Framework. RAND Corporation, Santa Monica, CA (2018)
2. Kalra, N., Paddock, S.: Driving to safety: How Many Miles of Driving Would It Take to Demonstrate Autonomous Vehicle Reliability? RAND Corporation, Santa Monica, CA (2018)
3. Blumenthal, M., Fraade-Blanar, L., et al.: Safe enough: Approaches to Assessing Acceptable Safety for Automated Vehicles. RAND Corporation, Santa Monica, CA (2020)
4. NHTSA Advanced notice of proposed rulemaking. https://www.federalregister.gov/documents/2020/12/03/2020-25930/framework-for-automated-driving-system-safety. Accessed 15 June 2021
5. UL 4600 Homepage. https://ul.org/UL4600. Accessed 15 June 2021
6. Doshi-Velez, F., et al.: Accountability of AI under the law: the role of explanation, arXiv (2017)
7. NHTSA Voluntary safety self-assessments. https://www.nhtsa.gov/automated-driving-systems/voluntary-safety-self-assessment. Accessed 15 June 2021
8. California department of motor vehicles disengagement. https://www.dmv.ca.gov/portal/vehicle-industry-services/autonomous-vehicles/disengagement-reports/. Accessed 15 June 2021

Levels of Autonomy and Safety Assurance for AI-Based Clinical Decision Systems

Paul Festor[1,2], Ibrahim Habli[3,4], Yan Jia[3,4], Anthony Gordon[5], A. Aldo Faisal[1,2,6], and Matthieu Komorowski[1,5,6](✉)

[1] UKRI Centre for Doctoral Training in AI for Healthcare, Imperial College London, London, UK
m.komoroswki14@imperial.ac.uk
[2] Department of Computing, Imperial College London, London, UK
[3] Assuring Autonomy International Programme, University of York, York, UK
[4] Department of Computer Science, University of York, York, UK
[5] Department of Surgery and Cancer, Imperial College London, London, UK
[6] Department of Bioengineering, Department of Computing, Imperial College London, London, UK

Abstract. Levels of Autonomy are an important guide to structure our thinking of capability, expectation and safety in autonomous systems. Here we focus on autonomy in the context of digital healthcare, where autonomy maps out differently to e.g. self-driving cars. Specifically we focus here on mapping levels of autonomy to clinical decision support systems and consider how these levels relate to safety assurance. We then explore the differences in the generation of safety evidence that exist between medical applications based on supervised learning (often used for prediction tasks such as in diagnosis and monitoring) and reinforcement learning (which we recently established as a way for AI-guided medical intervention). These latter systems have the potential to intervene on patients and should therefore be regarded as autonomous systems.

Keywords: Digital healthcare · Autonomy · AI safety

1 Introduction

Scales of increasing levels of autonomy have been proposed for AI applications in the healthcare domain [1,14]. This raises a concern, especially for safety engineers, about how to define safety requirements for such autonomous systems. There is general principle that higher levels of complexity and autonomy imply more stringent safety requirements [2].

Supervised learning forms the basis of most AI-based clinical applications that have been proposed in the literature for diagnostic and monitoring purposes, e.g. in the recognition of skin tumors from images [3]. In this paradigm, the purpose of the algorithm is to try to predict as accurately as possible a

I. Habli et al. (Eds.): SAFECOMP 2021 Workshops, LNCS 12853, pp. 291–296, 2021.
https://doi.org/10.1007/978-3-030-83906-2_24

defined prediction, labelling or classifications tasks using labelled data as training material - effectively making a single decision to the nature of patient's health state. In contrast, in reinforcement learning (RL), a agent learns a decision strategy (a so called policy) in a sequential decision making process. The policy is optimised so that it maximises some form of future expected total return [13]. Formally it is related to optimal control and model predictive control applications. RL differs from conventional prediction tasks used in most of the medical AI literature in that the model does not simply reproduce human behaviour, it attempts to improve and learn an optimal decision strategy from sub-optimal training examples acquired from humans. This is a highly appealing approach for many clinical scenarios with uncertainty, since the method would - in principle - be able to tease out the right decisions among a range of options selected by human doctors. We developed such an algorithm in previous research for the treatment of sepsis (severe infections with organ failure), which we called the "AI Clinician" system and is now being developed for prospective clinical evaluation [7]. As we discuss in the later sections, the performance and safety assessment of the output of RL algorithms is more difficult than conventional prediction tasks based on supervised learning.

In this article, our objectives are to define increasing levels of autonomy of AI systems in healthcare and discuss differences between supervised and RL based AI applications with regards to safety assurance, using the example of the AI Clinician for sepsis treatment.

In autonomous systems the nature of the action of the agent is an important concept to consider. In autonomous vehicles the agent has to steer, accelerate and brake the car, give signals etc. However, in clinical settings it is not necessarily essential that the system directly controls the intervention.

2 Levels of Autonomy of AI Applications in Healthcare

We define 5 increasing levels of autonomy, from zero (no AI involved at all) to four (fully autonomous AI), as shown in Table 1. The differences between levels essentially reside in how the AI system and human user interact to make a decision, and who bears the responsibility for the decision made. We illustrate those levels using two different scenarios: self-driving vehicles, for their ease of visualisation and understanding, and clinical decision support systems, which are at the core of our research interests and whose deployment represents a true challenge today.

Level 0 is used to serve as a reference: it designates a version of the setup in which no AI is involved. For self-driving cars, this means human drivers without any assistance, and for clinical decision support system, this is standard care. This level is the baseline for the other ones: the aim of introducing AI is to increase some aspects of performance of the system so the performance should be assessed by reference to the baseline. This level can also serve as a reference for measuring the safety of the system. This makes sense particularly when considering systems which operate in high-risk environments, such as self-driving cars and clinical decision support systems (CDSSs).

Table 1. Definition of the five increasing levels of AI autonomy

Level	Short definition	Illustration in self-driving cars	Illustration in an AI for drug administration	Who makes the final decision/burden of responsibility
0	No AI	No assistance	Standard care	Human
1	AI suggests decisions to human	GPS guidance system suggests direction	Clinicians can see the AI recommendation	Human
2	AI makes decisions, with permanent human oversight	Car following lanes on the motorway, driver has to keep hands on the wheel	AI changes the doses, with human doctors continually checking	Human
3	AI makes decisions, with no continuous human oversight but human backup available	Autonomous car with a human behind the wheel, asks for driver's help if the AI cannot deal with the situation	AI changes the doses, and can alert human users in case of high uncertainty. Continuous human oversight not required.	AI
4	AI makes decisions, with no human backup available	Autonomous self driving car with no driving cockpit	AI changes the dose with no human backup	AI

Level 1 is the first step towards AI autonomy. In level 1, the AI system is set up in its environment, it can produce outputs and these outputs can be seen by the human agent. The human has the choice to decide whether or not to look at the AI's output, and make their own decision accordingly. In this level, the requirements on the AI system are quite low as its dysfunction should have minimal impact on whether appropriate decisions are made or not. In the case of self-driving cars, consider GPS systems where the driver has the option to follow GPS guidance or not, and there is always a way for drivers to find their way without GPS, e.g. following signs and maps. Similarly, a level 1 AI-based CDSS would present treatment recommendations to the clinical team; however, the team should still be able to function without the AI.

Level 2 pushes the AI's autonomy one step further by letting it act directly on the environment. However, on level 2, the AI system is continuously monitored by a human expert who can take the lead at any time. An example of such an AI for self-driving cars is lane-following assistance where the car can change its steering and speed to stay in a given lane and maintain a reasonable distance from other vehicles. In the context of Clinical Decision Support Systems (CDSSs), a level 2 autonomous system would issue treatment recommendations which would be administered to the patient either directly or by a human. A human expert is continuously reviewing the system's output.

Level 3 represents what most people would call true "autonomy". Here, the AI acts directly on the environment, but it is not continuously monitored by humans anymore. Instead, the AI may request human input when needed. Note here that, because the interactions between human and AI are not as frequent or regular as in level 2, there can be delays in the human reaction, but these delays should be within an acceptable range for the application. This level introduces a new important requirement; the AI has to be uncertainty-aware, and be able to recognise states in which its output might not be appropriate and human input is required. In the case of lane-keeping for self-driving cars, such a system would be able to keep the car safely on the road in most of the situations, without continuous human supervision, but could ask for human help when the conditions prevent the AI from being confident in its decisions, e.g. at a complex junction. Similarly, in a drug administration scenario, the AI would normally be able to take care of drug administration with no human supervision, but acknowledge the states in which the decision is not clear and ask for the input of a human expert. A typical example in healthcare applications is represented by mechanical ventilators that autonomously adapt to patient characteristics to accelerate the weaning process, e.g. [9].

Level 4 stands at the top of the scale and represents the state of complete autonomy. An AI system of level 4 autonomy acts directly on its environment, and should be able to handle every situation within its defined scope of use. This level is unrealistic with the current state of the art in both self-driving cars and CDSSs for drug administration and is questionable if this would ever be desirable in the healthcare setting [14].

An essential aspect of our approach is that the level of autonomy of a system can only be assessed with respect to the environment in which it operates and the tasks it aims to address. Even though there is a correlation between risk and AI autonomy, the absolute amount of risk involved in acting in different environments can be very different (an AI failing to mow the lawn properly has very different consequences from failing to keep a car in line on the motorway).

3 Difference Between Supervised and Reinforcement Learning Applications

CDSSs based on supervised learning in the field of computer vision or closed-loops are already in use in clinical settings [14]. For example, the FDA has approved systems that detect strokes in brain CT scans and automatically alert physicians [11]. In the operating room, prototypes of AI closed-loop systems are being developed, to control the level of sedation (measured with real-time electroencephalography, EEG) or blood pressure during surgery [5,6,8], e.g. the amount of desired concentration of anesthetic in the brain (targeted controlled infusion, [12]). In our classification, such systems would correspond to level 1 (stroke detection) or 2 (drug dosing system). In the first example, the AI merely supports human clinicians in their decisions and improves workflow and care coordination. In the second example, the system controls the delivery of a drug that is being given to a human; it has latitude to increase or decrease the amount of drug flowing into the system. This is equivalent to a clinician ordering a certain amount of drug to be delivered and another human trying to control the amount injected accordingly. Yet, the clinician specifies the amount of drug in a part of the body, and the system is under the continuous supervision of a human expert (an anesthesiologist) who may take over the control of the system at any point.

Autonomy, however arises when the AI system is not directly working to precisely-defined human specification. An important distinction must be made between AI-based clinical systems based on supervised learning and reinforcement learning. Let us use the example of the AI Clinician system for sepsis resuscitation to illustrate this distinction [7]. Sepsis represents a global healthcare challenge, a leading cause of death and the most expensive condition treated in hospitals [15]. A cornerstone of the treatment of severe infections is the administration of intravenous fluids and vasopressors. However, there is much debate around the dosing of these treatments and what resuscitation targets should be used. Despite decades of research, resuscitation strategies in an individual patient remain mostly empirical. This is the clinical challenge that the AI Clinician attempts to address. The challenge of collecting safety evidence is much more complex for an CDSS based on RL (such as the AI Clinician) than for the medical applications based on supervised learning described above, for a number of reasons.

Firstly, there is no established "gold standard" for sepsis treatment [15]. Clinicians may have multiple objectives, which run in parallel and might be conflicting. For example, optimising blood pressure with large volumes of intravenous fluids may temporarily improve cardiac output whilst compromising organ perfusion and increasing the risk of renal failure at a later time point. In a conventional supervised learning setting, the equivalent task would be to train a model to replicate desirable human behaviours, which is in general more straightforward. In sepsis resuscitation, the desired effect of fluid resuscitation and vasopressors may not be clearly defined, which makes it difficult to use supervised learning.

Secondly, while the RL agent can in theory explore treatment strategies that have not been used in practice by clinicians, there is in reality limited opportunity with RL to learn the optimal policy using "on-policy learning" by trial-and-error, due to ethical and patient safety risks [4]. A major limitation is the lack of high fidelity human simulators that would enable safe exploration of various decisions without compromising safety. In many real-world applications of RL such as healthcare, the environment in which to train the model is not fully observable, which induces uncertainty about the state represented by the RL model. This is very different from simulation frameworks used in computer science research such as the Atari games [10], where the environment is fully observable at any time and provides all the information needed to make optimal decisions.

Thirdly, the effect of the decisions on outcomes represents a complex closed-loop with confounded causality. The effect of administering fluids and/or vasopressors is realised at multiple time horizons on multiple parameters. For example, the effect on cardiac output and blood pressure can be immediate, while the effect on the patient's kidney function can be delayed by a few hours or days, and the patient's final outcome can be weeks away but still influenced by a single early decision. This is reflected when defining the RL model reward, where researchers have to choose between immediate, intermediate or delayed rewards.

Finally, key areas of our current work focus on the development of a safety case for the AI Clinician system in which the safety evidence and its associated arguments vary with the intended level of autonomy of the system and the readiness of the wider clinical environment for the deployment of this novel technology.

Acknowledgements. PF was supported by PhD studentship of the UKRI Centre for Doctoral Training in AI for Healthcare (EP/S023283/1). AAF was supported by an UKRI Turing AI Fellowship (EP/V025449/1). All authors acknowledge support by the Assuring Autonomy International Programme (Lloyd's Register Foundation and the University of York; Project Reference 03/19/07).

References

1. Bitterman, D.S., et al.: Approaching autonomy in medical artificial intelligence. Lancet Digital Health **2**(9), e447–e449 (2020)
2. Catherine Menon, C.E., McDermid, J.: Defence standard 00–56 issue 4: towards evidence-based safety standards. In: Dale, C., Anderson, T. (eds.) Safety-Critical Systems: Problems, Process and Practice, pp. 223–243. Springer, London (2009). https://doi.org/10.1007/978-1-84882-349-5_15
3. Esteva, A., et al.: Dermatologist-level classification of skin cancer with deep neural networks. Nature **542**(7639), 115–118 (2017)
4. Habli, I., et al.: Artificial intelligence in health care: accountability and safety. Bull. World Health Organ. **98**(4), 251 (2020)
5. Joosten, A., et al.: Feasibility of closed-loop titration of norepinephrine infusion in patients undergoing moderate-and high-risk surgery. Br. J. Anaesth. **123**(4), 430–438 (2019)
6. Joosten, A., et al.: Automated closed-loop versus manually controlled norepinephrine infusion in patients undergoing intermediate-to high-risk abdominal surgery: a randomised controlled trial. Br. J. Anaesth. **126**(1), 210–218 (2021)
7. Komorowski, M., et al.: The artificial intelligence clinician learns optimal treatment strategies for sepsis in intensive care. Nat. Med. **24**(11), 1716–1720 (2018)
8. Lowery, C., Faisal, A.A.: Towards efficient, personalized anesthesia using continuous reinforcement learning for propofol infusion control. In: 2013 6th International IEEE/EMBS Conference on Neural Engineering (NER), pp. 1414–1417. IEEE (2013)
9. De Montfort University: Dräger smartcare®/ ps – the automated weaning protocol (2019)
10. Mnih, V., et al.: Human-level control through deep reinforcement learning. Nature **518**(7540), 529–533 (2015)
11. Murray, N.M., et al.: Artificial intelligence to diagnose ischemic stroke and identify large vessel occlusions: a systematic review. J. NeuroInterventional Surg. **12**(2), 156–164 (2020)
12. Struys, M.M., et al.: Comparison of plasma compartment versus two methods for effect compartment-controlled target-controlled infusion for propofol. J. Am. Soc. Anesthesiologists **92**(2), 399 (2000)
13. Sutton, R.S., Barto, A.G.: Reinforcement Learning: An Introduction. MIT Press (2018)
14. Topol, E.J.: High-performance medicine: the convergence of human and artificial intelligence. Nat. Med. **25**(1), 44–56 (2019)
15. Yealy, D.M., et al.: Early care of adults with suspected sepsis in the emergency department and out-of-hospital environment: a consensus-based task force report. Ann. Emergency Med. (2021)

Certification Game for the Safety Analysis of AI-Based CPS

Imane Lamrani(✉), Ayan Banerjee, and Sandeep K. S. Gupta

iMPACT Lab, Arizona State University, Tempe, AZ, USA
{ilamrani,abanerj3,skgupta}@asu.edu
https://impact.lab.asu.edu/

Abstract. Current certification procedures aim at establishing trust in manufacturers of artificial intelligence/machine learning based cyber-physical systems. The certification process usually requires the manufacturer to demonstrate excellence in following safety engineering standards and regulations throughout the holistic system's engineering process. This paper touches on the need for real-world performance monitoring performed by the certifier to ensure that the operational system does not deviate from the specifications. We propose an interactive cooperative process between the manufacturer and certifier which aims at verifying conformance and consistency between the specifications and the operational model while preserving the manufacturer's competitive advantage.

Keywords: Certification · Game theory · Operational safety · AI-based cyber-physical systems

1 Introduction

Recent cases of fatal failures of safety-critical cyber physical systems (CPS) have renewed the discussion on the certification process. Current certification procedures require manufacturers to demonstrate that the system is designed following safety engineering standards and regulations throughout the system's engineering process [5,11], as depicted in Fig. 1. The food and drug administration (FDA) introduced an action plan for the Digital Health Software Precertification (Pre-Cert) pathway which represents a least burdensome regulatory paradigm that aims at establishing trust in manufacturers of Artificial Intelligence/Machine Learning (AI/ML)-based[1] medical devices that are capable of leveraging transparency of product development and performance [5]. The proposed precertification pathway can be extended to other fields including autonomous driving. In the precertification pathway, manufacturers are required to demonstrate excellence in product development as well as real-world performance (RWP) monitoring and must extend a RWP analytics (RWPA) plan to FDA which aims at

[1] An (AI/ML)-based device is a system designed using AI and ML techniques to learn from and act on data [6].

I. Habli et al. (Eds.): SAFECOMP 2021 Workshops, LNCS 12853, pp. 297–310, 2021.
https://doi.org/10.1007/978-3-030-83906-2_25

actively monitoring RWP data for continued safety, effectiveness, security, and performance of (AI/ML)-based medical devices. However, manufacturers may tend to conceal some system's components due to trade secrets restrictions or if they are unable to demonstrate safety guarantees of these components. This has been witnessed in different incidents including the Volkswagen cheating software component that allowed vehicles to improperly pass the government emissions tests. We strongly believe that there is a crucial need to ensure that the specified RWPA, system's specifications, and RWP data (RWPD) are conformant and consistent with the operational system. This paper proposes an additive certification layer based on a cooperative game executed between the manufacturer and certifier, which can be performed periodically at the operation phase to verify consistency between the operational CPS and the specified CPS, also referred to as operational safety, while:

a) Protecting manufacturers' competitive advantage by allowing the manufacturer and certifier to analyse the possibility of system's certification using only partial system's specifications,

b) Detecting CPS' corruption scenarios and discrepancies between the specified system and the operational system before the occurrence of fatal accidents and,

c) Identifying accidents' root-causes during the accident analysis phase

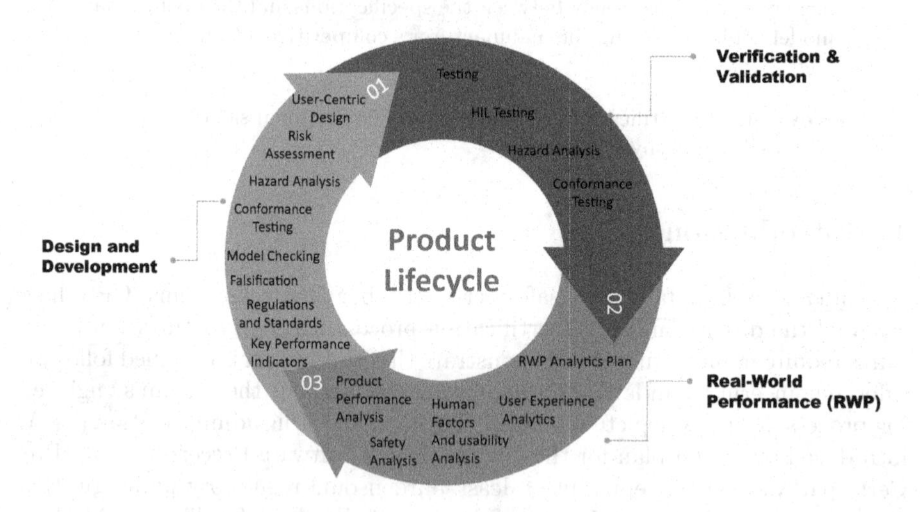

Fig. 1. Cyber-physical system lifecycle.

2 Definitions and Preliminaries

Cyber Physical System Model. A CPS model represents the interaction between the controller model, the user's behavioral model, and the physical system model, as shown in Fig. 2 (a). The CPS can be modeled as a hybrid automaton (HA) where the dynamics of the controlled physical system's variables are

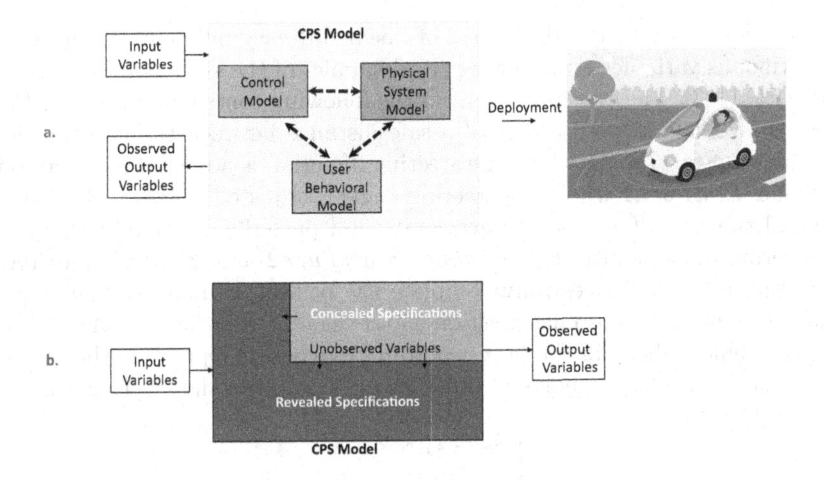

Fig. 2. Cyber-physical system.

represented by a set of ordinary differential equations (ODEs). The HA model includes a set of operational contexts (a.k.a control modes) where predefined initial and transient conditions are satisfied [7]. During real-world operations, a subset of the system's variables are observed and sampled data of these variables can be recorded and used during the certification process. As shown in Fig. 2 (b), the manufacturer reveals detailed information about some components of the CPS model while they may conceal information about other components. The manufacturer usually provides claims or partial information regarding the unspecified submodel of the system but does not reveal complete specifications of its inner workings. For example, Medtronic provides a model for the self-tuning component of Minimed 670 but conceals gain and window size parameters [9].

Automated Lane Change System. We consider the automated lane change (ALC) system as a running example in this paper. As shown in Fig. 3, the goal of the automatic lane change system of an autonomous vehicle (orange) is to automatically over take the vehicle in front (green) and return to the initial lane.

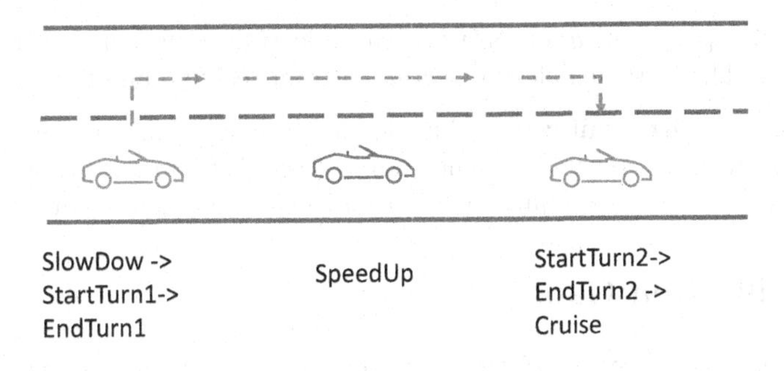

Fig. 3. Automated lane change system.

The vehicle can be modeled by a set of linear ordinary differential equations of its continuous variables representing the dynamics of the system in every control mode (Eq. 1) [3]. The system consists of the following control modes: *SlowDown* where the vehicle slows down until a safe distance between the two vehicles is maintained, *StartTurn*1 where the steering direction is adjusted to move to the next lane, *EndTurn*1 where the steering angle is adjusted to stay in the passing lane and the *SpeedUp* mode to accelerate and pass the green vehicle. Finally, the steering angle is adjusted in modes *StartTurn*2 and *EndTurn*2 to return the initial lane and successfully complete the passing maneuver. The vehicle's variables include the absolute position s and velocity v of the overtaking vehicle w.r.t the vehicle ahead in the (x, y) coordinates where x represents the direction of the road and y the orthogonal direction. w represents the steering angle and a the acceleration.

$$\begin{cases} \dot{s}_x = v_x - 2.5 \\ \dot{v}_x = 0.1a_x \\ \dot{s}_y = 0.1v_y \\ \dot{v}_y = f_1(v_y, s_y, w) \\ \dot{w} = f_2(w, s_y) \\ \dot{a}_x = f_3(s_x, v_x, a_x) \end{cases} \tag{1}$$

Observable Variables: These are the continuous variables of the process model. For the Lane Change (LC) system, the lane change controller has a_x, v_x, s_x, v_y, s_y, and ω as continuous variables that can be observed.

Data Contexts: The observable variables should be observed under certain initial, transient and final value conditions. In case of the LC system, to extract the full system, full observability condition is enough, since it is already a linear system. However, the manufacturer has not provided all observable parameters.

Sampling Requirements: For each observable parameter, a minimum sampling rate is required to ensure an error bound on the extracted constant parameters. In case of the LC system, to accurately extract the value of d, e, f, within an error bound of ϵ, the minimum sampling frequency required is given by

$$f_{\{d,e,f\}} = \frac{ds_x(0) - e + fv_x(0)}{\epsilon - a_x(1)}, \tag{2}$$

where $s_x(0)$, $v_x(0)$, and $a_x(1)$ are the variable values when $t = 0$ and $t = 1$, respectively. These frequencies are obtained from residual analysis of Eq. 1.

Duration Requirements: The duration requirements are dependent on the accuracy of the linear regression. The manufacturer is required to test the system until it satisfies the data requirements from the certification agent.

3 Certification Game

As shown in Fig. 4, we model the proposed certification process as a cooperative game executed between two types of agents: a) Certification agent, and b)

Manufacturer. Both agents share the same goal, that is the completion of the certification process in the least number of actions. The certification game is terminated if the certification agent is able to accurately extract an accurate approximative model of the unspecified/unrevealed component of the system to complete the certification process. However, the agents' reward functions differ. Usually, manufacturers have allocated budgetary resources and deadlines. Hence, the manufacturer's reward function is dependent upon providing a subset from the set of system's parameters S_1, a subset of operational data from the set of observed sampled variables S_2, and a subset of the system's specifications S_3 within a budget on time t and cost c while minimizing risk. The certifier's reward function is associated with completing the certification process in the least number of steps.

In the following, we define the rewards functions of the manufactuer (R_{man}) and certification agent (R_{cer}) at every time step n:

$$R_{man}(n) = R_{man}(n-1) - \frac{(Total\_Cost)_{Provide\_Information}}{Budgetary\_Resource} - \frac{Total\_Time}{Allocated\_Time} \tag{3}$$

$$R_{cer}(n) = R_{cer}(n-1) - 1 \tag{4}$$

where $Total\_Cost$ and $Total\_Time$ represents the sum of the cost and time correspondingly spent by the manufacturer at every $Provide\_Information$ action (every time step). Collecting the information and data for the proposed certification process may require the manufacturer to augment existing systems with new hardware or software components. Hence, the manufacturer decides the amount of information and resource allocation devoted at each time step. The certification agent may provide feedback to the manufacturer on the possibility of certification and the complementary material needed to finalize the process

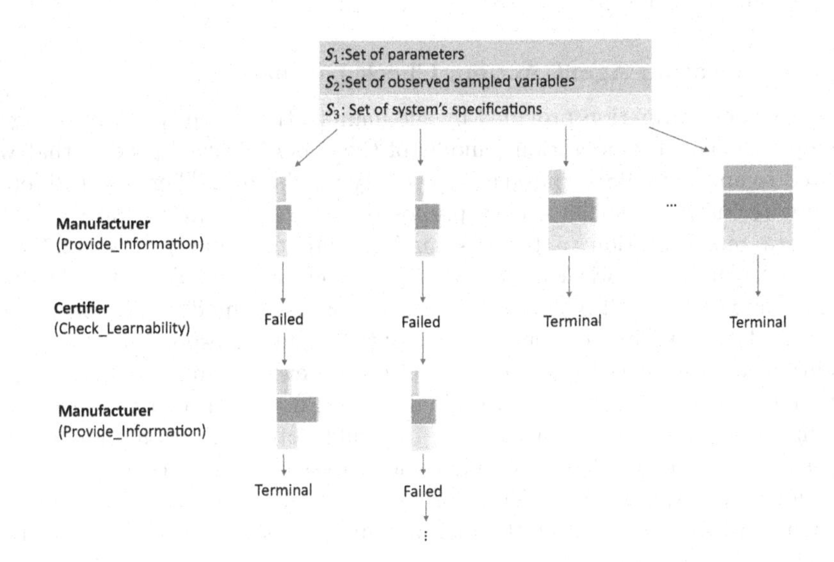

Fig. 4. Certification game.

following the previous *Provide_Information* actions taken by the manufacturer. At this point, the manufacturer has the possibility to provide the complimentary information or retract from the current path and explore a different path. The game involves the following actions alternatively taken by the manufacturer and the certification agent until a Terminal state is achieved.

3.1 Manufacturer Action: Provide_Information

In this step, the manufacturer performs a cost, time, and risk analysis to perform experiments and monitoring to provide the requested data. The manufacturer decides on a a subset of observed variables data, a subset of context-based operational conditions, and a subset of parameters to provide to the certifier using the following considerations:

- **Observability:** The possibility of monitoring a given variable during operation time is a function of: a) cost of sensors, b) availability of sensing at the required resolution, and c) intrusiveness of the sensor to the operation of the system.
- **Transparency:** Even if variables and parameters are observable, regulatory requirements and constraints may prevent sharing of such data among stake holders.
- **Confidentiality:** Internal parameters of the CPS processes often fall under trade secret rules and patent confidentiality agreements and may not be disclosed to the certification agent.
- **Risk to human subjects:** Often experiments under certain data contexts can involve undue risks to human subjects. This not only increases cost but also time required to perform complicated procedures.

For the LC system lets assume that the manufacturer only provides a_x and w at a certain frequency $f = \frac{1}{\tau}$.

3.2 Certification Agent Action: Check_Learnability

Based on the information provided by the manufacturer, the certification agent attempts to learn the operational model of the cyber/physical process that will be used to analyze CPS's compliance to safety regulations. There exist different learning techniques that have been proposed in the literature to learn a hybrid automata representation of the system from input/output traces [13,15,16]. Recurrent neural networks can also be applied to learn the ODEs at every operational mode of the CPS [2]. In Sect. 4, we discuss a model learning process that uses data from different operational contexts and corresponding operational conditions and extracts the parameters of the unrevealed submodel of the CPS using regression analysis. If the terminal state is not reached, the certifier can inform the manufacturer regarding the possibility of certification and the complimentary material needed to complete the process following the adopted course of actions. For example, completion of an operational safety verification of the Medtronic 670G requires that the manufacturer reveals the gain parameters of the self-tuning component [14].

3.3 Terminal State and Safety Analysis

The certification agent can utilize several safety analysis techniques on the derived model. One example of model-based safety analysis techniques is reachability analysis that provides the set of states, also called *reach set*, that the system can visit starting from a given set of initial conditions on the continuous variables [4,8]. The reachability analysis can be applied for safety check by checking intersection of the reach set with a safety threshold on any of the system variables.

4 Evaluation

Based on the information provided by the manufacturer, the certification agent attempts to recover the operational model of the cyber/physical process. The recovery process uses the data contexts and extracts the parameters of the process model.

For the LC system, the certification agent uses the following steps:

Action 1: Use the data collected for Mode 0. Compute the $\frac{da_x}{dt}$ using the Euler's method.

$$\frac{da_x}{dt}(t) = (a_x(t) - a_x(t-\tau))/\tau, \tag{5}$$

where $\tau = \frac{1}{f_{\{d,e,f\}}}$. Then a linear regression between $\frac{da_x}{dt}(t)$ and $a_x(t)$ gives the parameter g.

Action 2: Use the transient condition on the deceleration of the vehicle in Mode 0, to obtain a linear regression as follows:

$$\text{at } t = 0, a_x(0) = 0. \text{ which implies } a_x(1) = ds_x(0) + e + fv_x(0). \tag{6}$$

Here $s_x(0)$ and $v_x(0)$ are initial test conditions and hence are known as a part of test specification. a_x is provided by the manufacturer. Hence, the certification agent can utilize atleast three different test cases to extract d, e, and f.

Action 3: Another transient condition in Mode 0 is that $v_x(0) = v_x(1)$. Hence $\frac{ds_x}{dt}(0) = \frac{ds_x}{dt}(1)$. This implies the following linear regression:

$$\text{at } t = 0 \text{ and } t = 1, \frac{da_x}{dt}(2) = d*(s_x(0)+\tau(bv_x(0)-c))+e+f*v_x(1)+g*a_x(1). \tag{7}$$

Here d, e, f, g are known, while b and c are the only unknowns. Hence with at least two test cases with different $v_x(0)$ and $s_x(0)$, the parameters b and c can be obtained.

Action 4: Utilize initial transient conditions $v_x(2) = aa_x(1) * \tau + v_x(0)$. Again utilizing multiple initial conditions a linear regression can be obtained that relates $\frac{da_x}{dt}(1)$ with $a_x(1)$, as follows:

$$\frac{da_x}{dt}(2) = d*(s_x(0)+2*(b*v_x(1)+c)*\tau)+e+f*(v_x(1)+aa_x(1)\tau)+ga_x(1), \tag{8}$$

here the only unknown is a. With the given approach, the certification agent could estimate the parameter values using Eq. 9.

$$a = 0.1534, b = 1, c = -3.02, d = -0.0099, e = 0.737, f = -0.3, g = -0.506. \tag{9}$$

$$a = 0.1, b = 1, c = -2.5, d = -0.01, e = 0.737, f = -0.3, g = -0.5. \tag{10}$$

These are not the accurate parameters. The settings used in the simulink replication of the LC system is shown in Eq. 10.

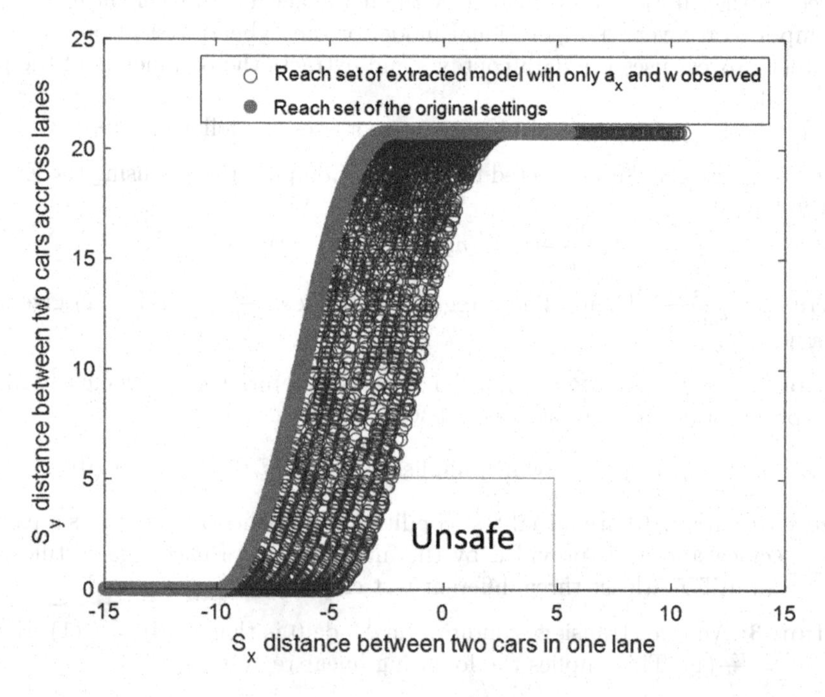

Fig. 5. Reach set for the derived LC system as compared to the original system.

For the LC system, the reach set can be obtained by using the *C2E2* reachability analysis tool [4]. Figure 5 shows the reach set for the derived system in Eq. 9 and the true system of Eq. 10. There can be three scenarios with the intersection of the reach set with the safety criteria as shown in the Fig. 6. The first case is when the extracted model reach set does not intersect with the unsafe set. The certification agent can then certify the system to be operationally safe and end the game. The second case is when reach set of the extracted model intersects with the unsafe set but the intersection area is more than error limits

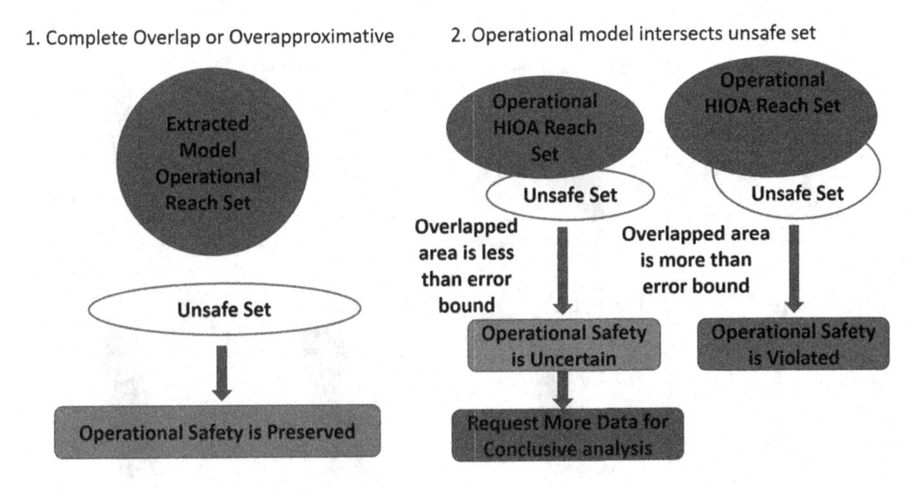

Fig. 6. Reach set comparison for operational safety assessment.

of the reach set computation due to inaccuracy in extracting the system parameters. The certification agent can stop the game and deem the system unsafe. The intersection can however be within the error limits of the analysis process. In such a case, the certification agent can go back to the original requirements, provide the safety analysis results to the manufacturer, and ask for reconsideration on the data sharing agreement. At this point the manufacturer can decide to repeat Step 2 and the game continues with certification agent performing Step 3. The example LC system, if the manufacturer only gives the a_x and w observations, then the system is deemed unsafe. The unsafe set is shown in the red box in Fig. 6 and it signifies safe distance between two cars while changing lanes. The reason for the unsafe categorization can be due to error in parameter extraction. The certification agent shows how the reach set varies with each of the parameter of the LC system. An example of such sensitivity analysis is shown in Fig. 7 that demonstrates how the reach set varies with changing parameter a of the LC system.

4.1 Step 1 Repeated by Manufacturer: New Sensor

The manufacturer can make the velocity or location of the car available to the certification agent. The velocity was already measured by the speedometer of the autonomous car. Hence, the manufacturer did not need to install any new sensor. However, to measure location, the manufacturer needs to install a proximity sensor. Thus, to save cost, the manufacturer provides access to the velocity parameter to the certification agent.

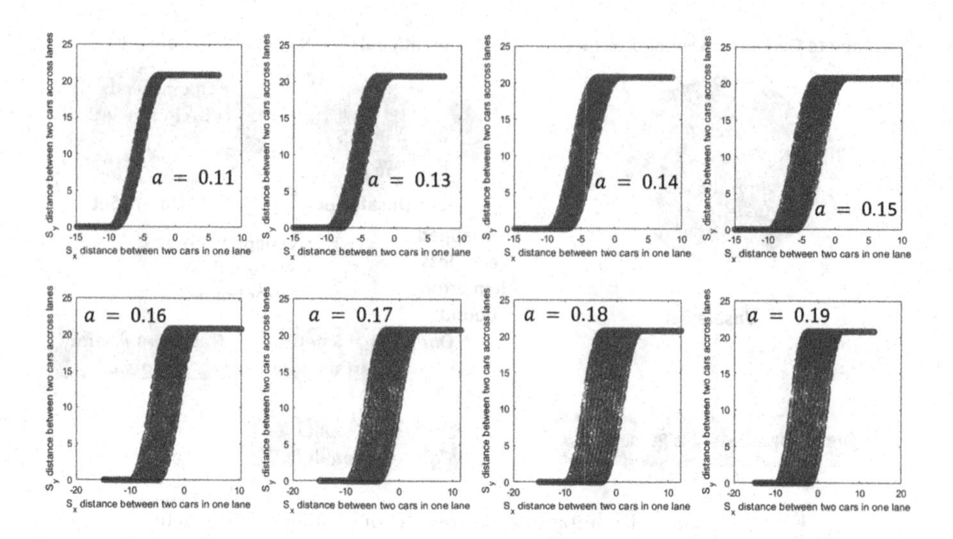

Fig. 7. Reach set variation with changing parameter a.

4.2 Step 2 Repeated by Certification Agent: Model Extraction and Update Safety Analysis

Model Extraction: If now v_x, a_x and w are available to the certification agent, then the agent can take the following steps to generate the system:

Action 1: Same as Action 1 in Sect. 3.2 to obtain g.

Action 2: Same as Action 2 in Sect. 3.2 to obtain d, e, and f.

Action 3: Utilize linear regression between $\frac{dv_x}{dt}$ computed using euler's method and a_x to obtain a.

Action 4: The transient condition for $\frac{da_x}{dt}$ can be utilized by considering two consecutive values and observing their difference.

$$\frac{da_x}{dt}(2) - \frac{da_x}{dt}(1) = d*(bv_x(1)-c)\tau + f*(v_x(2)-v_x(1)) + g*(a_x(2)-a_x(1)), \quad (11)$$

here all variables except b and c are known.

Utilizing the above-mentioned steps the certification agent obtains the parameter set of Eq. 12.

$$a = 0.102, b = 1, c = -2.23, d = -0.0099, e = 0.737, f = -0.3, g = -0.506. \quad (12)$$

Safety Analysis: The reachability analysis is again performed by the certification agent and is compared with the previous reach set when only a_x and w were observed. As shown in Fig. 8, the new derived system does not intersect with the unsafe set. At this point the certification agent concludes the game and certifies the safety of the LC system.

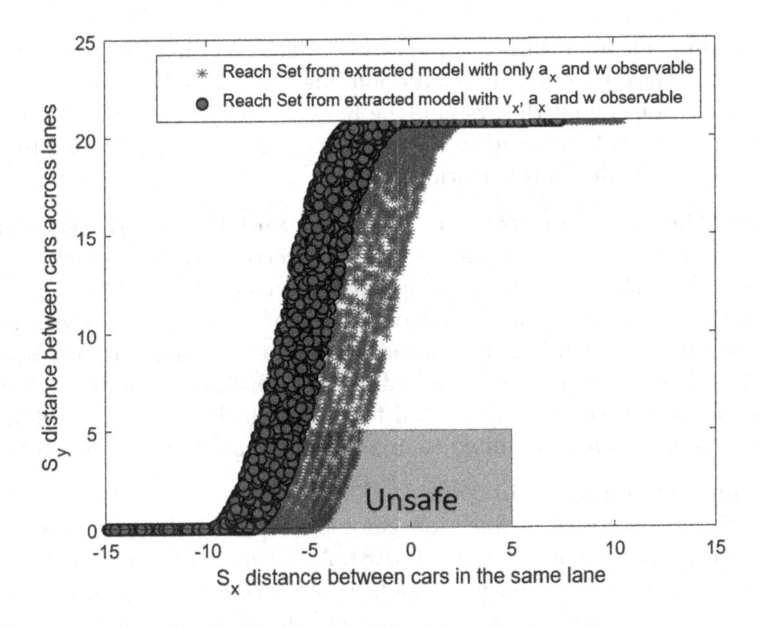

Fig. 8. Reach set variation with changing parameter a.

5 Discussion

Gaps in Traditional Safety Verification and Certification of Cyber-Physical Systems: The increasing number of incidents and accidents arising from the operation of safety certified cyber-physical systems bring with it a pressing need to re-evaluate the certification process of CPS of different domains including automotive, aviation, and healthcare. For example, the NHTSA vision for safety report encourages the adoption of an operational design domain (ODD) under which the system is guaranteed to be safe and capable of detecting and mitigating deviations from the defined ODD. The first concern is that the list of factors to consider in the ODD is extensive and difficult to generate without real-world experience [12]. In addition, to verify all possible behaviors of a system, there is a combinatorial explosion of test cases due to the interleaving variables of the physical system, human-in-the-loop, and the operational environment. As a result, real world performance (RWP) monitoring will eventually represent a crucial component of the certification process. In fact, the UL4600 specifically requires mechanisms for collecting operational data and ensuring safety after deployment [11]. Also, the Food and Drug Administration introduced a novel certification pathway (Pre-Cert), where manufacturers of medical devices are required to extend a RWP monitoring and analytics plan to demonstrate continued safety. The FDA requires full specifications of the RWP and access to operational data. However, it is crucial to verify that the specified system and the RWP conforms with the deployed system. Discrepancies between what is specified by the manufacturer and their third-party providers engender an accountability dilemma during accident investigation. This has been witnessed during the Toyota unintended acceleration bug that caused fatal crashes, where false specifications about the system (the RAM possesses an error detection and correction logic) and only partial access to the system were provided to NASA to conduct the accident investigation [10].

Bridging the Gap Between the Certified and Deployed System: This paper proposes a certification game between the certifier and manufacturer where the goal of the certifier is to learn an operational model and verify its consistency with the specifications while the goal of the manufacturer is to analyze the multiple possibilities of enabling operational model learning and the corresponding required budget on time and cost to find the most optimized solution. Other factors can be incorporated in the reward function, which can be product/domain specific including, but not limited to, market differentiation and liability.

Implementation and Validation: In this paper, we desmonstrated the applicability of the proposed certification game on the lane change module of the advanced driver assistance system (ADAS). As future work, we aim at conducting surverys and discussions with manufacturers to deduce the essential components of the reward function. We will also implement the approach as a tool that parses a file containing the specifications of the hybrid automata model of the system and mines the necessary attributes of the certification game including system variables, control modes, and corresponding ODEs. The manufacturer will also provide additional information including observability and sampling frequency. All this information along with the specifications file will be used to generate operational traces. The operational model learnability is assessed via the possibility of recovering the same specifications file using the generated traces. We also aim at enabling feedback to the manufacturer on the necessary specifications/traces that are needed to facilitate the certification. The experiments will be conducted using the example of heavy vehicle braking system [1].

6 Conclusions

The operation of cyber-physical systems (CPS) in the real world introduces several novel interaction scenarios between the environment and the CPS which manifest during in-the-field operation. In addition, recent cases of fatal failures from the operation of cyber-physical systems in the real world engendered a pressing need to reassess the certification process. This paper proposes an additive certification layer to verify consistency between the operational CPS and the specified CPS. The certification process is modeled as a cooperative game between the certification agent and the manufacturer. Both agents share the same goal which is the completion of the certification but their utility functions differ. The proposed approach can be implemented as a certification tool that can be used by manufacturers to analyze the certification outcome of their CPS using different amounts of information and data to provide to the certifier. The tool can also be implemented as a black-box where inner workings are kept confidential to the certification agency. This tool will enable a certification process with minimized resources through cooperation between the certifier and manufacturer while preserving the manufacturer's and certifier's confidentiality.

Acknowledgment. This work is partly funded by a DARPA AMP grant.

References

1. Banerjee, A., Lamrani, I., Gupta, S.K.: Faultex: explaining operational changes in terms of design variables in cps control code. In: Proceeding of the 4th IEEE International Conference on Industrial Cyber-Physical Systems (ICPS). IEEE Press (2021)
2. Chen, R.T., Rubanova, Y., Bettencourt, J., Duvenaud, D.: Neural ordinary differential equations. arXiv preprint arXiv:1806.07366 (2018)
3. Fan, C., Qi, B., Mitra, S., Viswanathan, M.: DryVR: data-driven verification and compositional reasoning for automotive systems. In: Majumdar, R., Kunčak, V. (eds.) CAV 2017. LNCS, vol. 10426, pp. 441–461. Springer, Cham (2017). https://doi.org/10.1007/978-3-319-63387-9_22
4. Fan, C., Qi, B., Mitra, S., Viswanathan, M., Duggirala, P.S.: Automatic reachability analysis for nonlinear hybrid models with C2E2. In: Chaudhuri, S., Farzan, A. (eds.) CAV 2016. LNCS, vol. 9779, pp. 531–538. Springer, Cham (2016). https://doi.org/10.1007/978-3-319-41528-4_29
5. FDA: Developing a software precertification program: a working model (2019)
6. FDA: Proposed regulatory framework for modifications to artificial intelligence/machine learning (AI/ML)-based software as a medical device (2019)
7. Henzinger, T.A.: The theory of hybrid automata. In: Inan, M.K., Kurshan, R.P. (eds.)Verification of Digital and Hybrid Systems, NATO ASI Series (Series F: Computer and Systems Sciences), pp. 265–292. Springer, Heidelberg (2000). https://doi.org/10.1007/978-3-642-59615-5_13
8. Henzinger, T.A., Ho, P.H., Wong-Toi, H.: Hytech: a model checker for hybrid systems. Int. J. Softw. Tools Technol. Transf. (1997)
9. Kanderian Jr, S.S., Steil, G.M.: Apparatus and method for controlling insulin infusion with state variable feedback, 5 October 2010. US Patent 7,806,886
10. Koopman, P.: A case study of Toyota unintended acceleration and software safety. Presentation (2014)
11. Koopman, P., Ferrell, U., Fratrik, F., Wagner, M.: A safety standard approach for fully autonomous vehicles. In: Romanovsky, A., Troubitsyna, E., Gashi, I., Schoitsch, E., Bitsch, F. (eds.) SAFECOMP 2019. LNCS, vol. 11699, pp. 326–332. Springer, Cham (2019). https://doi.org/10.1007/978-3-030-26250-1_26
12. Koopman, P., Fratrik, F.: How many operational design domains, objects, and events? SafeAI@ AAAI 4 (2019)
13. Lamrani, I., Banerjee, A., Gupta, S.K.: Hymn: mining linear hybrid automata from input output traces of cyber-physical systems. In: 2018 IEEE Industrial Cyber-Physical Systems (ICPS), pp. 264–269. IEEE (2018)
14. Lamrani, I., Banerjee, A., Gupta, S.K.: Operational data driven feedback for safety evaluation of agent-based cps. IEEE Trans. Ind. Inform. (2020)

15. Medhat, R., Ramesh, S., Bonakdarpour, B., Fischmeister, S.: A framework for mining hybrid automata from input/output traces. In: Proceedings of the 12th International Conference on Embedded Software, pp. 177–186. IEEE Press (2015)
16. Summerville, A., Osborn, J., Mateas, M.: Charda: causal hybrid automata recovery via dynamic analysis. arXiv preprint arXiv:1707.03336 (2017)

A New Approach to Better Consensus Building and Agreement Implementation for Trustworthy AI Systems

Yukiko Yanagisawa and Yasuhiko Yokote(✉)

Advanced Data Science Project, RIKEN Information R&D and Strategy Headquarters, RIKEN, Tokyo, Japan
{yukiko.yanagisawa,yasuhiko.yokote}@riken.jp

Abstract. We propose a system that focuses on consensus building and agreement implementation as the basis for establishing AI trustworthiness. Our approach is new, and we have called it consensus building with an assurance case: it is based on applying open systems dependability engineering to the objective of achieving stakeholder accountability. The experimental validation of the proposed system was conducted on the issue of feature engineering within the context of a project on data-intensive medicine. Our findings show that the online nature of the proposed system is effective in facilitating stakeholders participation and contribution, and logging the consensus building process in a useful manner. Using this approach, stakeholders can achieve a logical understanding of the substance of the consensus and the process by which it was reached, thereby providing assurance that AI safety-related requirements are being met.

Keywords: AI safety · Assurance case · Consensus building · Agreement implementation · Consensus building process

1 Introduction

As applications of AI systems expand, so do concerns about their safety. In response, guidelines for AI system development have been issued in Europe, the United States and Japan [1–3]. These guidelines cover both technical and ethical issues, including fairness, accountability, and transparency. The European Commission has recently issued a white paper on its approach to excellence and trust in AI [4] along with a proposal for AI regulation [5]. In this paper, we focus our discussion on trustworthiness in the field of data-intensive medicine that uses such AI systems. We believe it is not sufficient for AI systems to be able to deal with these issues autonomously in this field. The requirements for autonomous AI systems come from stakeholders, organizations and people; model acquisition for AI systems training requires a high level of medical expertise; interpretation of AI system results also requires a high level of expertise. Toulmin [6] writes in his book that the correctness of computational results and their appropriateness are two different things that must be pursued independently. Shneiderman [7] states in

I. Habli et al. (Eds.): SAFECOMP 2021 Workshops, LNCS 12853, pp. 311–322, 2021.
https://doi.org/10.1007/978-3-030-83906-2_26

his paper that humans should be considered in the highest priority for reliable, safe and trustworthy AI systems and positioned existing AI systems at the axis of human-control and automation. In this paper, we focus on the stakeholder relationships at the interface with AI systems, and propose a method to address correctness and appropriateness independently and to minimize the risks associated with trustworthiness, particularly regarding AI safety.

Applications of data-intensive medical science to personalized medicine are increasing, and concerns about its safety are simultaneously being discussed. Brundage et al. [8] state that it is important for stakeholders to be able to verify AI developers' claims so they can be aware of the safety of AI systems. Ashmore et al. [9] focus on the lifecycle of machine learning systems and investigate techniques that enable AI assurance to be built in. One international standard for building trust in the stakeholder relationship context is IEC 62853 [10], where four process views are defined: consensus building, accountability achievement, failure response and change accommodation. It incorporates the concept of the ISO/IEC/IEEE 15288 standard's process view [11], which is a group of relevant processes of specific interest to the stakeholder. Hernández-Orallo et al. [12] identify categories of issues on the AI safety, where those categories can be used as claims of AI developers to resolve those issues. As the ability of these stakeholders to achieve accountability for AI safety can minimize its risk in data-intensive medicine, it is important to determine how to implement the consensus building process views defined in IEC 62853.

According to Yoshitake [13], in healthcare "decisions of informed consent alone are not sufficient", and "a decision on medical treatment should be made based on the mutual consent of the various parties involved, taking into account the unclear wishes of the patient". On this basis, she argues that consensus building in healthcare should be defined as "a creative process of finding a convincing solution based on the values of the various parties involved." Susskind et al. [14] say the process of consensus building includes implementation of the agreement, and that such implementation is difficult in practice [15]. As Inohara [16] pointed out, it is necessary to develop "consensus theory and consensus building methods" to improve the quality of consensus building.

In this paper, we focus on consensus building and agreement implementation to ensure the safety of AI systems, and we propose a new approach along with a supporting tool to achieve it. In Sect. 2, an approach for "consensus building with an assurance case" (abbreviated as CBAC) is proposed as a new way of achieving high-quality accountability in open systems dependability engineering [17]. Section 3 uses proof-of-concept experimentation to show CBAC's effectiveness: we conducted experiments on feature engineering within the process of data-intensive medicine, and the results are discussed in this section. Section 4 sets out our conclusions.

2 Our Approach

To allow open systems dependability engineering to be understood by society as a whole, the assurance case has been adopted as standard notation without any special wording to describe the dependability of the system [18]. It is characterized using Toulmin's argumentation layout [6] in the argumentation framework, which makes the argumentation structurally expressible. An assurance case describing the arguments and evidence for consensus building and agreement implementation by stakeholders in a structured way enables stakeholders to understand the agreement's content, including its process, in a logical manner, and to be confident that AI safety requirements will be enforced. To achieve this, we have developed a new interactive consensus building approach that makes stakeholders' arguments transparent and makes them confident about agreement by presenting them with a structured document accompanied by the assurance case.

2.1 Consensus Building and Agreement Implementation with an Assurance Case

There are a wide range of issues involved in building a consensus and implementing its agreement including stakeholder selection, communication, processes, and decision-making rules. In this paper, we have identified the following three issues that are important in building a consensus and implementing its agreement among diverse stakeholders:

- The meeting place for easy participation and concentration for stakeholders
- The need for a facilitator
- The need to record the consensus building and agreement implementing process

On the first issue, it is proposed to use an online virtual space where stakeholders can participate when they are free to do so. Nakai [19] describes the premise of consensus building thus: "For consensus to be reached within a group, stakeholders must first come to the table to discuss it." In Japan, work reforms have increased levels of freedom in terms of work locations and working hours, and there is a trend towards fewer wasteful meetings and improved efficiency [20]. Due to the recent pandemic, it is difficult for stakeholders to meet physically at one place at a given time to form a consensus. Virtual meetings are part of the "new normal" and are becoming more and more common. Because unanimity is important to consensus building, all stakeholders must be involved. Discussing the same agenda for a long period can also be problematic. Straus [21] cites "loss of focus" as a typical problem with meetings. The CBAC approach proposes a consensus building process using the channel mechanism provided by Slack, an online business communication tool. This also limits participation to a single channel, which helps participants focus on the discussion.

In response to the second issue, it is proposed to structure consensus in assurance cases and let the chatbot function assume the role of facilitator. To facilitate good consensus building there has to be a commitment to various actions, such as clarifying the agenda to be agreed upon, providing relevant information for discussion, offering alternatives in the event of an impasse, and merging discussions when professionalism is needed. Everyone agrees on the importance of the facilitator in meetings, but it is not always possible to find a competent one. The CBAC approach therefore uses the chatbot function of Slack and, as far as possible, automates the facilitator's role. A proposed agreement is structured in an assurance case, and even if it is in standard notation a certain amount of expertise is needed to write and understand it. The Slack chatbot function makes it possible to present expert-written assurance cases as drafts for discussion and to display information that is relevant to what someone has said by searching previous discussions in a separately stored database. It can also encourage agreement among all stakeholders when the discussion has been concluded. However, it cannot negotiate emotional responses or provide alternatives in the event of an impasse. During actual consensus building and agreement implementation, the human facilitator and the chatbot function can be combined to facilitate consensus building.

On the third issue, it is proposed that all conversations (message exchanges) on the Slack channel be recorded in a database. This would help to address duplication of ideas in meetings, misunderstandings, and problems with post-meeting recall, while simultaneously improving accountability. Straus [21] calls the recording of a meeting "group memory", an effective method for solving typical problems in meetings. In open systems dependability engineering, the importance of documenting the consensus process to achieve accountability is recognized [22]. In the CBAC approach, within the Slack channel, participating stakeholders can type and send messages, and the dialogue proceeds in chronological order. Participants can also attach documents to substantiate their opinions. Those messages and documents are then recorded in a database, which also records participants' actions using the chatbot function. All records within the channel in which consensus is reached will be traceable so that all records will be linked with the assurance case on the agenda. Traceable records will be used effectively to identify the cause of any problems arising from the implementation of agreements that have been reached.

2.2 Consensus Building Process

This section describes the consensus building process using the CBAC approach (see Fig. 1).

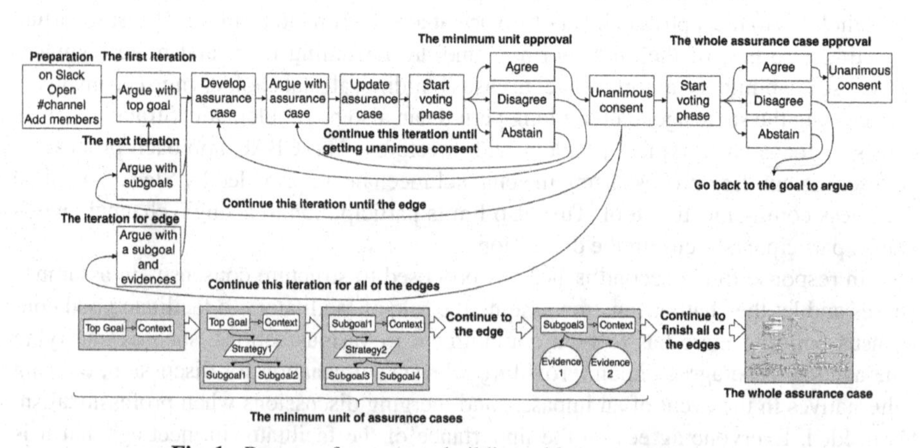

Fig. 1. Process of consensus building with an assurance case.

Preparation. Before the facilitator begins to build the consensus, she or he creates a dedicated channel in Slack, then selects the stakeholders needed for the consensus building process and registers them as members of the channel.

Description of the Top Goal. The process begins with a determination of the main requirements for ensuring AI safety. The assurance case expert writes the assurance case, with the top goal (or top claim) being the most important requirement defined by the stakeholders. We recommend that the assurance case expert writes the document, because there are several patterns in the strategy that affect the direction of the argument for consensus building. The top goal describes the context and the assumptions of the argument. This is where the first step towards consensus building is taken. It will have a significant impact on subsequent discussions, so stakeholders must agree on the subject and the extent to which they wish to ensure the AI safety requirement.

Initiating Consensus Building. Consensus building is initiated by the facilitator typing a message into the Slack channel and declaring it. Triggered by the message, the chatbot function searches for the assurance case needed for consensus building, parses it, and displays it in text that can be read in Slack. Stakeholders can read this case and, if necessary, view it as a diagram. Through channel messages, in an interactive discussion with other stakeholders, they can ask and answer questions, rationales and agree or disagree on the top goal and its context. The assurance case expert modifies the assurance case accordingly, in line with the content of the discussion, and presents it to the stakeholders via channel messages. The facilitator observes the progress of the discussion and, if there is a convergence, activates the "voting function" to confirm the stakeholders' agreement on the top goal of the current assurance case. Each stakeholder presses a button to indicate if they agree, disagree or wish to abstain. Consensus building is complete when all stakeholders select "agree". If any stakeholder selects "disagree" or "abstain", the facilitator continues the discussion until the top goal, rationale, and preconditions are unanimously agreed on by the stakeholders. All messages and documents submitted to the Slack channel up to this point are recorded in the database in a form that links to the assurance case.

Incremental Assurance Case Development. Once the top goal has been agreed upon, the facilitator starts a discussion about the strategy needed to achieve the top goal and how to break it down into subgoals by using the strategy. Based on the stakeholders' comments, the assurance case expert will develop the assurance case. When the facilitator wants to confirm that achieving these subgoals leads to the top goal, she or he must reactivate the voting function and obtain further stakeholder consensus. Stakeholders repeat the assurance case description and consensus building process until they finally identify evidence that will prove the subgoal has been achieved. Depending on the AI system involved, the consensus building discussion to ensure the AI safety requirement is likely to be extensive, as the assurance case that structures it will. A case may be hard to grasp immediately: CBAC uses an incremental assurance case development approach.

At the beginning of the process, the complete assurance case is not subject to consensus building. We start with development of the top goal and consensus building, then build consensus on the assurance case incrementally, i.e., 1 top goal + 1 strategy + multiple subgoals. In this approach, we define the 1 goal + 1 strategy + multiple subgoals as a minimum unit of assurance case. By forming a consensus in an assurance case with such a small set, we limit the scope of the case and minimize the number of stakeholders involved. This clarifies the topics stakeholders should be discussing and makes it easier for them to understand what they need to agree on, so improving the quality of consensus building, and also reduces the time needed to reach a consensus.

With each iteration of deconstruction of the goal, the content becomes more detailed. Eventually, a goal is tied to evidence that will confirm it. The unit of consensus building at the edge of an assurance case is 1 goal + context of the goal if necessary + multiple evidence to confirm the goal. Once consensus building up to the edge of the case is complete, we can say that the entire assurance case has been agreed upon and be confident of its implementation. Repeating consensus building with the minimum assurance case is also effective in advancing the case that has been developed. Building a consensus on the top goal of the already developed assurance case, then repeating this with a minimum case, enables us to develop a much deeper assurance case.

In addition, the edges of an assurance case describe the concrete actions to be taken, making it easier to understand and judge than would be the case with an abstract top goal. Thus, it is possible to start building a consensus from the smallest unit (which includes supporting evidence).

Agreement Implementation. With incremental assurance case development, the assurance case evolves through repeated consensus building at each project milestone of the project's progress as shown in Fig. 2 and Fig. 3. The initial assurance case at time t_1 is about argumentations on the objectives of the project (as the top goal) and on data preprocessing (part [A]). Then, at time t_2, evidence is collected for claims on data preprocessing (part [B]) and the project progresses to the phase of argumentations on model to be used for data analysis (part[C]).

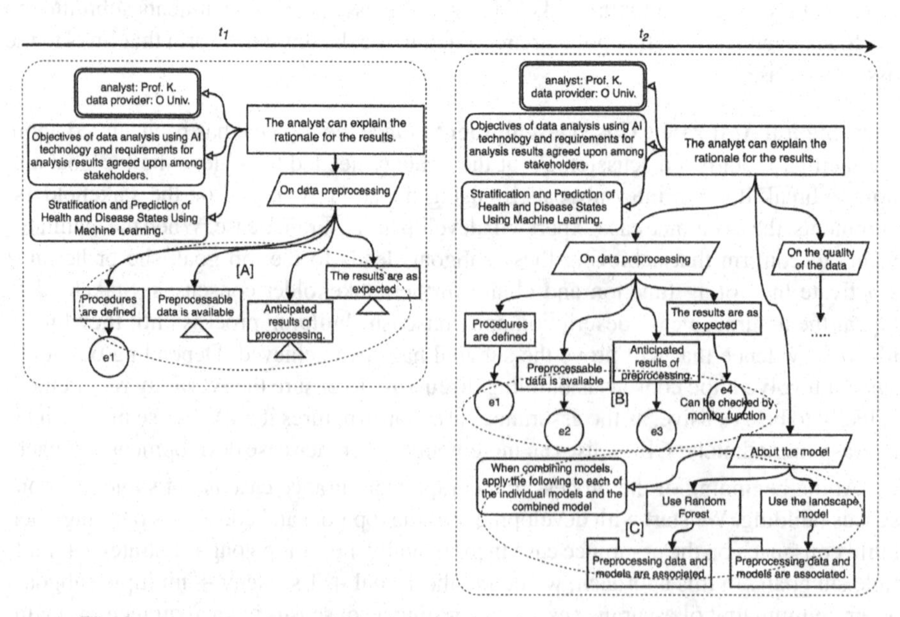

Fig. 2. Assurance case evolution in agreement implementation at time t_1 and t_2.

As the project progresses further, at time t_3 in Fig. 3, data analysis is performed (part [D]), where the argumentations when the results did not meet the requirements are also discussed. Stakeholders require new argumentations on use of a new model (part[E]). In this way, agreement implementation can be incorporated into the consensus building process, i.e., the CBAC approach.

We have developed a support tool that facilitates implementation of the CBAC approach, in the form of a graph-database that records the evolution of assurance cases over time. The support tool encourages multiple stakeholders to jointly develop and agree on assurance cases in a distributed environment and to collect objective evidence. An assurance case targeted by the consensus building is loaded into the database while its network structures are maintained, and the relationships between the supplemental information are also loaded into the database. This supplemental information may include documents that reinforce its context and evidence, the meanings of terms in the content, messages in the channels, results of actions and so on. The support tool uses these relationships to make the necessary decisions to support the facilitator by calculating the features of the assurance case, the comprehensiveness of the evidence, the possibility of misunderstandings, etc., and posting messages to the Slack channel where necessary.

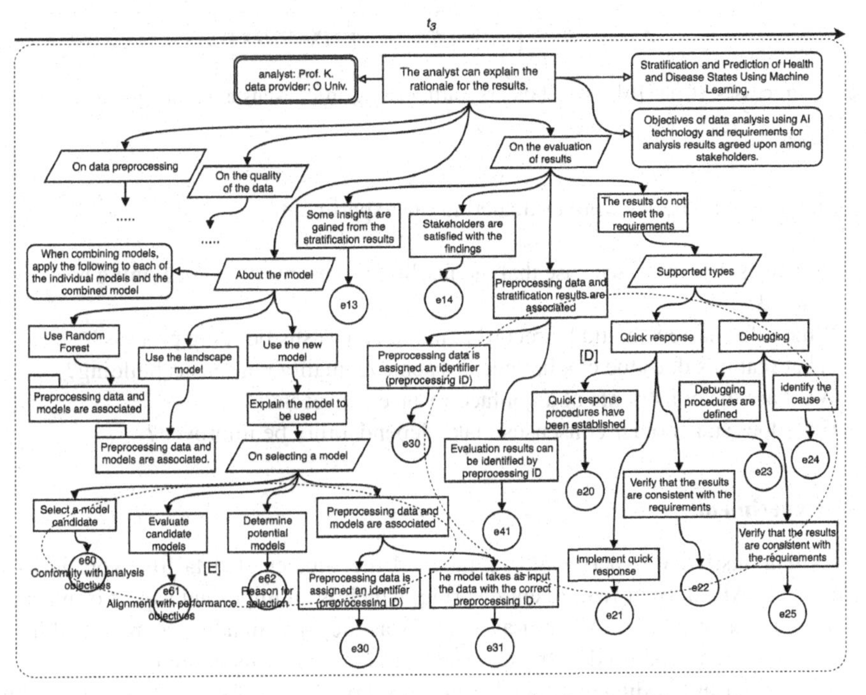

Fig. 3. Assurance case evolution in agreement implementation at time t_3.

3 CBAC Approach Experimentation

We have applied the effectiveness of the CBAC to a case study of data-intensive medical science. This consists of several processes as shown in Fig. 4. In this case study, four stakeholders are assigned: (dm) as the stakeholder responsible for data management, (md) as the stakeholder responsible for model development, (mi) as the stakeholder responsible for model implementation, and the subject. Stakeholders (dm), (md), and (mi) are assumed to establish one consensus (CP_1 in the figure), while (mi) and (subject) are assumed to establish another consensus (CP_2 in the figure). In the experiments in this section, we will focus on the formation of consensus CP_1.

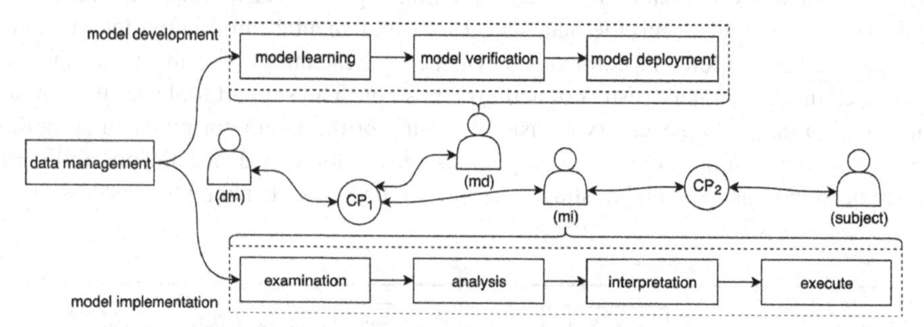

Fig. 4. Processes (from [9]), stakeholders participating in consensus building, and consensus points.

The following issues were evaluated in the experiment.

1. How does the assurance case that the facilitator presents at the beginning need to be detailed?
2. What information should be recorded in the consensus building process?
3. What features does the chatbot need to achieve quality consensus building?
4. What are the constraints of a nonface-to-face approach?
5. How does stakeholder consensus make dependability be improved?

3.1 Experiment

In this demonstrative experiment, we targeted the subject of data trustworthiness in terms of the AI safety of feature engineering, which is an important aspect of machine learning-based prediction. An assurance case was prepared in advance, recorded in the supporting system, and used as the first proposal for consensus building.

The consensus building individuals were stakeholders, one for dm and one for md/mi in Fig. 4, involved in feature engineering, an assurance case specialist, and a facilitator. The facilitator began the process with the message "Start building consensus". This experiment was designed to continue until the stakeholders were confident that feature engineering would be carried out in accordance with the terms of the agreement reached through consensus building. After the experiment, we distributed a questionnaire (22 questions) that could be answered on Slack by the stakeholders. They were asked about (1) the discussion in Slack; (2) how the discussion was conducted using the assurance case; and (3) the chatbot function.

3.2 Result

The total number of messages registered on the channel was 327, comprising 39 for the two stakeholders, 36 for the assurance case specialist, 90 for the facilitator, and 162 for the chatbot. The total number of nodes on the assurance case increased to 29 after consensus was reached, compared with the previous total of 20 nodes. The biggest cause was the increase in the number of context nodes. Before the start of the experiment, there were no context nodes; afterwards, there were eight. The basis of the top goal was clarified by adding two context nodes. The three subgoals with evidence also became clear after five context nodes were added. The biggest argument in the consensus building process concerned the boundary between dm on data cleansing and md/mi on analysis, because it involved the scope of responsibility. This argument led to a change in strategy. The messages related to this argument in Slack itself became the basis for the strategy, and the links to the messages were described in the strategy context node. In the post-experiment questionnaires, we found that all the stakeholders were routine Slack users. They also stated that consensus building in Slack was easier than in face-to-face meetings, and indicated that they had been able to reach agreement on the assurance case.

3.3 Discussion

We examine the five evaluation items listed in the previous section based on the results of the experiment.

How did the assurance case that the facilitator presents at the beginning need to be detailed? The initial assurance case does not have to cover all the necessary information, but we found it was easier to discuss if it had the evidence as the edge of the assurance case. This is because the following series of stories clarify the issues that stakeholders should consider for the top goal to be achieved: i.e., the top goal is proposed, the strategy to achieve it is presented, the specific actions to implement it are presented as subgoals, and the warrants to proof that the subgoals have been achieved. The assurance case we presented as the consensus proposal had a rough flow from the top goal to the evidence, but the warrants for the goal or evidence were not developed at all. The discussion began by reinforcing the missing warrants. As a result, we believe that stakeholders who were unfamiliar with the assurance case could subsequently understand it more easily. Also, we could refer to the subgoals and the evidence to discuss the necessary warrants for the top goal. The subgoals and the evidence were easily understood by stakeholders because they comprised the specific information that was related to the stakeholders' actual work. Understanding the whole picture based on this and examining the top goal from below made consensus building around the top goal at higher hierarchies of abstraction intelligible.

What information should be recorded in the consensus building process? The results showed that all dialogue (the messages on the Slack channel) and the meeting minutes needed to be recorded. During consensus building, there were a number of occasions when the stakeholders examined the minutes of the past meetings and presented them as the warrant or support for the issue. In the post-experiment questionnaire, the stakeholders indicated that being able to explore the discussion later on was beneficial. Records of the decision-making process were useful as confirmation of goals and evidence.

What features should the chatbot have to achieve quality consensus building? We expected the chatbot function to have support features to organize the multiple arguments that took place on the Slack channel. In this experiment, the number of stakeholders was quite small, as was the assurance case - less than 30 nodes. Nevertheless, multiple arguments arose simultaneously. Specifically, a pattern of multiple detailed arguments emerged during one major argument. The facilitator presented the list of current arguments periodically in the channel, both for the benefit of those stakeholders who would speak later rather than in real time, and to clarify which arguments the stakeholder statements related to. If the number of stakeholders and the size of the assurance case are large, multiple discussions could confuse the channel. The facilitator would then need to be replaced by a function that visualizes the argument in a timely manner to help participants focus on the discussion.

What were the constraints of a nonface-to-face approach? This issue did not arise during the experiment. In the post-experiment questionnaires, stakeholders stated that it was easier to talk in a virtual environment than in a face-to-face meeting. They also said that they had the opportunity to revisit difficult words that were occasionally missed in real time conversations by going back and verifying the content of the discussion. The stakeholders' responses indicated that the Slack channel, far from being a constraint, was an effective method of consensus building. However, in the post-experimental questionnaire, one of the stakeholders stated that she knew the other stakeholder and knew she was an easy person to question. She said that being able to imagine the faces of the other stakeholders made her feel psychologically secure. The effectiveness of online discussions depends on the nature of the participants, which is a point to be taken into account in support systems in any future CBAC approach.

How does stakeholder consensus make dependability be improved? In the experiment, consensus was reached when the stakeholders were assured that they could carry out the feature engineering in accordance with unanimous consent obtained. In future, as stakeholders' work progresses, we will review whether the evidence described in the assurance case has been obtained. We will also build consensus on whether the goals are reached with this evidence. Trustworthiness is achieved by building two points of consensus: before the evidence is gathered, and after.

4 Conclusions

We have proposed a process of consensus building and agreement implementation with an assurance case as an approach that makes the process of discussion between diverse stakeholders visible and improves the quality of the consensus they reach. With argumentations through assurance case development addressed appropriateness, and evidence to proof the goal's achievement addresses correctness. Thus, stakeholders are assured of the AI safety within the implementation. In considering the safety of AI in personalized medicine, as opposed to other complex systems, it is particularly important to build consensus among stakeholders on ethical goals. In addition, objective evidence is required to ensure the implementation of the agreement. This should define the process by adding a new Ethical Process View to four process views specified in 62853. In doing so, it is effective to describe the arguments for ethical goals in the assurance case and to implement the process in the CBAC approach.

However, there is still work to be done related to ensuring that stakeholders can understand the assurance case. In our experiment we attempted to handle this by having an assurance case expert participate in the discussion and make revisions to the assurance case, and by deploying a chatbot function to display the assurance case in a text format that could be read by non-experts, but we cannot say that these features fully resolved the issue. Since the ultimate purpose of an assurance case is to serve as a document which people decide whether or not they will agree to, there has to be a certain amount of explanation of its conventions in advance, as well as a format which they can consult for an overview of the whole structure at any time.

Another issue that still needs work is managing multiple detailed discussions. Using the CBAC approach, since the consensus building process extends beyond a single conference, stakeholders may be joining or leaving the discussion arbitrarily, which raises problems arising from the nature of Slack's UI. Due to Slack's linear timeline interface metaphor, when a stakeholder who has been absent from the discussion for a certain period returns to the Slack channel, it can be difficult for them to catch up with developments in the discussion of the assurance case that occurred in the interim, and what interactions between other stakeholders have resulted in what resolutions. This reveals the need for a means of clearly displaying the status of a pattern of multiple detailed arguments at any time.

To solve the problems set out above, we plan to conduct further experiments in consensus building among stakeholders across a larger number of projects using the support system. First, we will confirm by experiment whether presenting an assurance case as a consensus proposal is beneficial. Second, if an assurance case needs to be presented as a proposal for an agreement, we will have to determine indicators of what explanations are needed in advance and how complete they should be. We will also streamline the requirements specifications for displaying assurance cases and for displaying a pattern of multiple detailed arguments, improve the chatbot function, and assess the results. The CBAC-based support system developed in this paper is now being evaluated as a platform for disease prediction research and development using AI data analysis, as well as in broader AI systems. We anticipate that using this support system to address various issues in that field will allow research initiatives to achieve a high degree of trustworthiness from the outset and contribute to consensus building for AI systems in this role.

Acknowledgments. The authors would like to thank Naoko Takanashi for her technical assistance implementing the CBAC-based support system, along with Akiko Hanai and Manami Kato, members of RIKEN, as stakeholders in our experimentation. This paper is partially supported by Innovation Platform for Society 5.0 from Japan's Ministry of Education, Culture, Sports, Science and Technology.

References

1. The High-Level Expert Group on AI: Ethics Guidelines for Trustworthy AI, European Commission (2019)
2. Vought, R.T.: Guidance for Regulation of Artificial Intelligence Applications (2020)
3. Cabinet Secretariat of Japan: Social Principles of Human-centric AI (2019)
4. European Commission: White Paper on Artificial Intelligence – A European Approach to Excellence and Trust (2020)
5. European Commission: Proposal for a Regulation on a European Approach for Artificial Intelligence (2021)
6. Toulmin, S.: The Uses of Argument (Updated Edition), Cambridge University Press (2003)
7. Shneiderman, B.: Human-centered artificial intelligence: reliable, safe & trustworthy, arXiv.org. https://arxiv.org/abs/2002.04087v1. (Extract from forthcoming book by the same title), 23 February 2020
8. Brundage, M., et al.: Toward trustworthy AI development: mechanism for supporting verifiable claims. https://arxiv.org/abs/2004.07213v2 (2020)
9. Ashmore, R., Calinescu, R, Paterson, C.: Assuring the machine learning lifecycle: desiderata, methods, and challenges. https://arxiv.org/abs/1905.04223v1 (2019)
10. IEC 62853:2018 – Open Systems Dependability (2018)
11. ISO/IEC/IEEE 15288:2015 – Systems and Software Engineering - System Life Cycle Processes (2015)
12. Hernández-Orallo, J., et al.: AI paradigms and ai safety: mapping artefacts and techniques to safety issues. In: 24th ECAI (2020). https://ebooks.iospress.nl/volumearticle/55181
13. Yoshitake, K.: Reproductive Medicine on Consensus Building, in [16]. (in Japanese)
14. Susskind, L., McKearnan, S., Thomas-Larmer, J. (eds.): The Consensus Building Handbook - A Comprehensive Guide to Reaching Agreement. SAGE Publications (1999)
15. Potapchuk, W., Crocker, J.: Implementing Consensus-based Agreements, Ch.14 of [14]
16. Inohara, T. (ed.): Study on Consensus Building. Keiso Shobo, Tokyo (2011). (in Japanese)
17. Tokoro, M.(ed.): Open Systems Dependability: Dependability Engineering for Ever-Changing Systems, 2nd ed., CRC Press (2015)
18. Yamamoto, S., Matsuno, Y.: D-Case - Building Consensus and Achieving Accountability in [17]
19. Nakai, Y.: The creation and breakdown of mutual cooperation as a precondition for agreement, in [16]. (in Japanese)
20. Ministry of Health: Labour and Welfare, Act on the Arrangement of Related Acts to Promote Work Style Reform (2019)
21. Straus, D.A.: How to Make Collaboration Work: Powerful Ways to Build Consensus, Solve Problems, and Make Decisions. Berrett-Koehler (2002)
22. Yokote, Y., Nagayama, T., Yanagisawa, Y.: D-ADD - The Agreement Description Database, in [17]

Author Index

Printed in the United States
by Baker & Taylor Publisher Services